AF293749

Advances in Hypersonics

Modeling Hypersonic Flows

Volume 2

J. J. Bertin
J. Periaux
J. Ballmann
Editors

Springer Science+Business Media, LLC

John J. Bertin
Sandia National Laboratories
Albuquerque, NM 87111
USA

Josef Ballmann
Lehr-Und Forschungsbiet
für Mechanik der Rheinisch-Westfälischen
Technischen Hochschule Aachen
Templergraben 64
Germany

Jacques Periaux
Dept. of Aerodynamic Theory
Avions Marcel Dassault-Brequet Aviation
92214 Saint Cloud
France

Library of Congress Cataloging-in-Publication Data
Advances in hypersonics / edited by J. J. Bertin, J. Periaux, J.
   Ballmann
        p.   cm.
     Includes bibliographical references.
     Contents: v. 1. Defining the hypersonic environment -- v.
   2. Modeling hypersonic flows -- v. 3. Computing hypersonic flows.
     ISBN 978-1-4612-6730-0

     1.   Aerodynamics, Hypersonic.   I. Bertin, John J., 1938-   .
   II. Periaux, Jacques.   III. Ballmann, Josef.   IV. Title: Advances in
   hypersonics.
   TL571.A27   1992   92-26882
   629.132'306--dc20   CIP

ISBN 978-1-4612-6730-0   ISBN 978-1-4612-0371-1 (eBook)
DOI 10.1007/978-1-4612-0371-1

Camera-ready copy prepared by the Authorss.

9 8 7 6 5 4 3 2 1

Contents

Preface vii

List of Contributors x

Turbulence Modeling for Hypersonic Flows
J. G. Marvin and T. J. Coakley 1

Advanced Topics in Turbulence Theory
Marcel Lesieur 44

Different Levels of Air Dissociation Chemistry
and Its Coupling with Flow Models
J. Warnatz, U. Riedel, and R. Schmidt 67

Modeling of Hypersonic Reacting Flows
Chul Park 104

Modeling of Hypersonic Non Equilibrium Flows
F. Grasso and V. Bellucci 128

Wall Catalytic Recombination and Boundary Conditions
in Nonequilibrium Hypersonic Flows—With Applications
Carl D. Scott 176

Physical Aspects of Hypersonic Flow:
Fluid Dynamics and Non-Equilibrium Phenomena
Maurizio Pandolfi 251

Permissions 269

Preface

These three volumes entitled *Advances in Hypersonics* contain the Proceedings of the Second and Third Joint US/Europe Short Course in Hypersonics which took place in Colorado Springs and Aachen. The Second Course was organized at the US Air Force Academy, USA in January 1989 and the Third Course at Aachen, Germany in October 1990.

The main idea of these Courses was to present to chemists, computer scientists, engineers, experimentalists, mathematicians, and physicists state of the art lectures in scientific and technical disciplines including mathematical modeling, computational methods, and experimental measurements necessary to define the aerothermodynamic environments for space vehicles such as the US Orbiter or the European Hermes flying at hypersonic speeds.

The subjects can be grouped into the following areas: Physical environments, configuration requirements, propulsion systems (including airbreathing systems), experimental methods for external and internal flow, theoretical and numerical methods. Since hypersonic flight requires highly integrated systems, the Short Courses not only aimed to give in-depth analysis of hypersonic research and technology but also tried to broaden the view of attendees to give them the ability to understand the complex problem of hypersonic flight.

Most of the participants in the Short Courses prepared a document based on their presentation for reproduction in the three volumes. Some authors spent considerable time and energy going well beyond their oral presentation to provide a quality assessment of the state of the art in their area of expertise as of 1989 and 1991.

The development of the Short Courses was a large success due to close cooperation of the following people whose talents cover large and impressive areas in science and engineering, organization, management, and fund raising abilities among others:

Colorado Springs Organizing Committee:

R. Bec (CNES, France); J. J. Bertin (Univ. of Texas at Austin, USA); C. Dujarric (ESA, France); R. Glowinski (Univ. of Houston, USA); R. Graves (NASA, USA); E. Krause (Univ. of Aachen, Germany); S. Lekoudis (ONR. USA); P. Le Tallec (Univ. of Paris

Dauphine & GAMNI, France); B. Monnerie (ONERA & AAAF, France); H. Oertel (DLR, Goettingen, Germany); R. Pellat (CNES, France); J. Periaux (Dassault Aviation & GAMNI, France); O. Pironneau (Univ. of Paris 6 & INRIA, France); L. Sakell (AFOSR, USA); M. Smith (US Air Force Academy, USA); J. Stollery (Cranfield Institute of Technology, UK); B. Stoufflet (Dassault Aviation, France); T. Texduyar (Univ. of Minnesota, USA); J. Wendt (VKI, Belgium)

Aachen Organizing Committee:

J. J. Bertin (Sandia National Laboratories, USA); J. Ballmann (RWTH Aachen, Germany); R. Bec (CNES, France); M. Borsi (Alenia, Italy); K. H. Brakhage (RWTH Aachen, Germany); A. Dervieux (INRIA, France); C. Dujarric (ESA, France); R. Glowinski (Univ. of Houston, USA); W. Goodrich (AGARD/NASA, USA); R. Graves (NASA, USA); H. Grönig (RWTH Aachen, Germany); E. H. Hirschel (MBB & GAMM, Germany); B. Holmes (NASA, USA); R. Jeltsch (ETH Zurich, Switzerland); G. Koppenwallner (DLR Goettingen & Hyperschall Technologie Goettingen, Germany); W. Kordulla (DLR Goettingen, Germany); E. Krause (Univ. of Aachen, Germany); S. Lekoudis (ONR, USA); P. Le Tallec (Univ. of Paris Dauphine & GAMNI, France); B. Monnerie (ONERA & AAAF, France); R. Pellat (CNES, France); J. Periaux (Dassault Aviation & GAMNI, France); M. Smith (US Air Force Academy, USA); J. Stollery (Cranfield Institute of Technology, UK); J. Wendt (VKI, Belgium).

The members of the Organizing Committees would like to address their warmest thanks to those institutions and companies for their support of the programs, in particular AFSOR, AGARD, NATO, CNES, Cray Research, EOARD, ESA, GAMNI, NASA OAST and the ONR for the Colorado Springs Short Course and Dassault Aviation, CNES, Deutsche Aerospace, Deutsche Forschungsgemeinschaft, EOARD, ESA, Fakultat I RWTH, GAMNI MBB, MTU and US Air Force Academy for the Aachen Short Course.

We would like to express our particular thanks to the faculty and staff of the US Air force Academy who made major contributions to the success of the Second Joint Europe/USA Short Course in Hypersonics; special thanks is due to Col. M. L. Smith and the Aeronautics Department (especially Capt. D. S. Adams and S. Orlofsky) and also USAFA families who provided accommodation to young scientists with the warmest and most generous hospitality, Special thanks is also due T. C. Valdez of the University of Texas at Austin.

We also express our gratitude to the faculty and staff of the RWTH University of Aachen who made major contributions to the success of the Third Joint Europe/US Short Course in Hypersonics; special thanks is due to Prof. R. Jeltsch and Prof. Krause from Fakultät I Mathematik and Aerodynamics Institute respectively for their outstanding contributions to the success of the Course. Special thanks is also due Dr. K. H. Brakhage for his help and assistance in the preparation of the Course. He carried out this difficult organizational task with enthusiasm and professional care.

The editors would like to thank the staff of Birkhäuser and Sylviane Gosset for their help and patience with us during the processing of the full manuscript.

We hope that these volumes will be used frequently as a classic reference in the years to come.

John J. Bertin
Jacques Periaux
Josef Ballmann
July 1992

List of Contributors

V. Bellucci, Department of Mechanics and Aeronautics, University of Rome "La Sapienza", Via Eudossiana 18, 00184 Rome, Italy

T. J. Coakley, National Aeronautics and Space Administration, Ames Research Center, Moffett Field, California 94035, USA

F. Grasso, Department of Mechanics and Aeronautics, University of Rome "La Sapienza", Via Eudossiana 18, 00184 Rome, Italy

Marcel Lesieur, Institut de Mécanique de Grenoble et Université Joseph Fourier, Grenoble, Domaine Universitaire, B.P. 53X–38041, Grenoble, France

J. G. Marvin, National Aeronautics and Space Administration, Ames Research Center, Moffett Field, California 94035, USA

Maurizio Pandolfi, Dipartimento di Ingegneria Aeronautica e Spaziale, Politecnico di Torino, Corso Duca degli Abruzzio 24, 10129 Torino, Italy

Chul Park, National Aeronautics and Space Administration, Ames Research Center, Moffett Field, California 94035, USA

U. Riedel, Institut für Technische Verbrennung, Universität Stuttgart, Pfaffenwaldring 12, 7000 Stuttgart, Germany

R. Schmidt, Institut für Technische Verbrennung, Universität Stuttgart, Pfaffenwaldring 12, 7000 Stuttgart, Germany

Carl D. Scott, NASA Johnson Space Center, Houston, Texas 77058, USA

J. Warnatz, Institut für Technische Verbrennung, Universität Stuttgart, Pfaffenwaldring 12, 7000 Stuttgart, Germany

Turbulence Modeling for Hypersonic Flows

J. G. Marvin
T. J. Coakley, Ames Research Center, Moffett Field, California

National Aeronautics and
Space Administration

Ames Research Center
Moffett Field, California 94035

SUMMARY

Turbulence modeling for high-speed compressible flows is described and discussed. Starting with the compressible Navier-Stokes equations, methods of statistical averaging are described by means of which the Reynolds-averaged Navier-Stokes equations are developed. Unknown averages in these equations are approximated using various closure concepts. Zero-, one-, and two-equation eddy viscosity models, algebraic stress models, and Reynolds stress transport models are discussed. Computations of supersonic and hypersonic flows obtained using several of the models are discussed and compared with experimental results. Specific examples include attached boundary-layer flows, shock-wave boundary-layer interactions, and compressible shear layers. From these examples, conclusions regarding the status of modeling and recommendations for future studies are discussed.

INTRODUCTION

In this report we will discuss turbulence models that are used in numerical simulations of complex viscous flows. Although there are many applications and uses of turbulence models, we will restrict our attention primarily to high-speed compressible flows. The material covered constitutes a brief survey of the essential features of turbulence models and their status in applications, but does not include many details important in practice. For these, the reader is encouraged to consult the references.

Turbulence models are necessary in numerical simulations because of the impracticality of computing all scales of turbulent motion. Since these scales compose a range many orders in magnitude, the computer storage required to resolve all scales is much larger than the storage capacity currently available on the most powerful computers. Even if computers did exist with the required capacity, the computational speed of current computers is too slow to handle all but the simplest of problems. Thus approximate methods, or models of turbulence, are introduced to simplify and make the computations practical.

There are several approaches to turbulence modeling depending on how many of the turbulent scales are included in the modeling process. A more rigorous approach is to use subgrid-scale modeling (also known as large-eddy simulation) in which only turbulent eddies equal to or smaller than the numerical grid sizes are modeled. In this case the largest eddies are computed, and because they move and deform in time, the calculations are necessarily unsteady. This results in relatively large computing times and restricts the applicability of subgrid modeling to fundamental studies.

A more practical approach is to model all the scales of turbulent motion. The equations solved in this case are the Reynolds-averaged Navier-Stokes equations and the numerical solutions, which represent long time averages of the flow variables, are usually steady in time. This is the approach described here.

The report is organized into several main sections. It begins with a discussion of averaging procedures and the development of the Reynolds-averaged Navier-Stokes and related equations. After a brief discussion of the various types of turbulence models available, attention is directed to describing a representative sample of eddy viscosity models including explicit modifications to account for high speeds and compressibility. This is followed by a section on results in which representative computations are discussed and compared with experimental measurements. The paper concludes with a section on the current status of turbulence modeling for hypersonic flows with recommendations for future experiments and computation.

REYNOLDS-AVERAGED NAVIER-STOKES EQUATIONS

The basic differential equations used in numerical simulations are the Reynolds-averaged Navier-Stokes equations. These equations are derived from the compressible Navier-Stokes equations by an averaging process that will be described shortly. The time-dependent, compressible Navier-Stokes equations are written as follows:

$$\rho_t + (\rho u_j)_j = 0$$
$$(\rho u_i)_t + (\rho u_i u_j + \sigma_{ij})_j = 0 \qquad i = 1, 2, 3 \tag{1}$$
$$(\rho E)_t + (\rho E u_j + u_i \sigma_{ij} + q_j)_j = 0$$

where subscript notation has been used for partial derivatives, i.e., $(\ )_t = \partial/\partial t$, $(\ )_{,i} = \partial/\partial x_i$, and the summation convention is used for repeated indices. The molecular stress tensor and heat flux vector are expressed as

$$\sigma_{ij} = \delta_{ij} p - \mu(u_{i,j} + u_{j,i} - \frac{2}{3}\delta_{ij} u_{k,k})$$
$$q_j = -\kappa T_j = -\frac{\mu}{Pr} h_{,j} \tag{2}$$

In these equations, ρ is density, u_i are Cartesian velocity components, E is total specific energy, T is temperature, h is enthalpy, e is internal energy, Pr is the Prandtl number, and the Stokes hypothesis is imposed. Assuming a perfect gas with constant specific heats, these variables are related as follows:

$$
\begin{aligned}
p = (\gamma - 1)\rho e \qquad & e = C_v T \qquad & h = C_p T \\
\gamma = C_p/C_v \qquad & E = e + u_i u_i/2 \qquad & Pr = C_p \mu/\kappa
\end{aligned}
\tag{3}
$$

In most applications, the Sutherland relation is used for molecular viscosity, i.e., $\mu = AT^n/(B + T)$, where n, A, and B are constants that depend on the gas.

To derive the Reynolds-averaged Navier-Stokes equations an averaging operation is defined as follows:

$$\bar{\rho}(x_j, t) = \frac{1}{2T} \int_{t-T}^{t+T} \rho(x_j, s)\, ds \tag{4}$$

where $2T$ is the averaging interval, which is assumed to be large compared with the energy containing turbulent time scales, but small compared with the time scale of the mean or average motion. The mean density, $\bar{\rho}$, in this sense is a slowly varying function of time. Although $\bar{\rho}$ depends on the averaging interval, it is tacitly assumed that a range of values for T exists for which $\bar{\rho}$ is practically independent of T and it is this range that is applicable in the averaging operation.

An alternate form of averaging which may be used in place of time-averaging is ensemble-averaging in which the averaging is performed over a large number of records or experiments. In this case the difficulty of selecting a time-averaging interval is not present and therefore this form of averaging is superior in many respects to time-averaging. Because results obtained using the two averaging forms are identical, we will continue to use the term time-averaging even though the preferred form is ensemble-averaging.

The fluctuating density, ρ', is defined as the difference between the density and its average value, i.e.,

$$\rho' = \rho - \overline{\rho} \tag{5}$$

Averages and fluctuating quantities for other variables such as p and u_i are defined similarly. Although not strictly true unless $T \to \infty$, we assume here that $\overline{\rho'} = 0$ and $\overline{\overline{\rho}} = \overline{\rho}$, which is consistent with ensemble-averaging.

The time-averaged Navier-Stokes equations are obtained by averaging equation (1). The result involves averages such as $\overline{\rho u_i u_j}$ and $\overline{\rho h u_j}$, which can be split into averages of mean and fluctuation quantities, e.g.,

$$\overline{\rho u_i u_j} = \overline{(\overline{\rho} + \rho')(\overline{u}_i + u_i')(\overline{u}_j + u_j')}$$
$$= \overline{\rho}\,\overline{u}_i \overline{u}_j + \overline{\rho}\overline{u_i' u_j'} + \overline{u}_i \overline{\rho' u_j'} + \overline{u}_j \overline{\rho' u_i'} + \overline{\rho' u_i' u_j'} \tag{6}$$

For incompressible flows, where $\rho' = 0$, the last three terms in the above equation are absent. From this it is apparent that the compressible averaged equations will contain many more terms than the incompressible averaged equations. For this reason, an alternative form of averaging for velocity and energy variables has been developed, which leads to a form of the compressible averaged equations that is almost identical to the incompressible form. This is called mass-averaging (or Favre-averaging) in which the mean and fluctuating velocities and enthalpies are defined as follows:

$$\tilde{u}_i = \overline{\rho u_i}/\overline{\rho}, \qquad \tilde{h} = \overline{\rho h}/\overline{\rho}$$
$$u_i'' = u_i - \tilde{u}_i, \qquad h'' = h - \tilde{h} \tag{7}$$

It is important to note that averages of fluctuating quantities are no longer zero, but finite, i.e.,

$$\overline{u_i''} = -\overline{\rho' u_i'}/\overline{\rho}, \qquad \overline{h''} = -\overline{\rho' h'}/\overline{\rho} \tag{8}$$

but that mass-weighted averages of u'' and h'' are zero, i.e.,

$$\overline{\rho u_i''} = \overline{(\overline{\rho} + \rho')u_i''} = 0, \qquad \overline{\rho h''} = \overline{(\overline{\rho} + \rho')h''} = 0 \tag{9}$$

By introducing mean and fluctuating quantities into equation (1) and averaging, we obtain the mass-averaged form of the compressible Navier-Stokes equations. These equations are written below. For simplicity, the bar and tilde notations have been omitted from averaged variables.

$$\rho_t + (\rho u_j)_j = 0$$
$$(\rho u_i)_t + (\rho u_i u_j + \sigma_{ij})_j = 0 \qquad i = 1, 2, 3 \tag{10}$$
$$(\rho E)_t + (\rho E u_j + u_j \sigma_{ij} + q_j)_j = 0$$
$$\sigma_{ij} = \sigma_{ij}^M + \sigma_{ij}^T, \qquad q_j = q_j^M + q_j^T$$

where σ_{ij}^M, σ_{ij}^T, etc., are defined as

Reynolds stress tensor:
$$\sigma_{ij}^T = \overline{\rho u_i'' u_j''} = \overline{(\overline{p} + p') u_i'' u_j''}$$

Molecular stress tensor:
$$\sigma_{ij}^M = p\delta_{ij} - \mu(u_{i,j} + u_{j,i} - \frac{2}{3}\delta_{ij} u_{k,k})$$

Reynolds heat-flux vector:
$$q_j^T = \overline{\rho h'' u_j''} = \overline{(\overline{p} + p') h'' u_j''}$$

Molecular heat-flux vector:
$$q_j^M = -\frac{\mu}{Pr} h_{,j}$$

Total energy:
$$E = e + k + u_k u_k / 2$$

Turbulent kinetic energy:
$$k = \overline{\rho u_k'' u_k''} / 2\overline{p}$$

$$(11)$$

In equation (11) it is assumed that μ is independent of time in the averages leading to σ_{ij}^M, and q_j^M.

The goal of turbulence modeling is to relate the Reynolds stresses and heat fluxes to known mean-flow quantities such as velocity and temperature. This can be done in various ways which lead to different types of turbulence models. If the Reynolds stresses and heat fluxes are related algebraically to the mean-flow variables, the corresponding models are called algebraic stress models. The most important and simplest subclass of these models are eddy viscosity models which relate Reynolds stresses to strain rates (or velocity derivatives) in a manner identical with molecular stresses. Eddy viscosity models will be the primary focus of this paper.

The simplest eddy viscosity models are the zero-equation models in which the eddy viscosity is modeled algebraically in terms of flow geometry and mean flow variables. More complicated turbulence models have been developed in which the Reynolds stresses are defined by field equations. These equations are derived by manipulating the Navier-Stokes equations for mean and fluctuating quantities (ref. 1). The resulting equation for the Reynolds stress tensor is given below

Reynolds stress equation:
$$(\sigma_{ik}^T)_t + (\sigma_{ik}^T u_j + q_{ikj})_{,j} = \rho(P_{ik} - \epsilon_{ik})$$

Production tensor:
$$\rho P_{ik} = -(\sigma_{ij}^T u_{k,j} + \sigma_{kj}^T u_{i,j})$$

Dissipation tensor:
$$\rho \epsilon_{ik} = -(\overline{\sigma_{ij}^F u_{k,j}''} + \overline{\sigma_{kj}^F u_{i,j}''})$$

Reynolds flux tensor:
$$q_{ikj} = (\overline{\rho u_i'' u_k'' u_j''} + \overline{\sigma_{ij}^F u_k''} + \overline{\sigma_{kj}^F u_i''})$$

Fluctuating stress:
$$\sigma_{ij}^F = \delta_{ij} p - \mu(u_{i,j} + u_{j,i} - \frac{2}{3}\delta_{ij} u_{k,k})$$

$$(12)$$

In these equations the fluctuating stress tensor, σ_{ij}^F, is interpreted to include both mean and fluctuating quantities. The dissipation tensor, $\rho \epsilon_{ik}$, contains both mean and fluctuating pressures and velocities and in this sense is a more general or extended definition than the conventional ones, which contain only fluctuating velocities. The production tensor is given directly in terms of the Reynolds stresses and mean velocity components, and thus requires no modeling. However, the dissipation and Reynolds flux tensors involve unknown averages and must be modeled.

A simplified form of the Reynolds stress equation is obtained by taking its trace and is called the turbulent kinetic energy equation, or TKE equation. This equation is written below

$$\text{TKE equation:} \quad (\rho k)_t + \left(\rho k u_j + q_{kj}\right)_{,j} = \rho(P - \epsilon)$$

$$\text{Production:} \quad \rho P = -\sigma_{ij}^T u_{i,j}$$

$$\text{Dissipation:} \quad \rho \epsilon = -\overline{\sigma_{ij}^F u_{i,j}''} \tag{13}$$

$$\text{TKE flux:} \quad q_{kj} = \frac{1}{2}\overline{\rho u_k'' u_k'' u_j''} + \overline{\sigma_{ij}^F u_i''}$$

The TKE equation forms the basis of several classes of turbulence models, including the one- and two-equation models, and the algebraic stress models. In these models, the (square of the) velocity scale of turbulence is given by the TKE and a length scale of turbulence is given either algebraically or in terms of a field variable governed by an equation similar to the TKE equation.

EDDY VISCOSITY MODELS

Eddy viscosity models are the simplest turbulence models in the sense that they model turbulent stresses and fluxes by analogy to molecular stresses and fluxes. This approach is generally referred to as the Boussinesq approximation. The models may be expressed in terms of an eddy viscosity function, μ_T, and a turbulent Prandtl number, Pr_T, as follows:

$$\sigma_{ij}^T = \overline{\rho u_i'' u_j''} = \frac{2}{3}\delta_{ij}\rho k - \mu_T(u_{i,j} + u_{j,i} - \frac{2}{3}\delta_{ij} u_{k,k}) \tag{14}$$

$$q_j^T = \overline{\rho h'' u_j''} = -\frac{\mu_T}{Pr_T} h_{,j}, \qquad Pr_T = \frac{C_p \mu_T}{\kappa_T}$$

where κ_T is the turbulent conductivity. With these models, the whole problem of modeling is reduced to defining the eddy viscosity and turbulent Prandtl number. The turbulent Prandtl number is usually assumed to be a constant of the order of unity, but it may vary between classes of problems. (It is normally set equal to 0.9 for boundary-layer problems.) The eddy viscosity function may be expressed in terms of length and velocity scale functions, l and q, as follows:

$$\mu_T = \rho l q \tag{15}$$

The way l and q are determined defines the type of eddy viscosity model to be used. If l and q are determined algebraically from mean flow data, the models are referred to as zero-equation models. If l is determined algebraically, but q is determined from a field equation such as the TKE equation, i.e., equation (13), the model is referred to as a one-equation model. If both l and q are determined from field equations, the resulting model is called a two-equation model. For this report we will discuss zero- and two-equation models.

Eddy viscosity models were developed originally for incompressible flows and only later were extended to compressible flows. Aside from the use of mass-averaging instead of time-averaging, there is very little difference in form between the two types of models. This is because in the initial investigations, which were restricted to attached transonic and moderately supersonic flows, it was found that the incompressible forms were quite satisfactory. As we shall see, the extensions to higher-speed flows in some simple cases are satisfactory, but other more complex cases require specific corrections for compressibility effects.

Zero-Equation Models

Zero-equation models are the simplest of eddy viscosity models in the sense that they do not make use of additional field equations. In this section we will discuss two widely used models that are representative of most other zero-equation models and that will be discussed in the results section. Unless otherwise stated, it is assumed that these models are applied at solid walls using no-slip boundary conditions

Cebeci-Smith Model. - The model described here is a simplified version of the model described in reference 2. It is a two-layer model that uses Prandtl's mixing length model (ref. 3) for the inner layer and Clauser's model (ref. 4) for the outer layer. The model is expressed as follows:

$$\mu_T = min(\mu_{TI}, \mu_{TO})$$
$$\mu_{TI} = \rho l^2 s, \quad \mu_{TO} = 0.0168 \rho \delta_i^* u_e / I$$
$$l = 0.4 yd, \quad d = 1 - exp(-y^+/A^+)$$
$$u_\tau = \sqrt{\tau_w/\rho_w}, \quad y^+ = u_\tau y/\nu_w$$
$$\delta_i^* = \int_0^\delta (1 - u/u_e)\, dy, \quad I = 1 + 5.5(y/\delta)^6$$

$$(16)$$

the strain-rate parameter, s, is usually taken to be the shearing strain , $|u_y + v_x|$, for two-dimensional problems. The operation of taking the minimum in equation (16) is interpreted to mean using the inner eddy viscosity, μ_{TI}, until it first becomes larger than the outer eddy viscosity, μ_{TO}, beyond which point the outer formula is used exclusively. Parameter I is Klebanoff's intermittency factor, u_τ, is the friction velocity, τ_w, is the wall shear, $\nu_w = \mu_w/\rho_w$, and the subscript w indicates wall values. The nondimensional parameter A^+, from van Driest (ref. 5) generally depends on the streamwise pressure gradient and surface blowing and roughness characteristics (ref. 2). For boundary layers with smooth solid walls and zero or small pressure gradients, A^+ is a constant, i.e.,

$$A^+ = 26$$

A more complicated and general version of this model including transition modeling terms is given in reference 2.

Baldwin-Lomax Model. – The Baldwin-Lomax (B-L) turbulence model (ref. 6) is similar to the Cebeci-Smith (C-S) model, but it incorporates features that make it more advantageous for complex two- and three-dimensional flows. It is similar to the (C-S) model in that it uses nearly the same inner model, but it differs with respect to the outer model. This model can be expressed as follows:

$$\mu_T = min(\mu_{TI}, \mu_{TO})$$

$$\mu_{TI} = \rho l^2 s, \qquad s = \sqrt{\omega_i \omega_i}$$

$$\mu_{TO} = 0.027\, \rho y_{max}\, min(F_{max}, 0.25\, u_D^2/F_{max})/I \tag{17}$$

$$F = ysd, \qquad u_D = |\vec{u}|_{max} - |\vec{u}|_{min}, \qquad \delta = y_{max}/0.3$$

where y_{max} is the outermost value of y in the boundary layer where F has a local maximum, F_{max}. In these formulas l is the Prandtl mixing length given by equation (16), d is the Van Driest damping factor, and I is the intermittency factor of equation (16) in which δ is replaced by $y_{max}/0.3$ as indicated.

It should be noted that with the B-L model, in contrast with the C-S model, the strain-rate function, s, is defined as the magnitude of the vorticity vector and not the shearing strain. This makes the model directly applicable to three-dimensional problems where an invariant shearing strain is not well defined.

A basic advantage of this model over the C-S model is a result of how the outer model is defined. Referring to equation (16) we see that the C-S model requires both the displacement thickness, δ_i^*, and the boundary-layer thickness, δ, which is used in the intermittency function. Both of these thickness parameters frequently are not well defined and are difficult to compute, especially for separated flows. The advantage of the B-L model is that it uses a length scale, y_{max}, which is well defined and easily computed for a wide class of flows. This does not necessarily mean that the B-L model is superior to the C-S model on a physical basis, but it does mean that it is more convenient on a numerical basis.

From a physical standpoint, it has been found that the C-S and B-L models give similar predictions of both attached and separated boundary-layer flows for low to moderate supersonic flows. Predictions of attached flows are usually in good agreement with experiments, but predictions of separated flows are frequently deficient. At hypersonic speeds, the models also tend to give similar predictions, although there is some evidence that the B-L model may be more sensitive to Mach number than the C-S model is.

The procedure of applying no-slip boundary conditions is frequently referred to as the integration-to-the-wall procedure. To be applicable, the numerical mesh spacing normal to the wall must be chosen such that the value of y^+ at the first point off the wall is of the order of unity, placing it well within the viscous sublayer.

For some numerical algorithms, such as explicit methods, the procedure of integrating to the wall has detrimental effects on numerical stability because of the fine mesh spacing required. In this case an alternate approach called the law-of-the-wall procedure, or wall-function method, is used. In this approach a slip-type boundary condition, based on the logarithmic velocity law of turbulent boundary layers, is used. This law can be written (for incompressible flow) as

$$u = \frac{u_\tau}{\kappa}\, ln\, Ey^+$$

$$\kappa = 0.4, \qquad E = 9.128 \tag{18}$$

To apply this method, within the context of time-marching Navier-Stokes solvers, the above formula for velocity is solved (by a Newton-Raphson procedure) for the friction velocity, u_τ, using for u and y their values at the first point off the wall. The values of ν_w and ρ_w are obtained by extrapolating the temperature to the wall. Once u_τ is determined, the wall shear stress, τ_w, is obtained, which is then used directly in the boundary condition for wall velocity. More complicated formulas have been developed for compressible flows and flows involving wall heat transfer.[7]

Two-Equation Models

Zero-equation models are well adapted to simple attached flows where a single well defined shear layer is easily identified. There are many complex flows where this is not the case, however, and use of zero-equation models becomes difficult or unwieldy. Examples of such flows include separated flows behind bluff bodies and multiple intersecting shear layers. In these cases it is difficult to define appropriate velocity and length scales because several such scales are usually present in the flow. For this reason, more advanced models have been developed in which the velocity and length scales are determined from field equations. These are the two-equation eddy viscosity models.

The prototype field equation for the two-equation models is the turbulent kinetic energy equation, equation (13). In order to use this equation, averages or correlations in the dissipation and TKE flux terms must be modeled in terms of known or mean-flow quantities. It is beyond the scope of this paper to explain in detail how these terms are modeled. Instead, we will simply discuss the results of the modeling.

There are essentially two terms in the TKE equation that must be modeled. These are the TKE flux, q_{ki}, and the dissipation, ϵ. For the TKE flux, a gradient-diffusion approximation is used, i.e.,

$$q_{kj} = -\left(\mu + \frac{\mu_T}{Pr_k}\right)k_{,j} \tag{19}$$

where Pr_k is a modeling constant (Prandtl number) of the order of unity.

The absolute dissipation rate, ϵ, is obtained from a separate field equation similar to the TKE equation given below. The velocity and length scales, and the eddy viscosity, are expressed in terms of k and ϵ as follows

$$\mu_T = C_\mu f \rho q l = C_\mu f \rho k/\omega = C_\mu f \rho k^2/\epsilon \tag{20}$$

$$q = \sqrt{k}, \quad l = q/\omega, \quad \omega = \epsilon/k$$

where f is a damping function analogous to the van Driest damper, equation (16), C_μ is a modeling constant, and ω is the specific dissipation rate. With these approximations, the TKE equation can be written

$$(\rho k)_t + (\rho k u_j + q_{kj})_j = \rho(P - \epsilon) \tag{21}$$

$$\rho P = -\sigma_{ij}^T u_{i,j} = \mu_T S - \frac{2}{3}\rho k D \tag{22}$$

$$S = (u_{i,j} + u_{j,i})u_{i,j} - \frac{2}{3}u_{k,k}^2, \quad D = u_{k,k}$$

In these equations, P is the turbulent production that is reexpressed in terms of the eddy viscosity and the strain invariants S and D. For incompressible flows the dilatation, $D = u_{k,k}$, is zero, but for compressible flows this term is nonzero and can be an important modeling term in some cases.

The term on the right-hand side of equation (21) is a turbulence source function that may be expressed in terms of a nondimensional source function, h_k, using equations (20) and (22), i.e.,

$$\rho(P - \epsilon) = h_k \rho \omega k \qquad (23)$$

$$h_k = C_\mu f \frac{S}{\omega^2} - \frac{2}{3} \frac{D}{\omega} - 1$$

Equation (21) is the prototype field equation for all two-equation models. With general two-equation models, the variables k and ϵ (or ω) of equation (20) are expressed in terms of two auxiliary variables, s_1, and s_2, each of which is governed by field equations similar to equation (21). This general form of a two-equation model can be written as follows:

$$k = k(s_1, s_2) , \quad \epsilon = \epsilon(s_1, s_2)$$

$$(\rho s_i)_t + (\rho s_i u_j + q_{ij})_j = h_i \rho \omega s_i \qquad i = 1, 2$$

$$h_i = C_{i1} \left(C_\mu f \frac{S}{\omega^2} - \frac{2}{3} \frac{D}{\omega} \right) - C_{i2} \qquad (24)$$

$$q_{ij} = -\left(\mu + \frac{\mu_T}{Pr_i} \right) s_{i,j}$$

In these equations there is no summation on the index i.

The eddy viscosity damping function, f, is usually expressed in terms of a turbulence Reynolds number, R_T, which in turn is written in terms of k and ϵ (or ω). Typical expressions for f and R_T are

$$f = 1 - exp(-\alpha R_T) , \quad R_T = \frac{\rho k}{\mu \omega} \qquad (25)$$

where α is a constant.

For fully developed turbulent flows, the turbulent Reynolds number becomes very large and the damping function approaches unity. On the other hand, at low Reynolds numbers (e.g., in laminar regions or the viscous sublayer) f goes to zero. In general, the variables C_{i1} and C_{i2} of equation (24) are also functions of the turbulence Reynolds number, analogous to equation (25), although in some cases they may involve additional terms. The Prandtl numbers Pr_1, Pr_2 are usually taken to be constant. At large values of the turbulence Reynolds number the variables C_{i1} and C_{i2} generally approach constant values along with the damper f.

1. $k - \epsilon$ Model

One of the most widely used two-equation models is the $k - \epsilon$ model originated by Launder and Spalding (ref. 8). In this model $s_1 = k$ and $s_2 = \epsilon$. The high Reynolds number form of the constants C_{i1} and C_{i2}, as well as the other constants in equation (24), are given by

$$C_\mu = 0.09 , \quad C_{11} = C_{12} = f = Pr_1 = 1$$
$$C_{21} = 1.45 , \quad C_{22} = 1.92 , \quad Pr_2 = 1.3 \tag{26}$$

These constants have been obtained by comparing solutions of the governing equations with experimental results. For example, the constants C_{11} and C_{12} come directly from the TKE equation. The constant C_{22} is determined from experiments on the decay of isotropic turbulence in which case all production and diffusion terms are absent from the equations, and an exact solution is easily obtained. The other constants are determined by obtaining approximate solutions for the wall region of equilibrium boundary layers, where $P = \epsilon$, and the logarithmic law, equation (18), is applicable, and by numerically optimizing free shear flow solutions.

The high Reynolds number form of the $k - \epsilon$ model described above is applicable to fully developed turbulent flows and does not apply to the viscous sublayer. For such applications, the molecular viscosity is much smaller than the turbulent viscosity and usually is neglected in the diffusion fluxes. In these cases, however, special slip-type boundary conditions based on equation (18) must be applied to the velocity and turbulence variables because the first numerical grid point must be taken well outside the viscous sublayer (in the fully turbulent region) and no-slip conditions are inappropriate. This approach has been followed by Launder and Spalding (ref. 8) and others. The generalization to compressible flow is described by Viegas, Rubesin and Horstman (ref. 7). Although this approach is convenient for many problems it is not easily adapted to low Reynolds number flows where transition phenomena are important. In these cases, a more general low Reynolds number form of the model must be used in which C_{11}, C_{12}, etc., depend on R_T. Several such models have been developed, including those by Jones and Launder (ref. 9), Chien (ref. 10), and Wilcox and Rubesin (ref. 11). Because the formulas defining these models are relatively complicated, we will not give them here. Instead, we will describe an alternative low Reynolds number, two-equation model that is given below.

2a. $q - \omega$ Model a

The $q - \omega$ model was developed to overcome numerical stability problems encountered with several low Reynolds number, two-equation models (refs. 9, 10, and 11). A discussion of these problems, and the development of the $q - \omega$ model, is given in references 12, 13, and 14. For this model, the variables s_1 and s_2 of equation (24) are taken as

$$s_1 = q = \sqrt{k} , \quad s_2 = \omega = \epsilon/k \tag{27}$$

The parameters and constants in the equations are given by the following relations:

$$C_\mu = 0.09 \quad , \quad f = 1 - exp(-0.02\, qy/\nu) \quad , \quad Pr_1 = Pr_2 = 2$$

$$C_{11} = C_{12} = 0.5 \quad , \quad C_{21} = 0.055 + 0.5 f \quad , \quad C_{22} = 0.833 \quad , \quad C_{23} = \frac{2}{3} \tag{28}$$

$$h_2 = C_{21}\left(C_\mu \frac{S}{\omega^2} - C_{23}\frac{D}{\omega}\right) - C_{22}$$

Numerical boundary conditions to be applied with this model at solid walls are given by $u = v = q = \omega_y = 0$.

2b. $q - \omega$ Model b: Compressibility correction

The previous model was tested on an oblique shock-wave boundary-layer interaction flow for a separated case, but it failed to predict any separation (ref. 15). In reference 15, a correction to the model was introduced that led to substantially improved predictions. This modification was based on the work of Morel and Monsour (ref. 16) who observed that in a uniaxial compression, the standard $k - \epsilon$ model predicts that the turbulence length scale should increase, which runs counter to the physical expectation that it should decrease. Arguing that the product of the density and the turbulent length scale should remain constant in a uniaxial compression, they derived a correction to the source term of the ϵ equation. Translated to the ω equation, this modification results in a new value of the constant multiplying the dilatation term, i.e.,

$$C_{23} = 2.4 \tag{29}$$

In reference 17, Vandromme proposed a compressibility modification with some similarities to the modification described here. His modification was based on earlier work by Rubesin and included density gradient terms as well as dilatation terms. Results using this model will be reported in the section on compressible shear layers.

2c. $q - \omega$ Model c: Heat transfer correction

The previously described correction for compressibility improves the predictions of pressure distribution and separation, but the surface heat transfer remains relatively unaffected and too high in the region of reattachment. To remedy this difficulty, a modification or constraint on the turbulent length scale was imposed, following the work of Monsour reported in Kline, Cantwell, and Lilley (ref. 18). In this correction, an upper bound is placed on the length scale appearing in the eddy viscosity such that it can never be greater than a constant times the Prandtl length scale in the wall region. The result is

$$\ell = min\,(2.4\,y, \; q/\omega) \tag{30}$$

This correction generally does not change the predictions of the turbulence model exept near a reattachment point, and, to a lesser extent, near a separation point. This occurs because in equilibrium or attached flows, the turbulent length scale q/ω is approximately equal to 2.4y in the wall region. This model will be referred to as the $q - \omega$ model c.

RESULTS

The status of modeling for high-speed flows will now be described by comparing predictions with the results of experiments. Experimentation plays an important role in the development of turbulence models by providing data on the flow physics required to substantiate modeling assumptions and in verifying the performance of models in testing. Wherever possible, experimental data from low- and high-speed flows will be contrasted to illustrate similarities and differences. Emphasis will be on attached flows, shock-wave boundary-layer interaction flows, and shear layers. References 19, 20, and 21 cite data that have been used to evaluate turbulence models for aerodynamic flow predictions.

Attached Boundary-Layer Flows

Modeling for hypersonic attached flows is more mature than for the other flows we consider. Eddy viscosity models perform reasonably well, as our examples will show. This fact may not be surprising because the modeling has been founded on a rather substantial experimental data base used together with knowledge regarding the behavior of incompressible flows.

Figure 1 shows a composite sketch of a turbulent boundary layer constructed from a substantial incompressible data base. Velocity profile data can be collapsed onto a single curve using the friction velocity, u_τ, as a scaling parameter. Regions of the viscous sublayer, the logarithmic region, and the outer layer are depicted. The viscous sublayer is the region where molecular viscosity is important. It consists of a laminar sublayer region and a buffer region that blends with the logarithmic turbulent region. The logarithmic region is characteristic of all turbulent boundary layers and can be expressed as a function of the Reynolds number based on the friction velocity, or y^+. The outer region, which actually begins quite close to the wall (y/δ between 0.1 and 0.2), is characterized by a wake-like region whose shape and thickness depend on the pressure gradient imposed by the outer inviscid flow field and the Reynolds number.

At the high speeds associated with supersonic and hypersonic Mach numbers, similar experimental observations have been made. In these cases, however, it is necessary to introduce a compressiblity transformation (ref. 22) to adjust the profiles appropriately. Figure 2 shows the transformed velocity profiles taken in a very high Mach number helium wind tunnel. It should be noted that at very high Mach numbers, the pressure gradient must balance turbulent normal stresses arising form the normal momentum balance. Also, the sublayer becomes thicker as the Mach number increases.

Representing turbulent velocity profiles using log-law variables enabled integration of the mean momentum equation to determine such quantities as skin friction and heat transfer. But, with the advent of finite difference methods for solving the boundary-layer equations, the development of mixing length and eddy viscosity models was facilitated when it was experimentally observed that the shear stress distribution across a boundary layer changed little because of compressibility. Figure 3, taken from Sandborn (ref. 23), shows compressible data up to Mach 7, compared with similar data representative of incompressible flows. Earlier, Maise and McDonald (ref. 24), using a similar approach with adiabatic wall temperature data, showed that the mixing length and eddy viscosity, scaled by the boundary-layer thickness and the incompressible displacement thickness, respectively, were essentially independent of compressibility effects up to a Mach number of 5. See figures 4a and 4b.

The effects of compressibility and wall temperature on skin friction are shown in figures 5a and 5b. The solid line is the van Driest correlation based on the Karman-Schoener incompressible friction law and represents the available skin friction data to within 10% for the adiabatic wall data and to within 20% for the data with heat transfer. See Hopkins and Inouye (ref. 25). Computations using the boundary-layer equations are compared with the data in the figure. Aside from showing that eddy viscosity models predict the correct influence of compressibility on skin friction (fig. 5a), several other conclusions can be reached. The choice of mass-averaging or time-averaging has no significant effect on the predicted results. The zero-equation C-S model reproduces the van Driest result somewhat more accurately than do the other models; thus this model would have to be the choice for prediction, considering its simplicities. The effects of heat transfer are illustrated in figure 5b where a two-equation model prediction is compared with the van Driest correlation for M=5. The result, which is typical of most eddy viscosity predictions, deviates from the van Driest variation as total temperature ratio decreases and points to a caution regarding accurate prediction of cool-wall heat transfer trends, although the data are considerably scattered in these cases.

Shang (ref. 26) extended computations using the C-S model to higher Mach numbers. He incorporated the normal momentum equation to account for nonzero normal pressure gradients and, more importantly, accounted for triple correlations involving density fluctuations usually omitted at lower Mach numbers. Results are shown in figures 6 and 7. Data and computations from two models, one with density fluctuation terms and one without density fluctuation terms, are compared. The inclusion of these terms affects the heat-transfer predictions somewhat more than the skin friction, but either approach produces reasonably accurate results, considering the uncertainties in the data.

It is interesting to note that at lower Reynolds numbers the data tend to be underpredicted, particularly the heat transfer. Such results are common because boundary layer transition influences the region encompassed by the low Reynolds number. These influences also tend to affect data correlation and may explain why there is more scatter in the cold-wall data around the van Driest predictions at higher Mach numbers where transition lengths are substantial. Figure 7, from Shang (ref. 26), shows skin-friction measurements for low and high Reynolds numbers compared with computations obtained with and without accounting for density fluctuation effects. At high Reynolds numbers, where the turbulent flow is fully developed, the computations compare reasonably well with the data, although it is difficult to conclude whether it is necessary to include the density fluctuation effects because of the data scatter. At low Reynolds numbers, the data are scattered for all Mach numbers, but this is especially pronounced at the highest Mach numbers, probably because of transition effects. The computations show poorest agreement at the high Mach numbers, so a cautionary note is made for this regime.

The effects of low Reynolds numbers can be accounted for at low Mach numbers approximately by modifying either the maximum mixing length or the outer eddy viscosity used in the model formulations. (See McDonald, ref. 27). Bushnell (refs. 28, 29, and 30) investigated the low Reynolds number problem for high Mach numbers and provided a data analysis which indicated that such low-speed, low Reynolds number corrections could still be applied at high Mach numbers. However, it was necessary to define a different Reynolds number. He recommended δ^+, the Reynolds number based on the friction velocity, wall density, and boundary-layer thickness. Figure 8, taken from reference 30, shows the domain of importance for including low Reynolds number effects. For values of δ^+ below 3000, the effects become more important and, in particular, below 400, they are significant. Lines of constant Mach number indicate that low Reynolds number effects can become substantial at high Mach numbers even though the length Reynolds number is large.

Modeling for adverse pressure gradient flows at hypersonic Mach numbers is less well advanced because the data base is limited. Figure 9 presents a list of experiments and pertinent test variables. It represents a partial, but representative, list of benchmark flows available for model evaluation. Mach number is limited to 7 and the wall-to-total-temperature range is mostly adiabatic. For these representative flows, eddy viscosity models give reasonably accurate results. Typical comparisons between computations and experiments for the skin friction, taken from reference 19, are shown in figure 10. The C-S model with the pressure gradient correction (i.e., the p^+ term) and the higher-order eddy viscosity and Reynolds stress models all adequately predict the influence of pressure gradient over a wide range of Reynolds numbers.

Shock-Wave Boundary-Layer Interaction Flows

In this section we discuss several examples of shock-wave boundary-layer interaction flows, some of which are separated. Figure 11, taken from reference 21, summarizes the status of experiments and computation for a variety of compressible flows. We will discuss a limited number of these flows consisting of both supersonic and hypersonic cases.

Figure 12 illustrates the experimental geometry of two hypersonic flows to be discussed. The first is a Mach 7 flow about an ogive-cylinder geometry. Two subcases of this flow will be considered: the first is the flow over the clean body from the nose rearward, and the second consists of the shock-wave boundary-layer interaction on the cylinder produced by a $15°$ ring-shock generator. Figure 13 shows comparisons of predictions and measurements of surface properties for the clean-body case (ref. 15). Three predictions are shown, one corresponding to laminar flow and the other two corresponding to turbulent flow obtained using the C-S and $q - \omega, a$ models. Transition was enforced in the modeling at a location about 10 cm from the nose. It is apparent that both turbulence models accurately predict surface pressure, skin friction, and heat transfer (Stanton number) distributions.

Results of predictions and measurements of the shock-wave boundary-layer interaction flow on the cylinder are shown in figure 14. In this case, measurements of surface-pressure, skin-friction, and heat-transfer are compared with predictions made with the C-S, B-L, and the three versions of the $q - \omega$ model. It is apparent from these results that both the zero-equation models and the unmodified $q - \omega, a$ models strongly overpredict the peak pressure in the interaction (fig. 14a). This basically is the result of the inability of the three models to adequately predict the extent of separation, which is indicated by the plateau in the measured pressure distribution ahead of the interaction. Substantial improvement was obtained with the $q - \omega$ model b, which incorporated the compressibility correction. Results obtained with the $q - \omega, c$ model, which incorporated the modification for heat transfer, were similar to those for model b.

Skin-friction and heat-transfer distributions for this case are shown in figures 14b and 14c. It is apparent from these results that although the two zero-equation models give reasonable predictions of peak

15

heating and skin-friction, their predictions in the region of separation are less accurate. Predictions made with the unmodified $q - \omega, a$ model indicate no separation at all and, as a result, grossly overpredict both peak heating and skin-friction. The computation made with the $q - \omega, b$ model shows an improvement in skin-friction prediction, but still strongly overpredicts peak heating. Finally, the prediction made with the $q - \omega, c$ model produced results that were in reasonably good agreement with the measurements.

Computed and measured pressure contours of the flow are shown in figures 15 and 16. Figure 15 shows measured pressure contours compared with contours obtained with the $q - \omega, c$ model. It is apparent that the overall features of the flow are well predicted by the model. Figure 16 compares predictions obtained with the $q - \omega, c$ and the B-L models. This result illustrates that the differences in model predictions of surface characteristics are accurately reflected in predictions of flow-field variables as well.

Calculations and measurements of the $7.5°$ shock generator case are also discussed in reference 15. In this case the flow is attached and the predictions of the two zero-equation models and the $q - \omega, c$ model give similar results that are in good agreement with the experiment.

The second flow is a compression corner flow, also illustrated in figure 12. Results of measurements and computation are shown in figures 17 and 18, corresponding to attached and separated cases, respectively (ref. 15). Calculations of surface pressure and heat transfer for the attached flow case (fig. 17), made with the C-S, B-L, and $q - \omega, c$ models, indicate reasonably good agreement between computation and experimentation. In the separated case (fig. 18) the predictions are also in reasonable agreement, although the pressure plateau and extent of separation predicted by the $q - \omega$ model is better than that predicted by the B-L model. In the reattachment zone, both models underpredict overshoots in measured pressure and heat-transfer distributions. In this case, computations with the C-S model were unreliable because of difficulties in computing boundary-layer and displacement thickness distributions, and are therefore not shown.

It should be noted that the modifications made to the $q - \omega$ model were general in the sense that no arbitrary constants were introduced and then adjusted to improve predictions. Furthermore, the modifications introduced did not interfere with or change predictions of simple attached flows (e.g., the clean-body flow). This is the type of modification one seeks when improving turbulence models for complex flows.

The final shock-wave boundary-layer interaction to be discussed is a Mach-3 compression corner flow illustrated in figure 19. Calculations of this flow with a corner angle of $20°$ are compared with results of the experiment in figure 20, and are discussed in greater detail in reference 7. The turbulence model used in this case was the Jones-Launder $k - \epsilon$ model. Two wall treatments were investigated, namely, the integration-to-the-wall and wall-function procedures. In this ($20°$) case, the flow was mildly separated. When predictions and measurements are compared, it is clear that noticeable differences in predictions result from different wall treatments using the same model. From both skin-friction and pressure distributions it is clear that the wall function treatment gives better predictions of separation and surface pressure. In addition, it also gives much better agreement with downstream skin-friction distributions than does the integration-to-the-wall procedure. Results similar to these were also observed in the $16°$ and $24°$ cases (ref. 7).

The primary reason for the differences between model predictions in this case lies in the low Reynolds number (damping) terms of the Jones-Launder model that strongly influence results when the integration-to-the-wall procedure is used, but that are inactive when the wall function procedure is used. Although the low Reynolds number terms can produce accurate results for zero pressure gradient attached flows, it is apparent that unless they are chosen carefully they can lead to unreliable predictions of complex flows.

Compressible Shear Layers

Knowledge of the physics of high-speed shear layers is limited at present. Experiments have shown that the far-field spreading angle, a measure of mixing, is reduced considerably, compared to that for incompressible flows. Figure 21 illustrates the status. The inverse of the spreading angles obtained from various experiments on single-stream mixing layers are shown as a function of Mach number. Reduction in spreading angle by a factor of 3 occurs at Mach 5. Various postulates to explain this reduced mixing have been proposed, but experimental evidence to substantiate them is lacking.

Application of incompressible turbulence models, extended to account for compressibility as described in previous sections, fails to produce accurate predictions of the spreading rate. The line labeled $k - \epsilon$ represents such a prediction. In reference 31, Dash et al. proposed a new model to account for compressibility in free shear layers and predicted the spreading angle as shown by the symbols labeled $k - \epsilon, cc$. The compressibility correction (cc) involved the empirical function $K(M_\tau)$, shown in figure 22, where M_τ is the ratio of the square root of the turbulent kinetic energy to the speed of sound, $k^{1/2}/a$, at the point of maximum k in the layer.

It is noteworthy that the modification also gives reasonable predictions for two-stream supersonic mixing. Figure 23 shows a comparison of $k - \epsilon$ model predictions compared with the measurements of Chinzie (ref. 32). The inverse of the two-stream spreading rate, scaled by the spreading rate σ_0 for which one stream is stationary, compares reasonably well with the $k - \epsilon, cc$ model prediction. Deviation of the data from the modified model prediction for the higher second-stream Mach numbers may be a result of free-stream turbulence present in the experiment.

Vandromme (ref. 17) also reported successful predictions of the single-stream spreading rate. As mentioned in the section on modeling, he used the ideas of Rubesin to make compressibility corrections to the turbulent kinetic energy and dissipation equations of the $k - \epsilon$ model. The results of his predictions are shown compared with results of experiments in figure 24. Substantial agreement was achieved.

More work will be necessary before the compressible mixing layer problem can be considered solved. Current modeling modifications are, to a considerable extent, ad hoc and have not been verified for a wide range of cases. Futhermore, they are not based on an understanding of the physical mechanisms involved. To understand these mechanisms, more experimentation is needed. It should also be noted that research is underway at Ames Research Center to use full simulations of compressible shear layers using the time-dependent Navier-Stokes equations to provide more complete information on mixing phenomena. It is hoped that this research will lead to improved modeling of compressible shear flows.

CONCLUDING REMARKS

In the preceding paragraphs we have described the development and status of turbulence models used in the numerical simulation of complex hypersonic flows. In our discussion we emphasized eddy viscosity models which constitute the simplest but most widely used class of turbulence models. Two subgroups of models were discussed– zero-equation and two-equation models. Each of these models has theoretical advantages over the other. For example, two-equation models provide a more general specification of turbulent length and velocity scales than zero-equation models, but they often display numerical stability problems which are not common to zero-equation models.

The basic models discussed are similar to those developed originally for incompressible flow. This is because in many applications, especially to simple attached boundary-layer flows at low to moderate supersonic speeds, the incompressible forms give satisfactory results. As discussed in the text, however, there is evidence that these incompressible forms become unsatisfactory as the flow complexity and/or the Mach number increase. With respect to flow complexity, it was shown that compressibility corrections were necessary to give satisfactory predictions of several hypersonic shock-wave boundary-layer interaction flows. With respect to Mach number, it was shown that incompressible model forms are unsatisfactory for compressible free-shear flows. In this case, too, compressibility corrections could be found which lead to satisfactory predictions.

The status of turbulence modeling for hypersonic flows is still far from complete, however. More experimental data and computational comparisons will be necessary to verify and establish the compressibility corrections made to date. In addition, more experimental and computational work will be needed, especially at low Reynolds numbers, because this flow regime is more prevalent at hypersonic speeds, and because the available data base in this case is still quite limited.

REFERENCES

1. Rubesin, M. W.; and Rose, W. C.: The Turbulent Mean Flow, Reynolds-Stress, and Heat-Flux Equations in Mass-Averaged Dependent Variables, NASA TM X-62248, 1973.

2. Cebeci, T. and Smith, A. M. O.: Analysis of Turbulent Boundary Layers. Academic Press, 1974.

3. Prandtl, L.: The Mechanics of Viscous Fluids, in Aerodynamic Theory, Vol. III, Durand, W. F. (Ed.), Springer-Verlag, Pasadena, CA, 1943.

4. Clauser, F. H.: The Turbulent Boundary Layer. Advances in Applied Mechanics, Vol. IV, Academic Press, 1956.

5. van Driest, E. R.: On Turbulent Flow near a Wall. J. Aeronaut. Sci., vol. 23, 1956, pp. 1007-1011.

6. Baldwin, B. S.; and Lomax, H.: Thin Layer Approximation and Algebraic Model for Separated Turbulent Flows. AIAA Paper 78-257, Jan. 1978.

7. Viegas, J. R.; Rubesin, M. W.; and Horstman, C. C.: On the Use of Wall Functions as Boundary Conditions for Two-Dimensional Separated Compressible Flows. AIAA Paper 85-0180, Jan. 1985.

8. Launder, B. E.; and Spalding, D. B.: Mathematical Models of Turbulence. Academic Press, 1972.

9. Jones, W. P.; and Launder, B. E.: The Prediction of Laminarization with a Two-Equation Model of Turbulence. Intern. J. Heat Mass Transfer, vol. 15, 1972, pp. 301-304.

10. Chien, K. Y.: Predictions of Channel Boundary-Layer Flows with a Low-Reynolds-Number Turbulence Model. AIAA J., vol. 20, Jan. 1982, pp. 33-38.

11. Wilcox, D. C.; and Rubesin, M. W.: Progress in Turbulence Modeling for Complex Flow Fields Including the Effects of Compressibility. NASA TP-1517, 1980.

12. Coakley, T. J.: Turbulence Modeling Methods for the Compressible Navier-Stokes Equations. AIAA Paper 83-1693, July 1983.

13. Coakley, T. J.: A Compressible Navier-Stokes Code for Turbulent Flow Modeling. NASA TM-85899, 1984.

14. Coakley, T. J.; and Hsieh, T.: Comparison between Implicit and Hybrid Methods for the Calculation of Steady and Unsteady Inlet Flows, AIAA Paper 85-1125, July 1985.

15. Voung, S. T.; and Coakley, T. J.: Modeling of Turbulence for Hypersonic Flows with and without Separation. AIAA Paper 87-0286, Jan. 1987.

16. Morel, T.; and Mansour, N. N.: Modeling of Turbulence in Internal Combustion Engines. SAE Technical Paper Series 820040, Feb. 1982.

17. Vandromme, D.: Contribution to the Modeling and Prediction of Variable Density Flows. Ph.D. thesis presented at the University of Science and Technology, Lille, France, 1983.

18. Kline, S. J.; Cantwell, B. J.; and Lilley, G. M.: Proceeding of the 1980-81 AFOSR-HTTM Stanford Conference on Complex Turbulent Flow, Stanford University, Stanford, Calif., 1981.

19. Marvin, J. G.: Turbulence Modeling for Computational Aerodynamics. AIAA J., vol. 21, no. 21, July 1983, pp. 941-955.

20. Marvin, J. G.: Modeling of Turbulent Separated Flows for Aerodynamic Applications. Recent Advances in Aerodynamics, Springer-Verlag, ed. by A. Krothapalli and C. A. Smith, Proceedings of an International Symposium held at Stanford University, Aug. 22-26, 1983.

21. Delery, J.; and Marvin, J. G.: Shock-Wave Boundary Layer Interactions. AGARDograph No. 280, Feb. 1986.

22. van Driest, E. R.: Problem of Aerodynamic Heating: Aeronaut. Engin. Rev., vol. 15, no. 10, Oct. 1956, pp. 26-41.

23. Sandborn, V. A.: A Review of Turbulence Measurements in Compressible Flow. NASA TM X-62337, 1974.

24. Maise, G.; and McDonald, H.: Mixing Length and Kinematic Eddy Viscosity in a Compressible Boundary Layer. AIAA J., vol. 6, no. 1, Jan. 1968, pp. 73-79.

25. Hopkins, E. J.; and Inouye, M.: An Evaluation of Theories for Predicting Turbulent Skin Friction and Heat Transfer on Flat Plates at Supersonic and Hypersonic Mach Numbers. AIAA J., vol. 9, no. 6, Jun. 1971, pp. 993-1003.

26. Shang, J. S.: Computation of Hypersonic Turbulent Boundary Layers with Heat Transfer. AIAA J., vol. 12, no. 7, July 1974, pp. 883-884.

27. McDonald, H.: Mixing Length and Kinematic Eddy Viscosity in Low Reynolds Number Boundary Layer. Rep. J214453-1, Res. Lab., United Aircraft Corp., Sept. 1970.

28. Bushnell, D. M.; and Morris, D. J.: Shear-Stress, Eddy-Viscosity, and Mixing-Length Distributions in Hypersonic Turbulent Boundary Layers. NASA TM X-2310, 1971.

29. Bushnell, D. M.; Cary, A. M.; Jr. and Holley, B. B.: Mixing Length in Low Reynolds Number Compressible Turbulent Boundary Layers. AIAA TN, AIAA J., vol. 13, no. 8, Aug. 1975, pp. 1119-1121.

30. Bushnell, D. M.; Cary, A. M., Jr.; and Harris, J. E.: Calculation Methods for Compressible Turbulent Boundary Layers-1976. NASA SP-422, 1977.

31. Dash, S.; Weilerstein, G.; and Vaglio-Laurin, R.: Compressibility Effects in Free Turbulent Shear Flows. AFOSR TR-75-436, Aug. 1975.

32. Chinzie, N.; Masuya, G.; Komuro, T.; Murakami, A.; and Kudou, K.: Spreading of Two-Stream Supersonic Mixing Layer. Phys. Fluids, vol. 29, no. 5, May 1986.

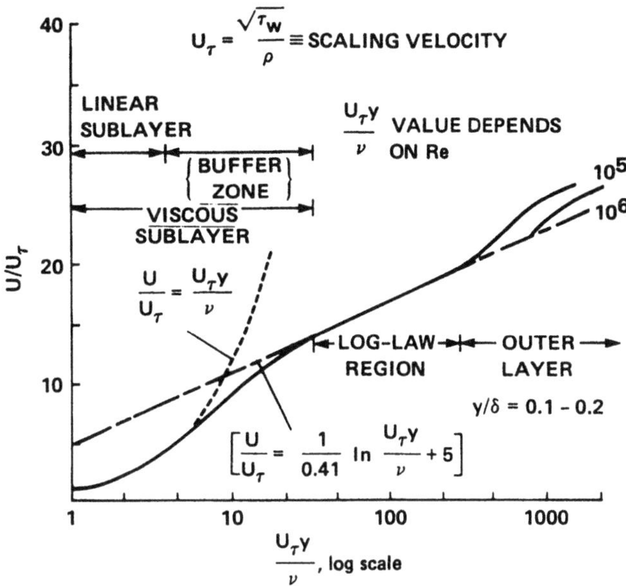

Fig. 1. Composite sketch of a turbulent boundary layer.

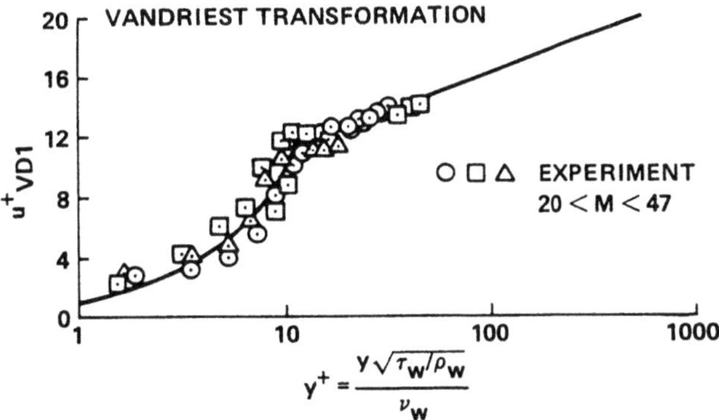

Fig. 2. A log-law representation of hypersonic boundary layer profiles.

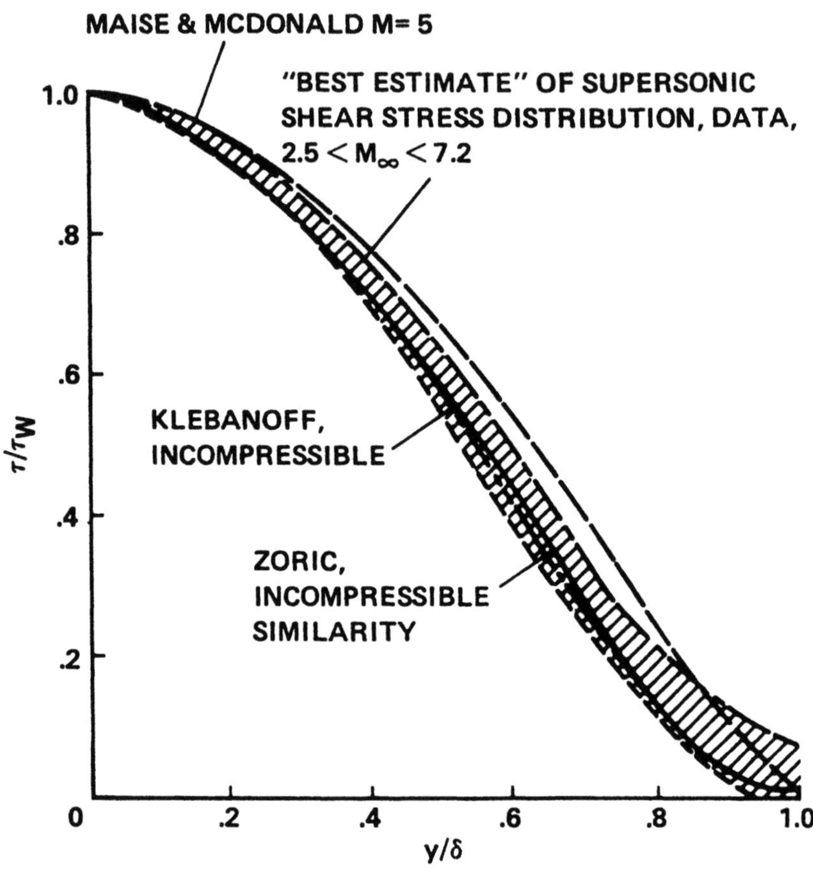

Fig. 3. A comparison of the best estimate of shear stress data with incompressible measurements and with Maise and McDonald's compressible approximation.

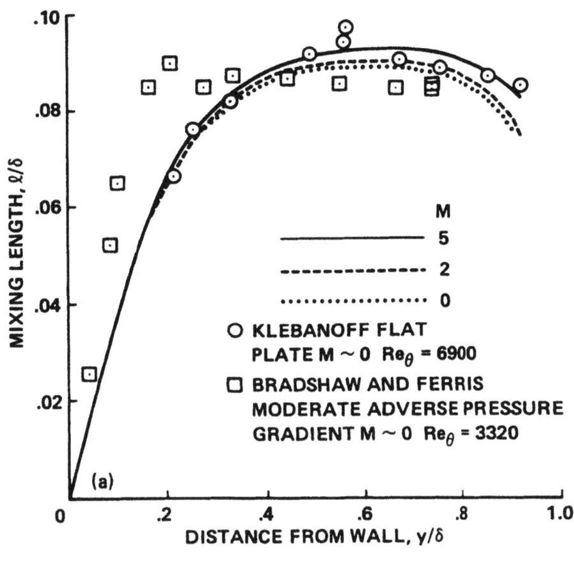

(a) Mixing length.

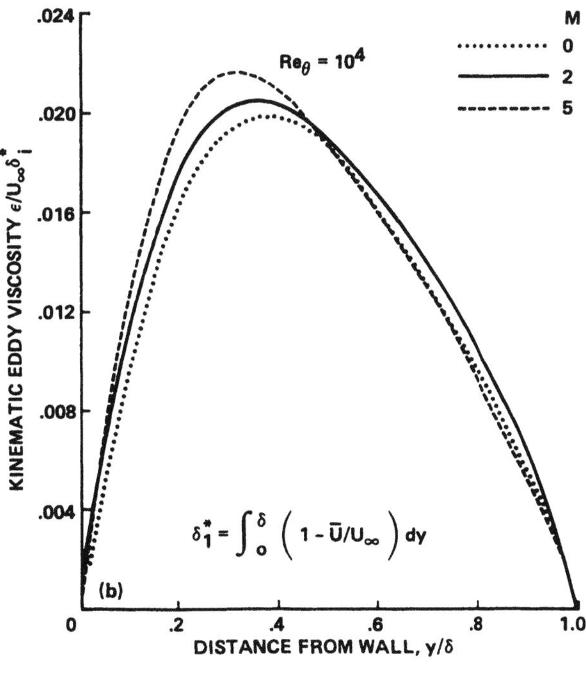

(b) Scaled eddy viscosity.

Fig. 4. Mixing length and scaled eddy viscosity from Maise and McDonald.

23

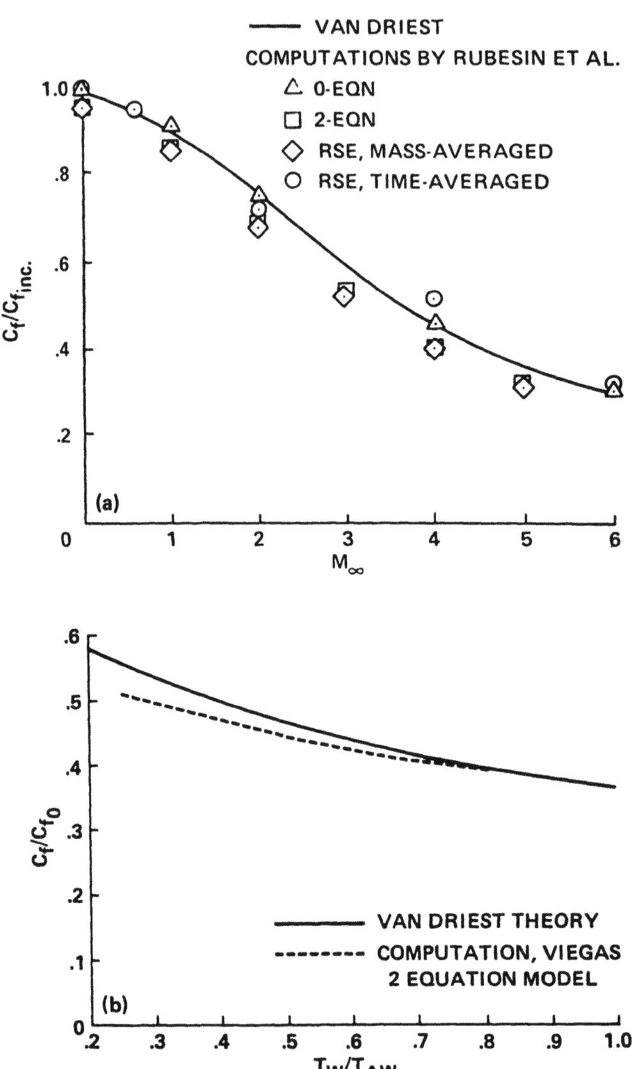

Fig. 5. Ability of turbulence models to predict compressibility effects: (a) adiabatic wall temperature; (b) Mach 5 and variable wall temperature.

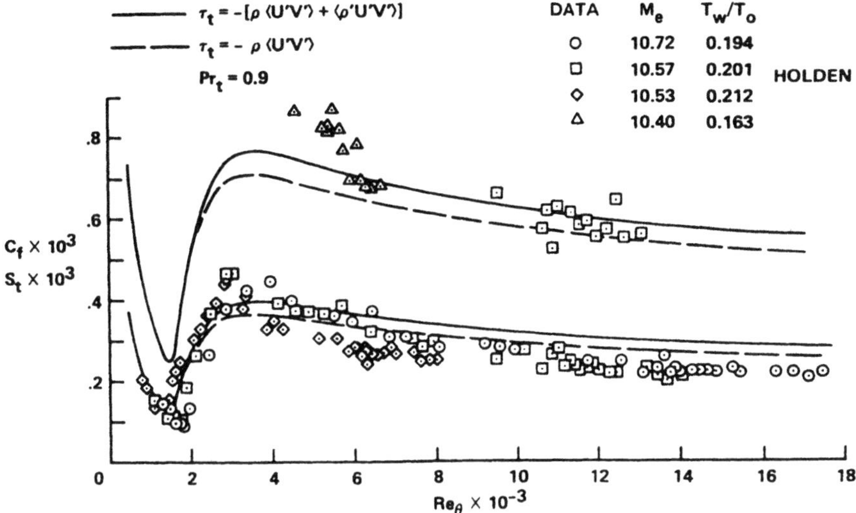

Fig. 6. A comparison of experimental skin friction and heat transfer with computation using a turbulence model with and without corrections for density fluctuations.

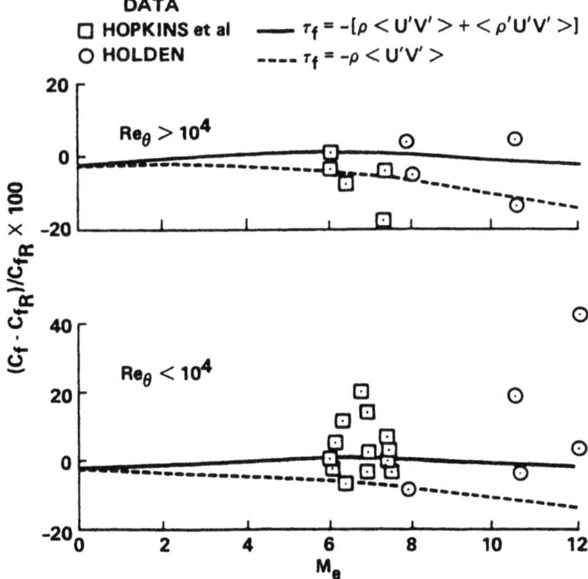

Fig. 7. Error in skin friction prediction for two ranges of Reynolds numbers using a turbulence model with and without correction for density fluctuations.

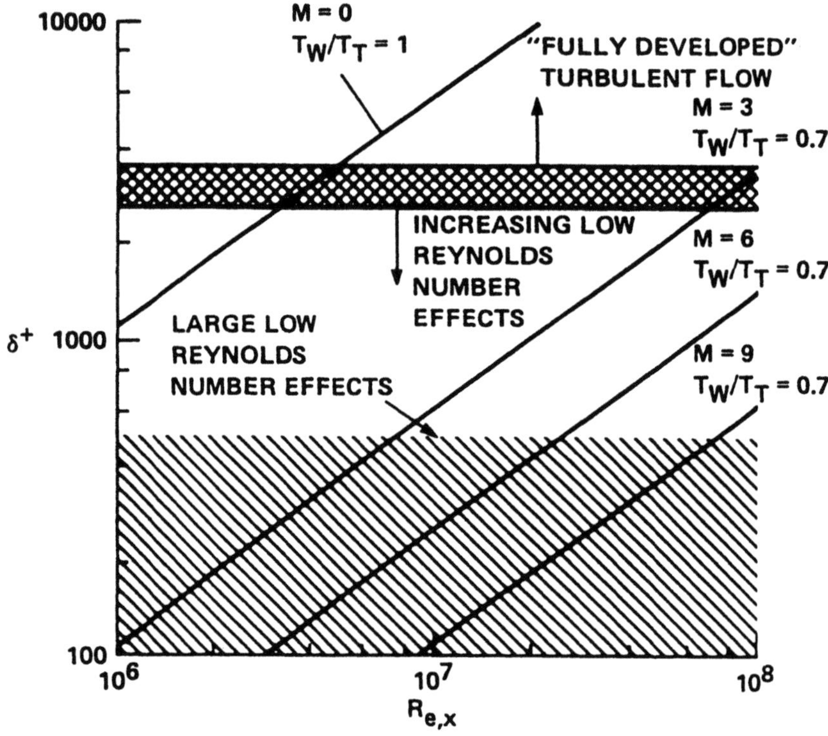

Fig. 8. Increasing importance of Low Reynolds number effects with Mach number.

2-D ATTACHED – PRESSURE GRADIENT

REFERENCE	EXPERIMENTAL CONFIGURATION	M_∞	$Re_{\theta_\infty} \times 10^{-4}$	T_w/T_o	P^+_{max}
ZWARTS	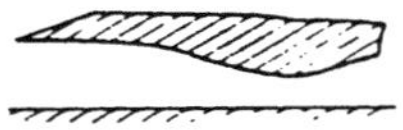	4.02	3.5	1	0.004
PEAKE, BRAKMANN AND ROMESKIE	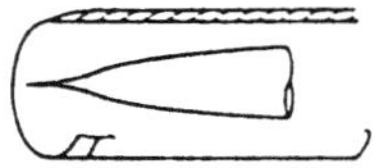	3.93	1.1	1	0.006
STUREK AND DANBERG		3.54	2.0-2.8	1	0.0085-0.0125
LEWIS, GRAN AND KUBOTA	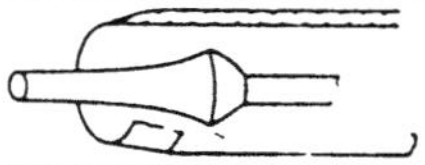	3.98	0.5	1	0.011
KUSSOY AND HORSTMAN	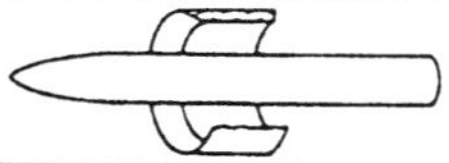	6.7	0.8	0.43	0.07
KUSSOY AND HORSTMAN	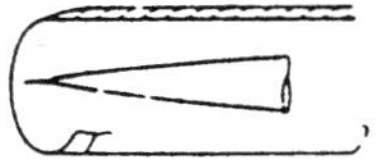	2.3	104-2270	1	0.12

Fig. 9. Benchmark flows with pressure gradient effects. See reference 19.

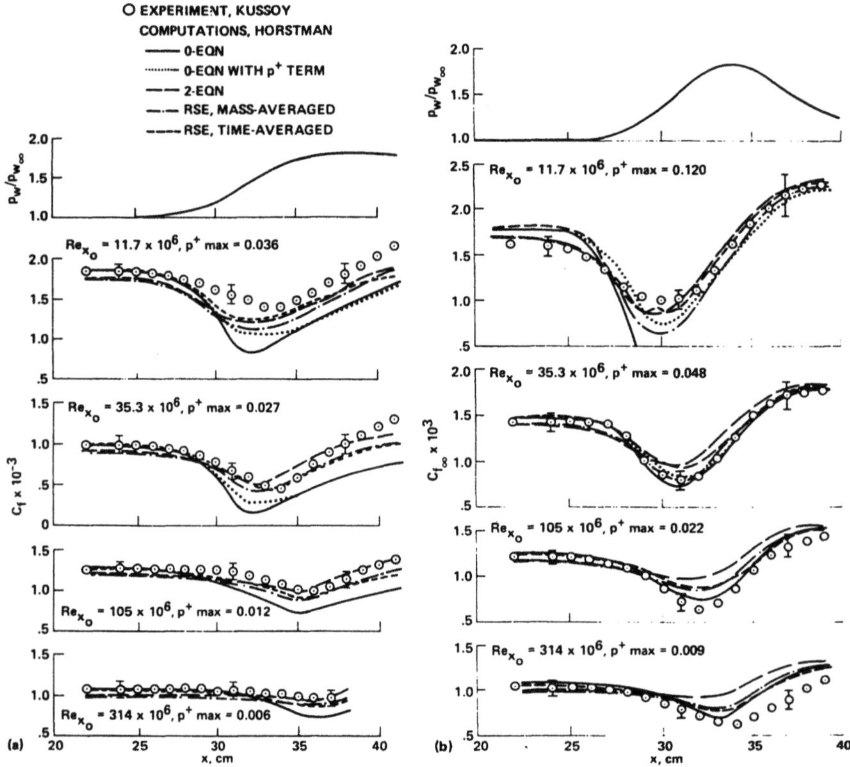

Fig. 10. A comparison of computations using various models with experiment.

28

FLOW	EXPERIMENTS	M_∞	$Re \times 10^{-6}$	COMPUTATIONS	GRID x, y	MODEL[a]
2-d M > 1 M < 1 STRAIGHT WALLS	SEDDON (1960)	1.5	0.6	VIEGAS AND HORSTMAN (1979)	40, 35	0; 1; 2
				VIEGAS AND RUBESIN (1983)	20, 35	2 (MOD.)
	MATEER et al. (1976, 1979)[b]	1.3-1.5	10-200	VIEGAS AND HORSTMAN (1979)	40, 35	0; 1; 2
				VIEGAS AND HORSTMAN (1982)	38, 40	2
					76, 40	2
				VIEGAS et al. (1983)	38, 20	2 (MOD.)
	OM et al. (1982)	1.3-1.5	0.5-1	McDONALD (1982)	31, 41	2
				OM et al. (1982)	170, 40	2
				VIEGAS et al. (1985)	85, 20	2 (MOD.)
2-d M > 1 M < 1 CURVED/DIVERGED WALL(S)	DELERY (1983)[b]	1.3-1.5	2-4	CAMBIER et al. (1981)	61, 21	0
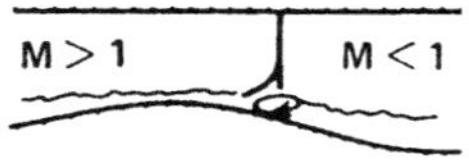				HORSTMAN (PVT. COMM.)		0; 2
	SALMON et al. (1981)	1.3-1.4	4	HA MINH et al. (1985)	79, 39	0; 2; RSE
				LIOU et al. (1981)	80, 50	2

[a]ZERO-, ONE-, TWO- AND REYNOLDS STRESS EQUATION MODELS

[b]SELECTED FOR AFOSR/HTTM STANFORD CONFERENCE (KLINE et al., 1981)

(a) Transonic flows with impinging shock waves.

Fig. 11. Benchmark flows with separation. Taken from reference 20.

FLOW	EXPERIMENTS	M_∞	$Re \times 10^{-6}$	COMPUTATIONS	GRID x, y, ϕ	MODEL[a]
2-d 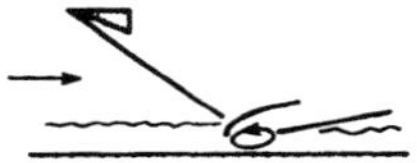	HOLDEN (1972)	8.5	22	BALDWIN AND MacCORMACK (1974)	40, 32	0; 2
	REDDA AND MURPHY (1973)	3	57	BALDWIN AND ROSE (1975)	40, 32	0 (MOD.)
				BALDWIN AND LOMAX (1978)	40, 32	0 (MOD.)
				VIEGAS AND HORSTMAN (1979)	40, 35	0; 2
	KUSSOY et al. (1975)[b]	7.2	13	MARVIN et al. (1975)	40, 78	0; 0 (MOD.)
				COAKLEY et al. (1977)	29, 45	0; 1; 2
				VIEGAS AND HORSTMAN (1982)	89, 50	2
				COAKLEY (PVT. COMM.)	81, 81	0; 2; 2 (MOD.)
3-d 	KUSSOY et al. (1980)[b]	2.2	36	KUSSOY et al. (1980)	47, 20, 20	0
				VIEGAS AND HORSTMAN (1982)	47, 20, 20	0; 2
3-d 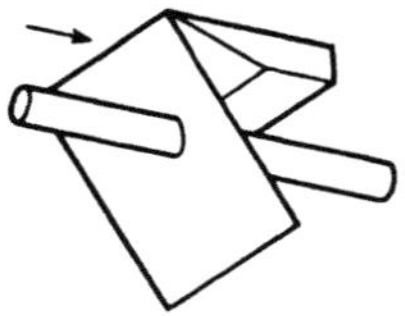	BROSH et al. (1983)	3	18	BROSH et al. (1983)	45, 34, 38	0 (MOD.)

[a]ZERO-, ONE-, AND TWO-EQUATION MODELS

[b]SELECTED FOR AFOSR/HTTM STANFORD CONFERENCE (KLINE et al., 1981)

(b) Supersonic flows with impinging shock waves.

Figure 11. Continued.

FLOW	EXPERIMENTS	M_∞	Re x 10^{-6}	COMPUTATIONS	GRID x, y, z	MODEL[a]
2-d 	SETTLES et al. (1976, 1979)[b]	2.9	63-200	HORSTMAN et al. (1977)	50, 32	0; 1
				VIEGAS AND HORSTMAN (1979)	50, 32	0; 1; 2
				VIEGAS AND HORSTMAN (1982)	80, 36	2
					160, 40	2
				VIEGAS et al. (1985)	80, 25	2 (MOD.)
				HUNG (1982)	77, 46	0 (MOD.)
	HOLDEN (1972)	8.7	22	HUNG AND MacCORMACK (1977)	63, 31	0; 0 (MOD.)
	LAW (1974)	3	10	SHANG AND HANKEY (1975)	64, 30	0; 0 (MOD.)
				HUNG AND MacCORMACK (1977)	63, 31	0; 0 (MOD.)
	DUNAGAN et al. (1985)	3	18	HORSTMAN (1985)	129, 45	2
	BROWN et al. (1985)	3	18	BROWN et al. (1985)	95, 95	2
3-d 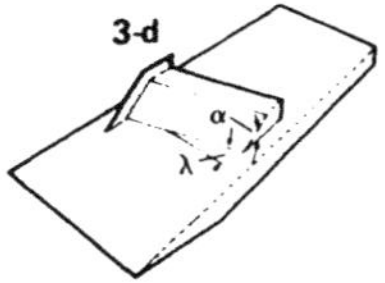	TENG AND SETTLES (1982)	2.9	6	SETTLES AND HORSTMAN (1984)	40, 35, 20	0; 2
				HORSTMAN (1984)	40, 35, 20	2
				HORSTMAN (1985)	40, 26, 27	2 (MOD.)
	KUSSOY AND LOCKMAN (1985)	3	18	HORSTMAN (1985)	64, 33, 38	2 (MOD.)
	BROWN (1986)	3	18	HORSTMAN (1986)	64, 33, 38	2 (MOD.)

[a]ZERO-, ONE-, AND TWO-EQUATION MODELS

[b]SELECTED FOR AFOSR/HTTM STANFORD CONFERENCE (KLINE et al., 1981)

(c) Supersonic flows with compression surfaces.

Figure 11. Continued.

31

FLOW	EXPERIMENTS	M_∞	Re x 10^{-6}	COMPUTATIONS	GRID x, y, z	MODEL[a]
3-d 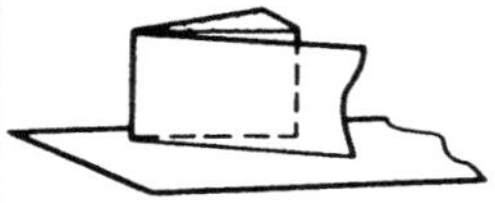	PEAKE (1976)[b]	2	10	HORSTMAN AND HUNG (1979)	21, 28, 31	0 (MOD.)
				VIEGAS AND HORSTMAN (1982)	21, 28, 31	0 (MOD.)
				HUNG AND MacCORMACK (1978)	32, 36, 32	0 (MOD.)
	OSKAM et al. (1975)	3	5	HORSTMAN AND HUNG (1979)	21, 28, 31	0 (MOD.)
	WEST et al. (1972)	3	0.8	SHANG et al. (1979)	17, 33, 33	0 (MOD.)
	GOODWIN et al. (1984)	3	10	KNIGHT AND HORSTMAN (1986)	32, 32, 48	0 (MOD.)
	SHAPEY et al. (1986)	3			64, 32, 44	2 (MOD.)
3-d 	DOLLING AND BOGDONOFF (1982)	3	0.8	HUNG AND KORDULLA (1983)	40, 32, 32	0 (MOD.)

[a]ZERO-, ONE-, AND TWO-EQUATION MODELS

[b]SELECTED FOR AFOSR/HTTM STANFORD CONFERENCE (KLINE et al., 1981)

(d) Supersonic flows with glancing shock waves.

Figure 11. Concluded.

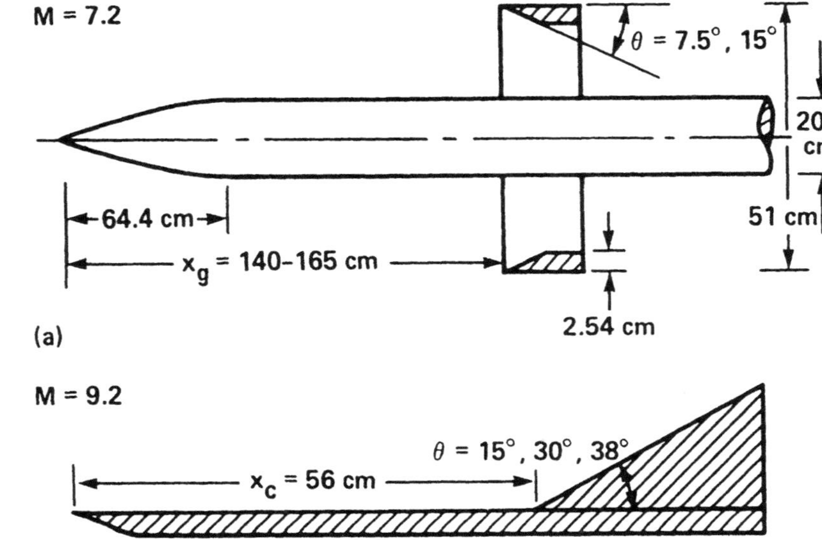

Fig. 12. Experimental arrangements for two flows used to assess models with compressibility corrections.

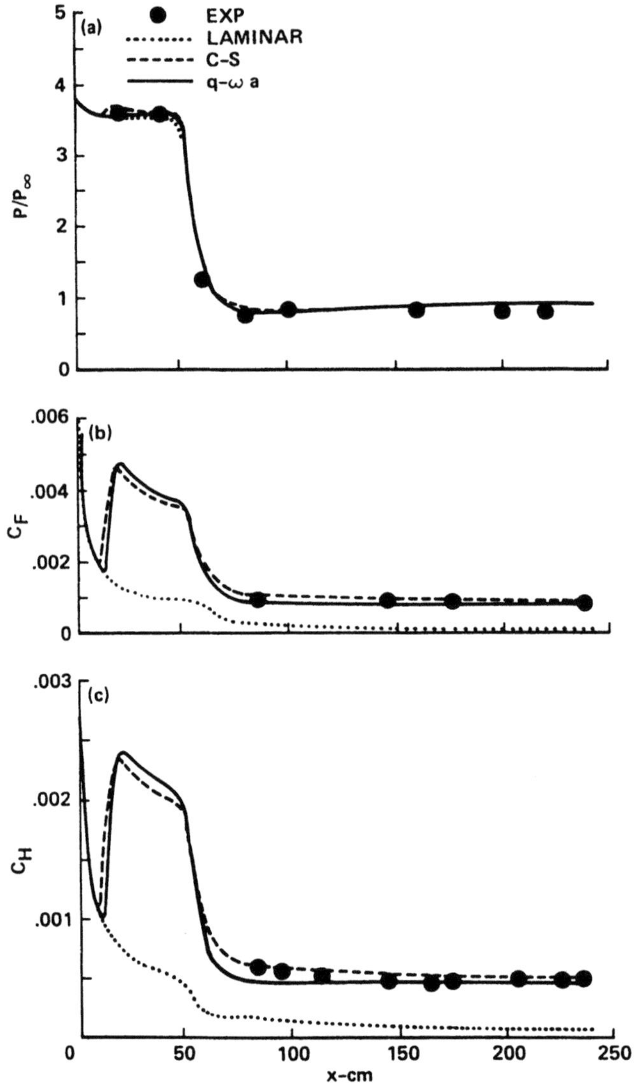

Fig. 13. Ogive-cylinder flow without the shock generator ring: (a) surface pressure;
(b) skin friction; (c) surface heat transfer.

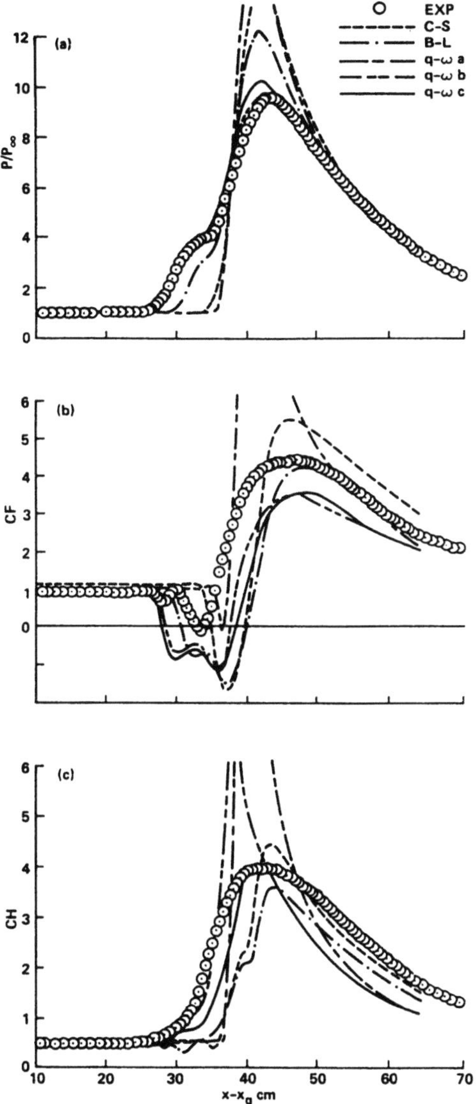

Fig. 14. Ogive-cylinder flow with 15^0 shock-generator ring: (a) surface pressure; (b) skin friction; (c) surface heat transfer.

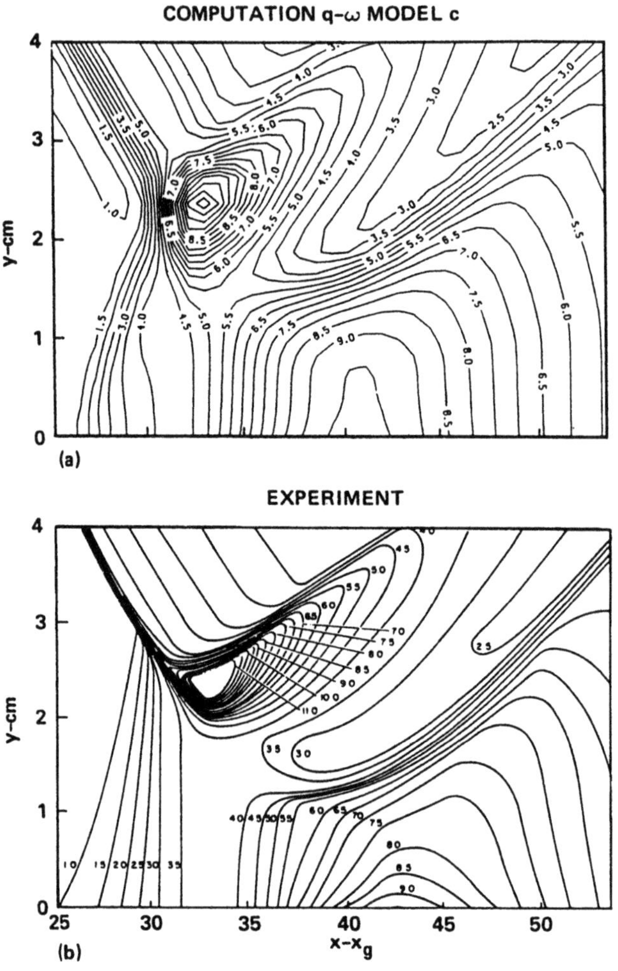

Fig. 15. Ogive-cylinder flow with 15^0 shock generator ring: (a) computed pressure contours; (b) experimental pressure contours.

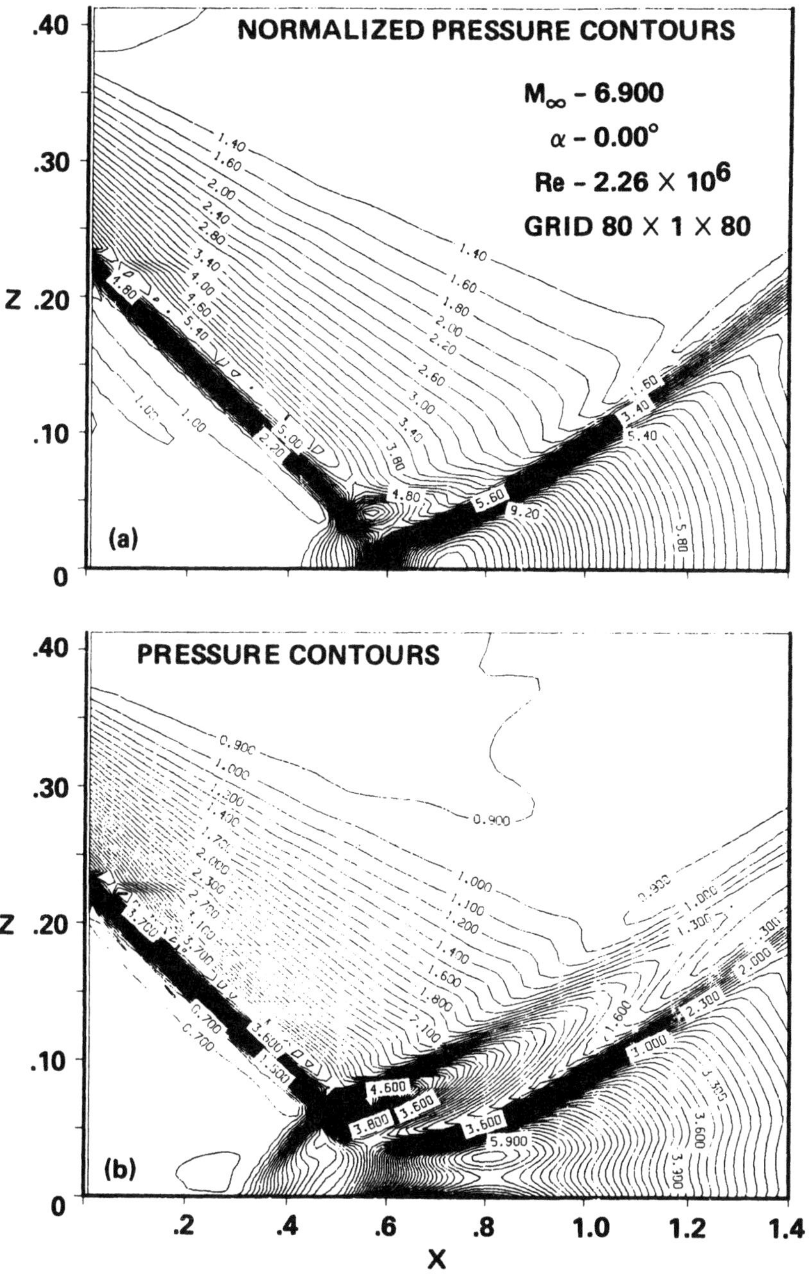

Fig. 16. Ogive-cylinder flow with 15⁰ shock generator ring: (a) zero-equation Baldwin-Lomax model; (b) $q - \omega$ model c

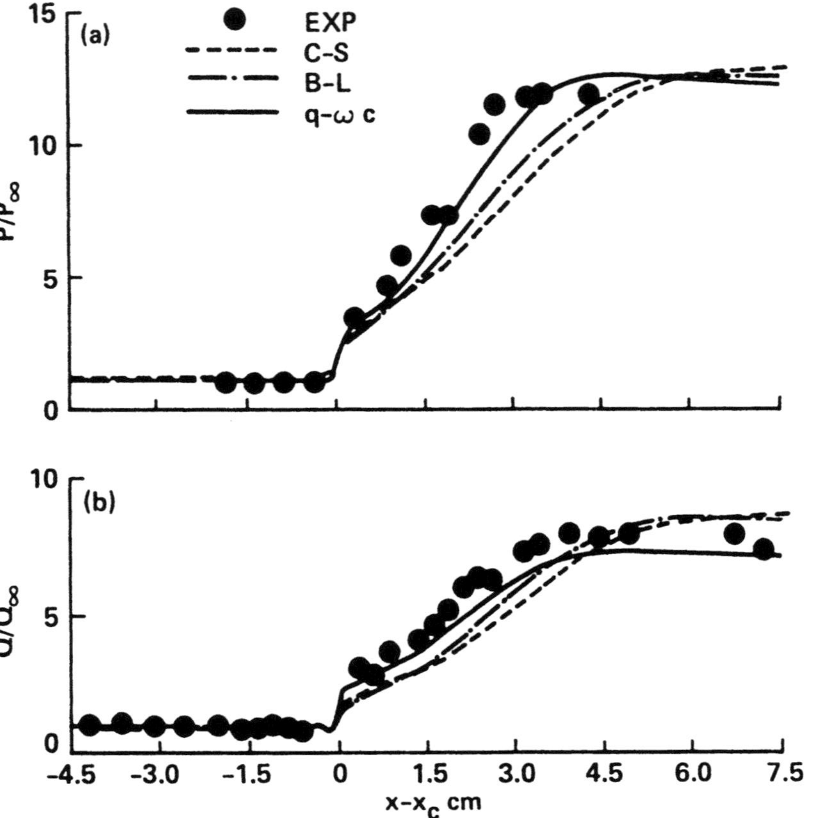

Fig. 17. Compression Corner Flow, 15º corner angle (a) surface pressure; (b) surface heat transfer.

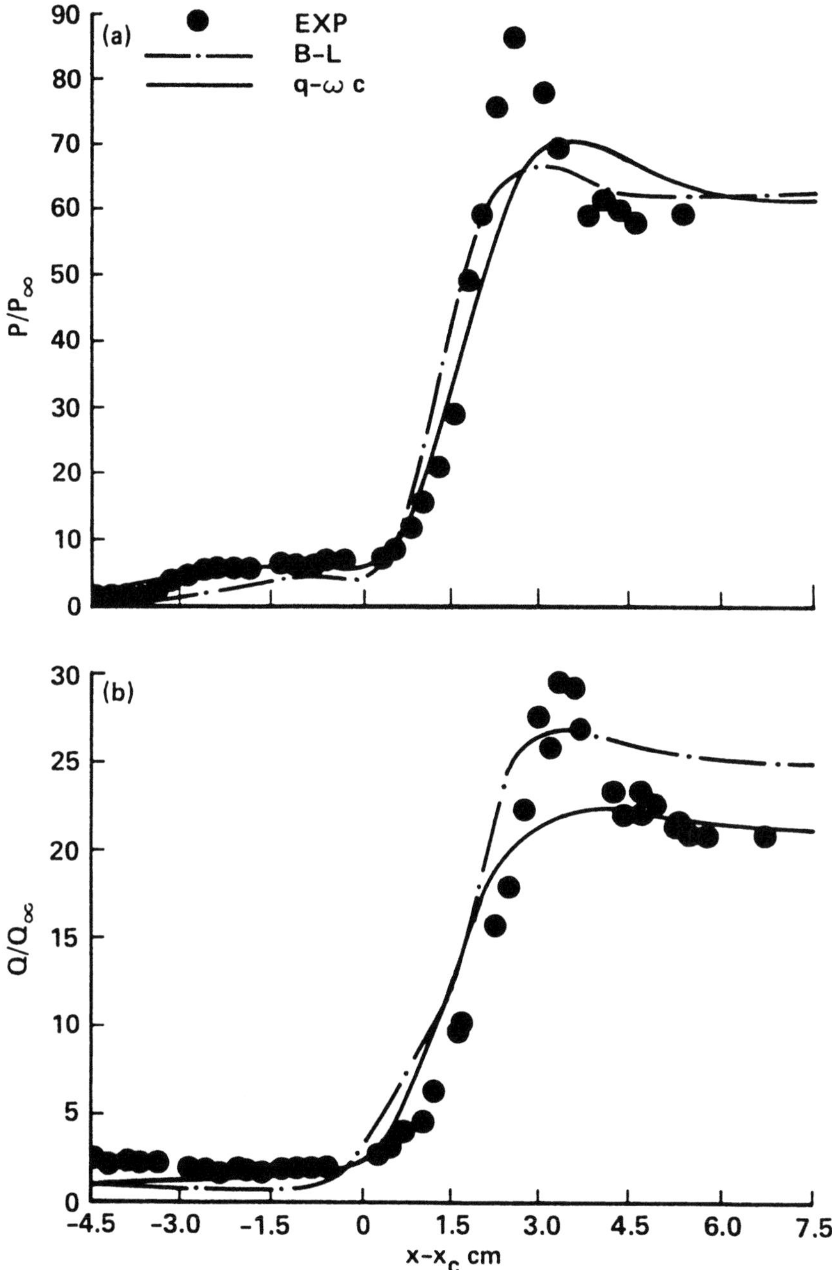

Fig. 18. Compression Corner Flow, 38^0 corner angle: (a) surface pressure; (b) surface heat transfer.

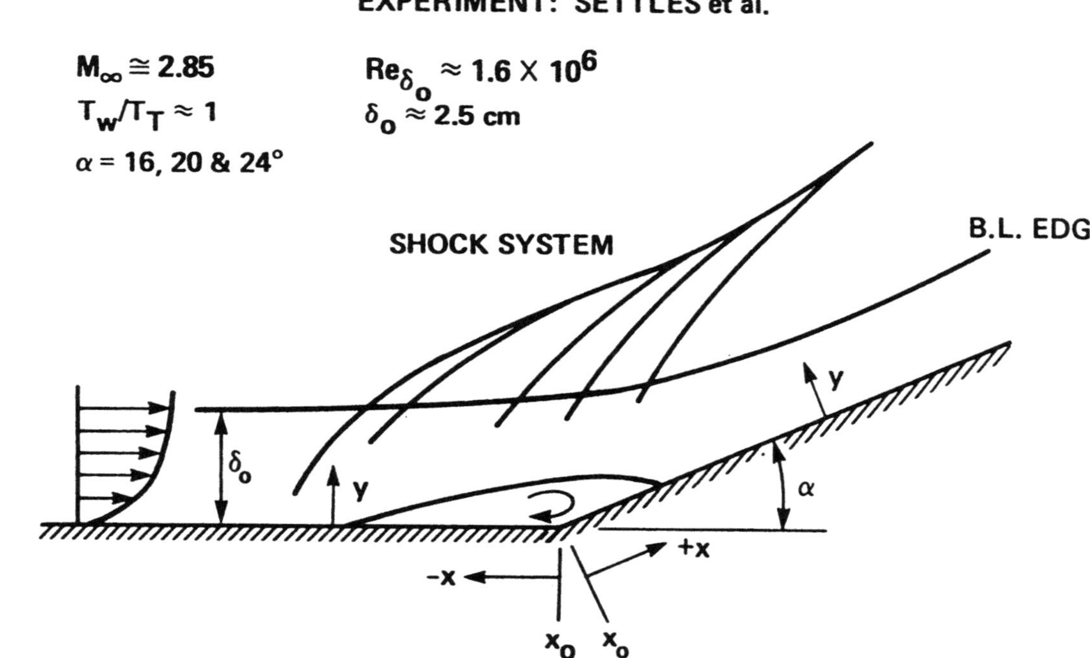

Fig. 19. Geometry and conditions of a compression corner experiment.

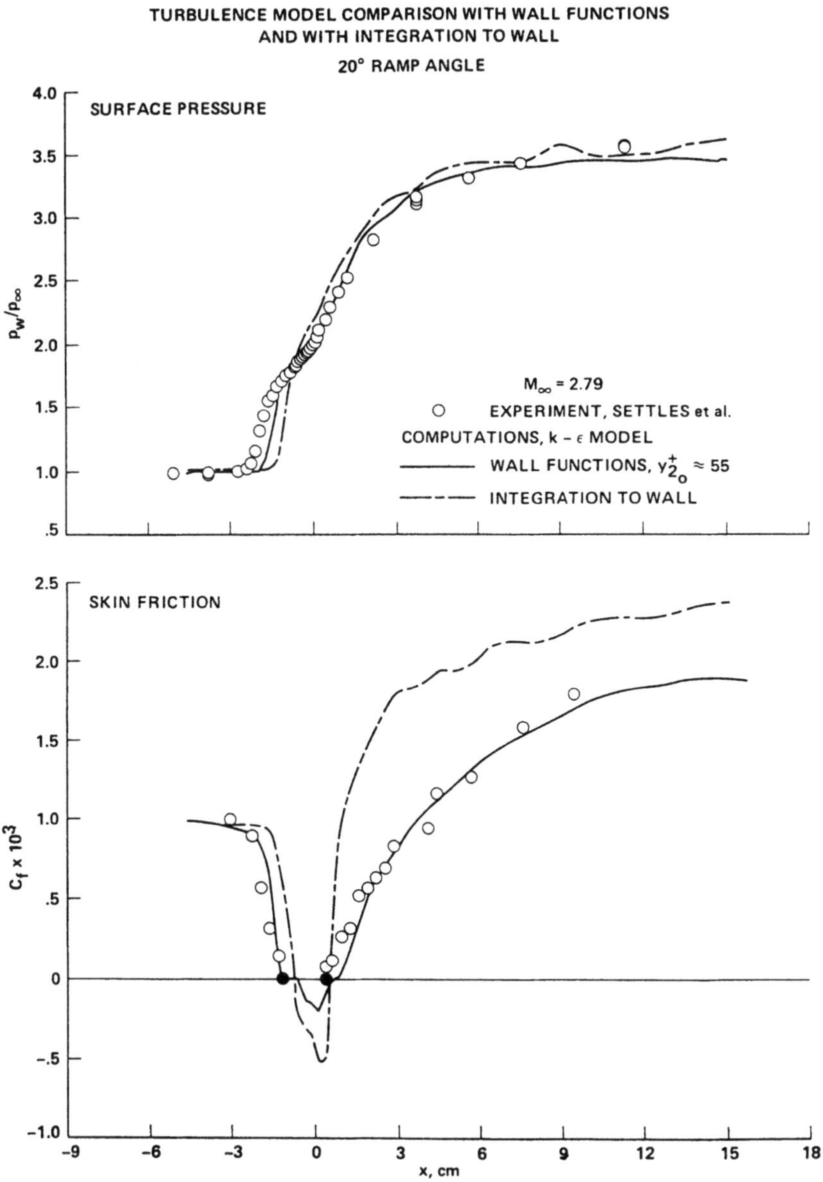

Fig. 20. Comparison of computations using $k - \epsilon$ model with experiment.

41

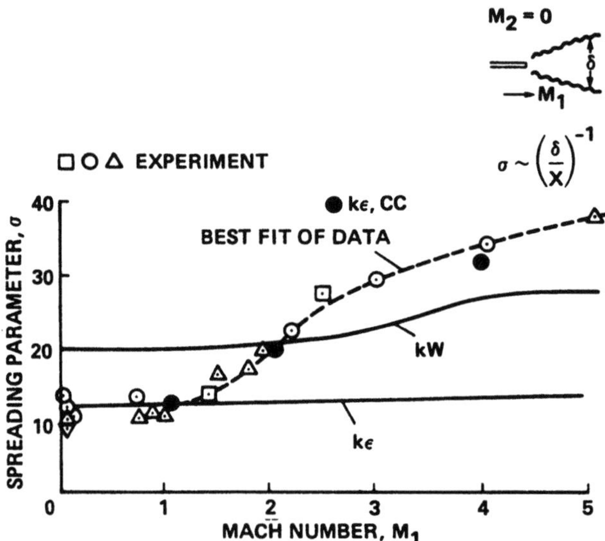

Fig. 21. Spreading rates for compressible 2-D shear layers. Data fits from NASA SP-321, 1972.

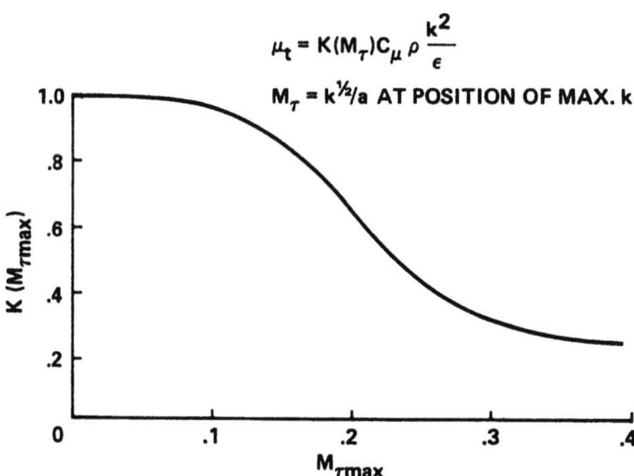

Fig. 22. Compressibility correction for $k - \epsilon$ model proposed by Dash, et. al[31].

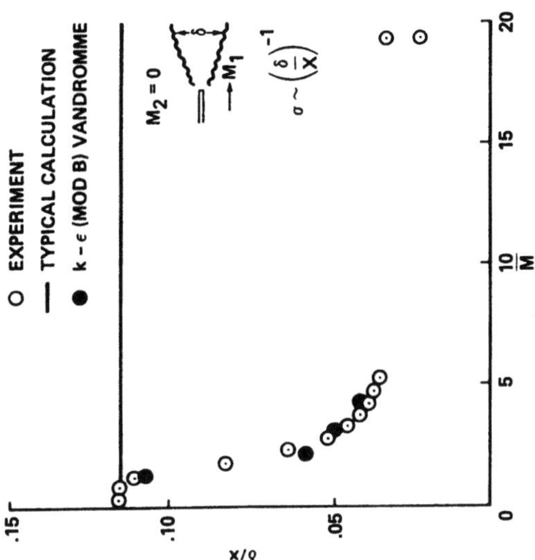

Fig. 24. Spreading rates for compressible 2-D shear layers.

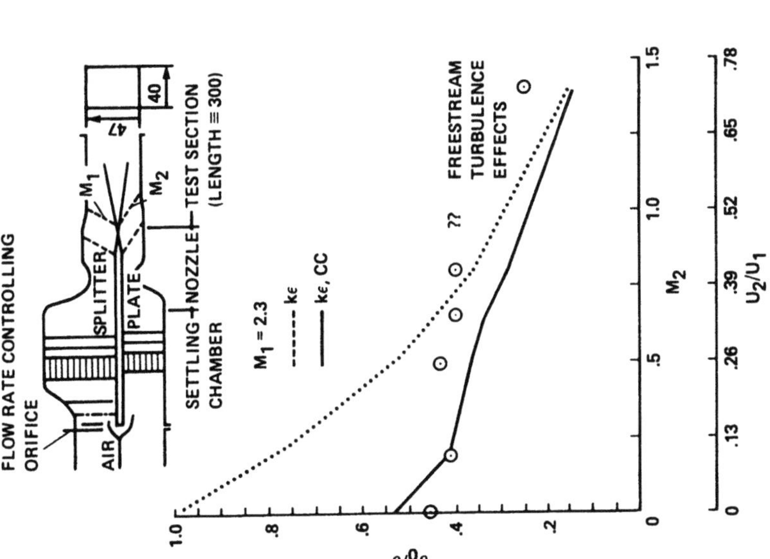

Fig. 23. Spreading rates for a two-stream compressible 2-D shear layer.

43

ADVANCED TOPICS IN TURBULENCE THEORY

Marcel **Lesieur**

Institut de Mécanique de Grenoble (L.E.G.I.)*

Institut National Polytechnique de Grenoble
et
Université Joseph Fourier, Grenoble
BP 53 X - 38041 Grenoble-Cedex, France.

Presented at the

SECOND JOINT EUROPE/US SHORT COURSE IN

HYPERSONICS

Colorado Springs, 16-20 January 1989

---

* Unité associée CNRS

ABSTRACT

One presents the spectral two-point closure point of view of turbulence theory. In the three-dimensional incompressible isotropic case, it is shown how these closures lead, in the inviscid limit, to singularities of the vorticity field at a finite time. A theory of non local interactions is developed, which allows to derive a parameterization of the subgrid scales for large-eddy simulations.

Results of direct or large-eddy simulations utilizing the above subgrid-scale parameterization are presented for three-dimensional isotropic turbulence or for a temporally growing mixing layer forced initially with a random perturbation. Two-dimensional calculations of various incompressible and compressible free shear flows are also presented.

Finally, the role of compressibility on the dynamics of the coherent structures in free shear flows is examined.

1 Introduction

When studying fully-developed turbulence in an incompressible flow at high Reynolds number, it is of interest to make a distinction between the small-scale turbulence and the large-scale coherent structures, which are generally driven by instabilities of the mean velocity profile. When the latter is inflectional, such as in mixing layers (Brown and Roshko, 1974), jets or wakes, coherent structures are Kelvin-Helmholtz-like vortices. In wall-bounded flows, coherent structures at the wall are formed of "hairpin" vortices, inclined approximately 45^0 from the wall, and which lift slow fluid from the boundary, resulting into alternate longitudinal streaks of slow and fast fluid (Moin and Kim, 1982). In both cases, the coherent structures can be highly intermittent and unpredictable, as shown for instance numerically by Lesieur et al. (1988) in the mixing-layer case. Contrary to the coherent structures, the small-scale turbulence is generally not far (at least locally) from statistical homogeneity and isotropy, and displays at large wave numbers a $k^{-5/3}$ kinetic energy spectrum, extending sometimes quite close to the coherent structures dominant mode at low wave numbers.

Therefore, the concept of three-dimensional fully-developed isotropic turbulence provides a good starting point for the study of small-scale turbulence, even though it has no physical reality at all scales of motion.

As will be seen below, hypersonic compressibility effects may modify drastically the coherent structure geometry, enhancing the formation of longitudinal Λ-shaped vortices, both in free and wall-bounded turbulent shear flows. Compressibility might affect much less small-scale three-dimensional turbulence. In this respect, it is of interest for the study of hypersonic turbulence to develop theories of incompressible isotropic turbulence. These theories will help to understand the effects of non linearity, and to improve the subgrid-scale procedures for large-eddy numerical simulations, if the smallest resolved scale lies in the Kolmogorov $k^{-5/3}$ range.

In the second section, we will present the two-point closure statistical modelling of turbulence, focusing on the Eddy- Damped Quasi-Normal Markovian approximation (EDQNM). In three dimensions, this theory will be shown to yield enstrophy singularities in the inviscid limit. In the third section, we will present a theory of non local interactions, allowing to solve the kinetic energy and temperature fluctuations decay problem, and to provide spectral eddy-viscosities and conductivities for large-eddy simulations. In the fourth section, three-dimensional large-eddy simulations of incompressible isotropic turbulence and mixing-layers will be reported. Finally, the influence of compressibility on a three-dimensional mixing layer will be looked at in the fifth section.

2 Two-point statistical closures

2.1 The closure problem

When studying statistically non homogeneous flows, the closure problem of turbulence arises at the level of the Reynolds equations for the mean flow $< \underline{u} >$, where the Reynolds stresses $-\rho < u'_i u'_j >$ have to be evaluated in terms of the mean velocity. This is done for instance with the aid of an eddy-viscosity assumption, calculated either with the aid of the Prandtl mixing-length theory, or using methods such as the so-called $K - \epsilon$ modelling method. For statistically homogeneous turbulence whithout any mean velocity, the closure problem is posed when looking at the infinite hierarchy of the statistical moments. Here, we will work in Fourier space: let us write Navier-Stokes equation

$$(\frac{\partial}{\partial t} + \nu k^2)\hat{u}_i(\underline{k}, t) = -ik_m P_{ij}(\underline{k}) \int_{\vec{p}+\vec{q}=\vec{k}} \hat{u}_j(\underline{p}, t)\hat{u}_m(\underline{q}, t)d\underline{p}$$

whith $P_{ij}(\underline{k}) = \delta_{ij} - k_i k_j / k^2$, and where $\hat{u}_i(\underline{k}, t)$ is the Fourier transform of $u_i(\underline{x}, t)$. Symbolically, it can be written in the following manner

$$\frac{\partial \hat{u}(\underline{k})}{\partial t} = \hat{u}\hat{u} - \nu k^2 \hat{u}(\underline{k}) \qquad (2-1)$$

where the $\hat{u}\hat{u}$ term stands for the non linear terms (projection of $u_j \partial u_i / \partial x_j$ on the plane perpendicular to $\underline{k}$). Notice that the pressure gradient has been eliminated. The same equation is written for a second wave vector $\underline{k}'$

$$\frac{\partial \hat{u}(\underline{k}')}{\partial t} = \hat{u}\hat{u} - \nu k'^2 \hat{u}(\underline{k}') \quad . \qquad (2-2)$$

Multiplying (2-1) by $\hat{u}(\underline{k}')$ and (2-2) by $\hat{u}(\underline{k})$ and adding leads to the spectral tensor equation

$$[\frac{\partial}{\partial t} + \nu(k'^2 + k^2)] < \hat{u}(\underline{k}')\hat{u}(\underline{k}) >=< \hat{u}\hat{u}\hat{u} > \quad . \qquad (2-3)$$

Similar procedures allow to obtain the evolution equation for the third order velocity moments, namely

$$[\frac{\partial}{\partial t} + \nu(k^2 + p^2 + q^2)] < \hat{u}(\underline{k})\hat{u}(\underline{p})\hat{u}(\underline{q}) >=< \hat{u}\hat{u}\hat{u}\hat{u} > \quad . \qquad (2-4)$$

In (2-4), the fourth-order velocity moment can be expressed as the sum of the value it would take where the velocity a gaussian function, and of a cumulant:

$$< \hat{u}\hat{u}\hat{u}\hat{u} >= \sum < \hat{u}\hat{u} >< \hat{u}\hat{u} > + < \hat{u}\hat{u}\hat{u}\hat{u} >_c \quad . \qquad (2-5)$$

2.2 QN, EDQN and EDQNM approximations

Within the Quasi-Normal approximation (Q.N., Millionshtchikov, 1941), the fourth-order cumulant is set equal to zero, which has the unpleasant effect of producing negative kinetic energy spectra (Ogura, 1963). In the Eddy-Damped Quasi-Normal approximation (EDQN), it is assumed that fourth-order cumulants relax linearly the triple correlations (Orszag, 1970), in such a way that (2-4) becomes

$$[\frac{\partial}{\partial t} + \nu(k^2 + p^2 + q^2) + \mu_{kpq}] < \hat{u}(\underline{k})\hat{u}(\underline{p})\hat{u}(\underline{q}) >$$
$$= \sum < \hat{u}\hat{u} >< \hat{u}\hat{u} > \qquad (2-6)$$

with

$$\mu_{kpq} = \mu_k + \mu_p + \mu_q \quad .$$

The choice of the relaxation rate proposed by Orszag as

$$\mu_k \sim [k^3 E(k)]^{1/2} \qquad (2-7)$$

may be improved in:

$$\mu_k = a_1 (\int_0^k p^2 E(p,t)dp)^{1/2} \qquad (2-8)$$

(see Lesieur, 1987, for details). A so-called "Markovianization" of the equations leads to the Eddy-Damped Quasi-Normal Markovian (EDQNM) equation for the kinetic energy spectrum $E(k,t)$ of three-dimensional isotropic turbulence (without helicity)

$$(\frac{\partial}{\partial t} + 2\nu k^2)E(k,t) = \iint_{\Delta_k} dp\, dq\, \theta_{kpq}(t)E(q,t)$$

$$[\frac{k^3}{pq}\, a(k,p,q)E(p,t) - \frac{kp}{q}\, b(k,p,q)E(k,t)] =$$

$$\iint_{\Delta_k} dp\, dq\, \theta_{kpq}(t)\frac{k}{pq}\, b(k,p,q)\, E(q,t)[k^2 E(p,t) - p^2 E(k,t)] \quad (2-9)$$

where

$$b(k,p,q) = \frac{p}{k}\, (xy + z^3)$$

and

$$a(k,p,q) = \frac{b(k,p,q) + b(k,q,p)}{2}$$

are geometric coefficients depending upon the sides k, p, q and cosines of the interior angles x, y, z of the interacting wave vector triad $\underline{k}, \underline{p}, \underline{q}$. The time θ_{kpq}, characteristic of the triple velocity correlations relaxation through non linear interactions, is given by

$$\theta_{kpq} = \frac{1 - \exp -[\mu_{kpq} + \nu(k^2 + p^2 + q^2)]t}{\mu_{kpq} + \nu(k^2 + p^2 + q^2)} \quad (2-10)$$

(Leith, 1971). The integration domain Δ_k is in the $[p, q]$ plane, and such that p and q should be the sides of a triangle of third side k (see Figure 1). Notice finally that the constant a_1 in (2-8) has to be adjusted on the value of the Kolmogorov constant C_k in the $E(k) = C_k \epsilon^{2/3} k^{-5/3}$ inertial range.

2.3 Kraichnan's stochastic models

An alternate way for obtaining these spectral evolution equations consists in replacing the original Navier-Stokes equation by a system of equations coupling non linearly, through gaussian phases, N fictitious velocity fields $\hat{u}_i^\alpha$. The following stochastic model, proposed by Kraichnan (1961, see also Herring and Kraichnan, 1972):

$$(\frac{\partial}{\partial t} + \nu k^2)\hat{u}_i^\alpha(\underline{k},t) = -\frac{i}{N}k_m P_{ij}(\underline{k})\cdot\int_{\vec{p}+\vec{q}=\vec{k}} \Phi_{\alpha\beta\sigma}\hat{u}_j^\beta(\underline{p},t)\hat{u}_m^\sigma(\underline{q},t)d\underline{p}$$

$$< \Phi_{\alpha\beta\sigma}(t)\Phi_{\alpha\beta\sigma}(t') >= 2\, \theta_{kpq}(t)\delta(t - t') \quad (2-11)$$

possesses the same inviscid quadratic invariants as Navier-Stokes, and same absolute equilibrium ensembles for inviscid truncated systems (see Lesieur, 1987 for details). In the limit $N \to \infty$, the $\hat{u}^\alpha$ become *gaussian* and *independant*. Their kinetic energy spectrum

$$E^\alpha(k,t) = 2\,\pi\,k^2\ <\hat{u}_i^\alpha(-\underline{k},t)\hat{u}_i^\alpha(\underline{k},t)> \qquad (2-12)$$

satisfies the same equation as the EDQNM equation.

2.4 Passive scalar EDQNM equation

Let us consider a passive scalar (for instance a passive temperature of zero mean), diffused by the velocity field. The evolution equation for the temperature spectrum $E_T(k,t)$ is, in three-dimensional isotropic turbulence:

$$\frac{\partial}{\partial t} E_T(k,t) = \iint_{\Delta_k} dp\ dq\ \theta_{kpq}^T \frac{k}{pq}(1-y^2)E(q,t)[k^2 E_T(p) - p^2 E_T(k)]$$
$$- 2\kappa k^2 E_T(k,t) \qquad (2-13)$$

with

$$\theta_{kpq}^T = \frac{1 - \exp-[\kappa(k^2+p^2) + \mu'(k) + \mu'(p) + \nu q^2 + \mu''(q)]t}{\kappa(k^2+p^2) + \mu'(k) + \mu'(p) + \nu q^2 + \mu''(q)} \qquad (2-14)$$

$$\mu'(k) = a_2(\int_0^k p^2 E(p,t)dp)^{1/2}$$

$$\mu''(k) = a_3(\int_0^k p^2 E(p,t)dp)^{1/2} \qquad . \qquad (2-15)$$

a_2 and a_3 are adjustable constants in the theory. κ is the molecular conductivity.

2.5 Inviscid enstrophy blow up

The enstrophy is defined as

$$D(t) = \int_0^{+\infty} k^2\ E(k,t)\ dk \qquad . \qquad (2-16)$$

A simplified version of the EDQNM, the MRCM (corresponding to $\theta_{kpq} = \theta_0$), provides a very simple form of the enstrophy evolution equation at zero viscosity:

$$\frac{dD}{dt} = \int \frac{k^2}{q}(k^2-p^2)(xy+z^3)\theta_{kpq}E(p)E(q)dpdqdk = \frac{2}{3}\theta_0 D^2 \quad (2-17)$$

which leads to:

$$D(t) = \frac{3}{2\theta_0} \frac{1}{t_* - t} \qquad (2-18)$$

Starting with a kinetic energy spectrum decreasing rapidly when $k \to \infty$, and hence of finite initial enstrophy $D(0)$, the enstrophy blows up at $t_* = [3/2D(0)\theta_0]$.

Within the EDQNM approximation, it may be shown that the enstrophy follows approximately the equation

$$\frac{dD}{dt} \sim D^{3/2} \quad , \qquad (2-19)$$

leading to

$$D(t) \sim \frac{1}{(t_* - t)^2} \qquad (2-20)$$

which blows up at

$$t_* = 5.6 \ D(0)^{-1/2} \qquad (2-21)$$

(see André and Lesieur, 1977, and Lesieur, 1987)[1] In the limit $\nu \to 0$, the kinetic energy dissipation rate

$$\epsilon = -\frac{1}{2} \frac{d}{dt} < \underline{u}^2 > = 2\nu \ D(t) \qquad (2-22)$$

behaves in the following manner:
- For $t < t_*$, $\epsilon \to 0$ (since $D(t)$ is finite).
- For $t > t_*$, ϵ is *finite*. The enstrophy behaves like ϵ/ν . Let us remark finally that the idea of an enstrophy blow up at a finite time when $\nu \to 0$ is consistent with the existence of a Kolmogorov $k^{-5/3}$ inertial range extending to ∞ (indeed the Kolmogorov dissipation wave number $k_D = (\epsilon/\nu^3)^{1/4}$ tends to infinity): starting with a kinetic energy decaying rapidly to infinity, a very simple phenomenology permits to show that it takes a finite time to build up this $k^{-5/3}$ spectrum, and hence to get a diverging enstrophy. Since the enstrophy is proportional to the r.m.s. vorticity, this implies that vorticity should blow up in some places of the flow at least, which corresponds to singularities. Among the various possible types of singularities responsible for this enstrophy blow up, one has to quote intense localized vorticity spots, discovered by Wray (private communication, 1988) in numerical simulations of the Euler equation, and whose existence was confirmed by Métais (private communication, 1988) in large-eddy simulations of isotropic or stratified turbulence. These spots might result from an intense vortex stretching

---

[1] This is consistent with vorticity majorations done by Leray (1934) on Navier-Stokes equation.

in regions of local reconnections between vortex tubes, as occurs in the mixing layer for the helical-pairing case (see section 4).

It is of interest to look at the evolution of the passive scalar enstrophy

$$D_T(t) = \int_0^{+\infty} k^2 \; E_T(k,t) \; dk \qquad (2-23)$$

within the EDQNM approximation: it is found, for $\nu \to 0$, $\kappa \to 0$, $P_r = \nu/\kappa$ finite.

$$\frac{d}{dt} D_T(t) = \frac{8}{3} a_3 D_T(t) D(t)^{1/2} \qquad (2-24)$$

(see Lesieur et al., 1987), showing that the scalar enstrophy blows up with the velocity enstrophy $D(t)$. Numerical large-eddy simulations performed by Lesieur and Rogallo (1988) show that the velocity enstrophy peaks at $t_* = 4.5 \; D(0)^{-1/2}$, which can be compared with (2-21). But the temperature enstrophy peaks much before, at $t_* = 2.7 \; D(0)^{-1/2}$. This may be explained by comparing the critical enstrophy divergence exponents of the velocity and passive scalar, defined by

$$D(t) \propto (t_* - t)^{-d_e} \quad , \quad D_T(t) \propto (t_* - t)^{-d_T} \quad .$$

Within the MRCM approximation, it is found $d_e = 1$ and $d_T = 2$, while the EDQNM yields $d_e = 2$ and $d_T = (8/3)5.6 \; a_3 \approx 19$ (see Lesieur, 1987, for details). At any rate, and whatever the model used, the scalar enstrophy diverges much faster than the velocity enstrophy, indicating that the passive scalar cascades faster than the velocity towards small scales. This is due to the absence of pressure term in the scalar equation.

3 Theory of non local interactions for 3D isotropic turbulence

3.1 Evaluation of non local transfers and fluxes

In the spectral evolution equation, written as

$$(\frac{\partial}{\partial t} + 2\nu \; k^2) \; E(k,t) \; = \; T(k,t) \quad , \qquad (3-1)$$

the r.h.s. is called the *kinetic energy transfer*. The kinetic energy flux through the wave number k is defined by

$$\Pi(k,t) = \int_k^{\infty} T(p,t) \; dp \quad . \qquad (3-2)$$

Now, let $a << 1$ be a small parameter, and consider triadic interactions in Fourier space such as:

$$inf(k,p,q)/sup(k,p,q) \leq a \qquad (3-3)$$

Let $T_{NL}(k,t)$ be the corresponding non local transfer. Following Kraichnan (1966), and Lesieur and Schertzer (1978), the corresponding non local flux

$$\Pi_{NL}(k,t) = \int_k^\infty T_{NL}(p,t)\, dp \qquad (3-4)$$

may be written as

$$\Pi_{NL}(k,t) = \Pi_{NL}^+(k,t) - \Pi_{NL}^-(k,t)$$

where, to the leading order in a:

$$\Pi_{NL}^+(k,t) = \frac{2}{15} \int_0^{ak} \theta_{kkq}\ q^2 E(q) dq [kE(k) - k^2 \frac{\partial E}{\partial k}]$$

$$+ \frac{2}{15} \int_0^{ak} \theta_{kkq}\ q^4 dq\ \frac{E^2(k)}{k} \qquad (3-5)$$

$$\Pi_{NL}^-(k,t) = -\frac{2}{15} \int_0^k k'^2 E(k') dk' \int_{sup(k,k'/a)}^\infty \theta_{k'pp}\ [5E(p) + p\frac{\partial E}{\partial p}] dp$$

$$+ \frac{14}{15} \int_0^k k'^4 dk' \int_{sup(k,k'/a)}^\infty \theta_{k'pp}\ \frac{E(p)^2}{p^2} dp \qquad (3-6)$$

By derivation with respect to k, it is found for

$$T_{NL}(k,t) = -\frac{\partial \Pi_{NL}(k,t)}{\partial k} \qquad :$$

$$T_{NL}(k,t) = \frac{2}{15} [4E(k) + 2k\frac{\partial E}{\partial k} + k^2 \frac{\partial^2 E}{\partial k^2}] \int_0^{ak} \theta_{kkq}\ q^2 E(q) dq$$

$$- \frac{4}{15} [3E(k) + k\frac{\partial E}{\partial k}] \frac{E(k)}{k^2} \int_0^{ak} \theta_{kkq}\ q^4 dq$$

$$- \frac{2}{15} \theta_{k,k,ak}\ a^5 k^3 E(k)^2$$

$$- \frac{2}{15} \theta_{k,k,ak}\ a^3 k^3 E(ak)[E(k) - k\frac{\partial E}{\partial k}]$$

$$- \frac{2}{15} k^2 E(k) \int_{k/a}^\infty \theta_{kpp}[5E(p) + p\frac{\partial E}{\partial p}] dp$$

$$+ \frac{14}{15} k^4 \int_{k/a}^\infty \theta_{kpp} \frac{E(p)^2}{p^2} dp \qquad (3-7)$$

Let us consider first the case $k << k_i$, wave number where the kinetic energy spectrum is maximum. Setting $a = k/k_i$, the non local transfer, approximately equal to the total transfer[2], reduces to

$$T_{NL}(k,t) = -\frac{2}{15}k^2 E(k) \int_{k_i}^{\infty} \theta_{kpp}[5E(p) + p\frac{\partial E}{\partial p}]dp$$

$$+ \frac{14}{15}k^4 \int_{k_i}^{\infty} \theta_{kpp}\frac{E(p)^2}{p^2}dp \qquad (3-8)$$

There is a k^4 *backscatter transfer*, which is dominant[3] with respect to the $k^2 E(k)$ *eddy-viscous drain*. This k^4 positive transfer is responsible for the formation of a k^4 kinetic energy spectrum in low k, when $E(k,0)$ is a sharp peak at the initial wave number $k_i(0)$. This last point was verified in various large-eddy simulations done, e.g., by Lesieur and Rogallo (1989), Lesieur, Métais and Rogallo (1989), Chollet and Métais (1989), Batchelor et al. (1992) or Métais and Lesieur (1992). For $k >> k_i$ (in the inertial-range), the *local transfer* is preponderant. There is no *backscatter transfer* in this case.

Now, let us consider the passive scalar non local transfer $T_{NL}^T(k,t)$ and flux

$$\Pi_{NL}^T(k,t) = \Pi_{NL}^{T+}(k,t) - \Pi_{NL}^{T-}(k,t) \quad ,$$

which can be calculated in the same way: it is found (Lesieur, 1987)

$$\pi_{NL}^{T+}(k,t) = \frac{2}{15} \int_0^{ak} \theta_{kkq}^T q^2 E(q)dq \ [2kE_T(k) - k^2\frac{\partial E_T}{\partial k}]$$

$$+ \frac{1}{4} \int_0^{ak} \theta_{kkq}^T q^3 E_T(q)dq \ E(k)$$

$$- \frac{1}{4} \int_0^{ak} \theta_{kkq}^T q^5 dq \ E(k)\frac{E_T(k)}{k^2} \qquad (3-9)$$

$$\pi_{NL}^{T-}(k,t) = -\frac{4}{3} \int_0^k k'^2 E_T(k')dk' \int_{sup(k,k'/a)}^{\infty} \theta_{k'pp}^T E(p)dp$$

$$+ \frac{4}{3} \int_0^k k'^4 dk' \int_{sup(k,k'/a)}^{\infty} \theta_{k'pp}^T \frac{E(p)}{p^2} E_T(p)dp \qquad (3-10)$$

The scalar non local transfer for $k << k_i^T$, wave number where the temperature spectrum is maximum, approximately equal to the total

---

[2] Indeed, the local transfer is negligible within this range.

[3] Or at least of the same order

transfer[4], is given by

$$T_{NL}^T(k,t) = -\frac{4}{3}k^2 E_T(k) \int_{k_i^T}^{\infty} \theta_{0pp}^T E(p)dp$$

$$+ \frac{4}{3}k^4 \int_{k_i^T}^{\infty} \theta_{0pp}^T \frac{E(p)}{p^2} E_T(p)dp \qquad (3-11)$$

3.2 Kinetic energy and passive temperature decay

The theory of non local interactions allows to solve the problem of decay of kinetic energy and temperature variance in three-dimensional isotropic turbulence. In the experiment of Comte-Bellot and Corrsin (1966), it was found

$$\frac{1}{2} < \underline{u}^2 >= \frac{1}{2}v^2 \propto t^{-1.26}$$

Warhaft and Lumley (1978) find

$$\frac{1}{2}v^2 \propto t^{-1.34} \propto \frac{1}{2} < T^2 > \quad .$$

From the preceding non local expansions, we have

$$E(k,t) = C_s(t)k^s, \qquad k << k_i$$

$$\frac{dC_s}{dt} = 0, s < 4; \qquad \frac{dC_s}{dt} = \infty, s > 4$$

$$\frac{dC_4}{dt} = \frac{14}{15} \int_{k_i}^{\infty} \theta_{0pp} \frac{E(p)^2}{p^2} dp$$

Looking for self similar solutions of the form

$$E(k,t) = v^2 l\ F(kl), l = \frac{v^3}{\epsilon}, \epsilon = -\frac{1}{2}\frac{dv^2}{dt} \quad ,$$

where l is the integral scale, it is found:
• for $s < 4$:

$$v^2 l^{s+1} = \text{constant} \quad ,$$

$$\frac{1}{2}v^2(t) \propto t^{-2(s+1)/(s+3)}$$

$$l(t) \propto t^{2/(s+3)} \qquad (3-12)$$

---

[4] As for the kinetic energy

The particular case $s = 2$ corresponds to Saffman's law

$$\frac{1}{2}v^2(t) \propto t^{-6/5}$$

• for $s = 4$, we have

$$v^2 l^{s+1} \propto t^\gamma; \gamma = \frac{1}{C_4} \frac{dC_4}{d\ln t}$$

$$\frac{1}{2}v^2(t) \propto t^{-2(s+1-\gamma)/(s+3)}$$

$$l(t) \propto t^{(2+\gamma)/(s+3)} \tag{3-13}$$

The EDQNM theory allows to determine $\gamma = 0.16$, and hence $\frac{1}{2}v^2(t) \propto t^{1.38}, l \propto t^{0.31}$.

The same analysis carried out for the passive scalar shows that, if $E_T(k,t) \propto k^{s'}, k \to 0$, a self similar decay implies $l_T \approx l$ (l_T is the scalar integral scale) and

$$\frac{1}{2} < T^2 > \propto t^{-(s'+1)/(s+3)} \quad \text{for} \quad s' < 4$$

$$\frac{1}{2} < T^2 > \propto t^{-5(2+\gamma)/(s+3)-\gamma'} \quad \text{for} \quad s' = 4$$

. For $s = s' = 4$,

$$\frac{1}{2} < T^2 > \propto t^{-1.48}$$

(see Herring et al., 1982, and Lesieur, 1987).

4 Large-eddy simulations of three-dimensional turbulence

4.1 Isotropic turbulence

Let us first introduce the concept of spectral eddy-viscosity and diffusivity for large-eddy simulations, first proposed by Kraichnan (1976): let k_c be a cutoff wave number in a spectral large-eddy simulation. For a given wave number $k < k_c$, the kinetic energy and temperature variance transfers from k to subgrid scales (p or $q > k_c$) may be expressed as:

$$T_{p|q>k_c}(E, E) = -2\nu_t(k|k_c) \, k^2 E(k) \tag{4-1}$$

$$T^T_{p|q>k_c}(E, E_\Theta) = -2\kappa_t(k|k_c) \, k^2 E_T(k) \tag{4-2}$$

where ν_t and κ_t are a spectral eddy-viscosity and diffusivity. If k_c is in the inertial range ($k_c > k_i$), the above non local expansions show that, for $k << k_c$

$$\nu_t(k|k_c,t) = \frac{1}{15} \int_{k_c}^{\infty} \theta_{kpp}[5E(p) + p\frac{\partial E}{\partial p}]dp \qquad (4-3)$$

$$\kappa_t(k|k_c,t) = \frac{2}{3} \int_{k_c}^{\infty} \theta_{kpp}E(p) \, dp \qquad (4-4)$$

These asymptotic values are independant of k , which shows that in this case ($k << k_c$), the damping action of subgridscale velocity fluctuations on the large scale kinetic energy and temperature variance is well described by these eddy-coefficients. In the general case when k can be arbitrarily close to k_c , the expression of the spectral eddy-coefficients is, when the kinetic energy spectrum follows a $k^{-5/3}$ law when $k \geq k_c$ (Chollet and Lesieur, 1981, Chollet 1983):

$$\nu_t(k|k_c,t) = \nu_t^+(k/k_c) \, [\frac{E(k_c,t)}{k_c}]^{1/2} \qquad (4-5)$$

$$\nu_t^+(k/k_c) = (0.267 + 9.21e^{-3.03(k_c/k)}) \qquad (4-6)$$

With a proper choice of the constants arising in (2-15) and discussed in Lesieur (1987)[5], the eddy conductivity writes:

$$\kappa_t(k|k_c,t) = \kappa_t^+(k/k_c).[\frac{E(k_c,t)}{k_c}]^{1/2} \qquad (4-7)$$

where the eddy-Prandtl number

$$P_r^t(k) = \nu_t^+/\kappa_t^+$$

is approximately constant and equal to 0.6. These spectral eddy coefficients are schematically shown on Figure 2. The "cusp-like" shape in the neighbourhood of k_c corresponds to local interactions. The latter are not taken into account by subgrid scale techniques based on an assumption of separation of scales, such as the RNG (Renormalization Group theory, see e.g. Forster et al., 1977, and Fournier, 1977), or the homogenization theory (see McLaughlin et al., 1985).

---

[5] The choice $a_2 = 0$ allows to recover Kraichnan's LHDIA passive temperature spectrum equation.

These spectral eddy-viscosities and diffusivities have been used in large-eddy simulations of decaying isotropic turbulence using pseudo-spectral methods developed by Orszag and Patterson (1972). We assume periodic boundary conditions. Calculations are carried out at a resolution of 128^3 grid points, with initial spectra given by:

$$E(k,0) = E_T(k,0) = Ak^8 e^{-4[k/k_i(0)]^2} \qquad (4-8)$$

$(k_c = 60, k_i(0) = 8 \text{ or } 20)$. Since $\nu_t >> \nu$, molecular viscosity can be neglected, hence we deal with infinite Reynolds number calculations. The main results of these calculations found by Lesieur and Rogallo (1989) and Lesieur, Métais and Rogallo (1989), are the following:

• The scalar spectrum possesses a k^{-1} inertial range in the large energetic scales.

• The eddy-diffusivity decreases logarithmically in this range, while the eddy-viscosity displays a plateau.

• Temperature variance decreases faster ($t^{-1.85}$ after 60 initial turn-over times instead of the $t^{-1.48}$ predictions of the EDQNM). The kinetic energy decay is in good agreement with closure predictions.

The k^{-1} range seems to be characteristic of large Reynolds numbers, with a Prandtl number of the order of 1. It is due to the scalar shearing by velocity gradients at k_i, and is well described throughout the calculation by the law:

$$E_T(k,t) = 0.1 \, \eta \, \frac{<u^2>}{\epsilon} \, k^{-1} \qquad (4-9)$$

Figure 3 shows the evolution in time of the kinetic energy and temperature spectra for a calculation with $k_i(0) = 8$.

4.2 Periodic mixing layer

The same LES code has been applied to simulate a temporally growing mixing layer (periodicity in the x longitudinal and in the z spanwise directions, free-slip boundary conditions in the y transverse direction). The initial velocity field is a hyperbolic tangent velocity profile $U \tanh y/\delta_0$ to which is superposed a small random three-dimensional perturbation of broad-band spectrum.

Figure 4, taken from Comte et al. (1989), shows the roll up of the Kelvin-Helmholtz vortices, whose formation is in good agreement with two-dimensional calculations carried out by Lesieur et al. (1988). One sees also the formation of longitudinal vortex filaments of vorticity superior to or of the order of the maximum basic spanwise initial vorticity U/δ_0. The latter have been observed in several experiments (see e.g. Bernal and Roshko, 1986). These vortices are the consequence of the straining by the large billows of vortex filaments undulating initially in the neighbourhood of the stagnation line: the result is a hairpin-shaped longitudinal vortex structure, superposed to the primary Kelvin-Helmholtz coherent eddies. We have checked that these stretched vortex filaments can carry a maximum longitudinal vorticity of $2\,U/\delta_0$. A similar secondary structure has been found in large-eddy simulations of the backwards-facing step (Siveira-Neto et al., 1991), using the *structure-function model*. This model, developed by Métais and Lesieur (1992), is derived from the above spectral eddy-viscosity, but is adapted in order to take into account the spatial intermittency and inhomogeneity of turbulence: the kinetic energy spectrum arising in eq. (4-5) is calculated in physical space with the aid of a local second-order velocity structure function.

In fact, mixing layer direct-numerical simulations involving a larger number of primary vortices (4) and forced randomly in the large scales show another type of interaction of the *helical-pairing type* between the large vortices: the latter oscillate in opposition of phase in the spanwise direction, undergoing pairings locally, and giving rise to a *vortex-lattice structure*: this is shown on Figure 5 (taken from Comte et al., 1992), representing a top view of a three-dimensional low-pressure chart. Indeed, in incompressible turbulence, strong vortices result into pressure troughs. Notice also on Figure 5 the presence of an asymetric hairpin vortex stretched between the large vortex patches, and whose right leg only is marked by low pressures. In this calculation, the maximum vortex stretching (about $4U/\delta_0$) occurs in zones of reconnection between the large vortices, with in particular creation of spanwise vorticity of opposite sign with respect to the primary vorticity. This is an indicator for local three-dimensional isotropy and build up of a Kolmogorov ultra-violet cascade.

5 Effects of compressibility on free shear flows

Calculations of compressible free shear flows have been performed by numerous researchers. Let us quote for instance Lele (1988) for a two-dimensional spatially-growing mixing layer; Soestrino et al. (1989), Sandham and Reynolds (1991) and Fouillet (1991) for a three-dimensional periodic mixing layer, Fouillet (1991) for a three-dimensional spatially-growing mixing layer. Lele (1988) confirms the importance of the *convective Mach number* M_c whose importance has been emphasized by Papamoschou and Roshko (1988), that is, a Mach number defined in a frame of reference moving downstream with the Kelvin-Helmholtz vortices: for $M_c \leq 0.4$, the flow is practically not affected by compressibility. In two dimensions, and for higher values of M_c, compressibility starts inhibiting the roll up and pairings. At $M_c = 0.7$, shocklets appear alternatively upwards and downwards of the vortices, as on a transsonic wing: the flow is accelerated around the vortices. Fouillet (1991) has noticed that shocklets disappear at higher Mach numbers, where compressibility flatens the large vortices. At any rate, these two-dimensional simulations for $M_c > 6$ are not relevant physically, since it has been shown (see e.g. Sandham and Reynolds, 1991) that three-dimensional instabilities (oblique waves) are then more amplified than their two-dimensional counterparts. The same authors find, by direct-numerical simulations, the occurrence of a structure of staggered Λ-shaped longitudinal vortices, stretched by the flow. The same vortex topology has been found by Fouillet (1991), in a three-dimensional direct-numerical simulation of a hyperbolic-tangent periodic compressible mixing layer forced initially by a random white-noise perturbation: for $M_c = 0.3$, the vortex-lattice structure found by Comte et al. (1992, see Figure 5) in the incompressible case is recovered. This may be interpreted as a staggered array of Λ vortices undergoing a pairing at their tip. At $M_c = 0.8$, the Λ vortices still exist, but the pairing has been inhibitted by compressibility. The same type of structure has been found by Normand and Lesieur (1992) in the numerical simulation of a transitioning boundary-layer on a flat plate at Mach 5: this calculation was rendered possible by the use of the structure-function subgrid-scale model described above. Returning to the compressible mixing layer, Fouillet (1991) has shown that, at $M_c = 1$, the low-pressure tubes do not correspond anymore to the vortex tubes: they reconnect between the two Λ vortices, giving rise to longitudinal pressure troughs. This is shown on Figure 6.

6 Discussion and perspectives

First, a few words about the validity and usefulness of the two-point closure approach: these closures are very useful to understand the phenomenology of isotropic turbulence, in three dimensions as well as in two dimensions. They give valuable informations about questions like the direction of the transfers or the existence of inertial ranges. They provide analytical models of singularities for infinite Reynolds number flows (in 3D). They allow to calculate explicitly non local interactions, and to predict decay laws for the kinetic energy. They permit to study turbulent diffusion and predictability. Finally they allow to derive efficient subgridscale parameterizations for large-eddy simulations, even in non homogeneous situations like the mixing layer.

The more we *explore* the structure of turbulence, the more we discover *coherent structures* at all scales, such as Kelvin-Helmholtz eddies, longitudinal vortices and streaks, horseshoes, mushrooms, hairpins, bananas, worms, hot spots,...). These structures are *unpredictable* and part of the turbulence. They might be compatible with a *statistical approach*, and hence be described by *closures*. Closures cannot of course describe strong departures from gaussianity, and hence are unable to predict quantities like structure functions of high order. But calculations carried out in Fourier space might not depend very much of the intermittency in physical space.

The intermediate position of coupling large-eddy simulations with closures in the subgrid scales could be the answer to the double constraint of describing accurately the coherent structures and statistically the Kolmogorov kinetic energy cascade towards the small unresolved scales. Another important perspective of the large-eddy simulations is to serve as a tool to assess the validity of turbulence models and to improve them. Attempts of incorporating compressibility effects in the closures have been developed by Marion (1988) for an isentropic gas. This could lead to compressible versions of the subgrid scale parameterizations developed above. The immediate perspectives concerning turbulent flows which are of interest for hypersonics fluid dynamics, that is free or wall bounded flow, consist in developing high resolution full or large-eddy three-dimensional simulations of these flows, both in temporal or spatially growing situations, in order to answer the following questions:
• What is the Mach number effect on three-dimensional turbulence developing in a hypersonic mixing layer, wake or jet?
• How does compressibility affects the transition to turbulence and its structure in a boundary layer?

Other important perspectives concern the coupling of real gaz effects with turbulence in a boundary layer.

References

André, J.C. and Lesieur, M., 1977, J. Fluid Mech., **81**, pp 187-207.

Batchelor, G.K., Canuto, V.M., and Chasnov, J.R., 1992, J. Fluid Mech., ?, pp ??-??

Bernal, L.P. and Roshko, A., 1986, J. Fluid Mech., **170**, pp 499-525.

Brown, G.L. and Roshko, A., 1974, J. Fluid Mech., **64**, pp 775-816.

Chollet, 1983, Thèse de Doctorat d'Etat, Grenoble University.

Chollet, J.P. and Lesieur, M., 1981, J. Atmos. Sci., **38**, pp 2747-2757.

Chollet, J.P. and Métais, O., 1989, European Journal of Mechanics B/Fluids, **8**, pp 523-548.

Comte, P., Lesieur, M. and Fouillet, Y., 1989, in *Topological Fluid Dynamics*, Cambridge University Press, H.K. Moffatt ed., pp 649-658.

Comte, P., Lesieur, M. and Lamballais, E., 1992, submitted to Phys. Fluids.

Comte-Bellot, G. and Corrsin, S., 1966, J. Fluid Mech., **25**, pp 657-682.

Forster, D., Nelson, D. and Stephen, M.J., 1977, Phys. Rev. A, **16**, pp 732-749.

Fouillet, Y., 1991, Thèse, University of Grenoble.

Fournier, J.D., 1977, Thèse, Nice University.

Herring, J.R. and Kraichnan, R.H., 1972, in *Statistical models and turbulence*, Springer-Verlag, **12**, pp 148-194.

Herring, J.R. et al., 1982, J. Fluid Mech., **124**, pp 411-437.

Kraichnan, R.H., 1961, J. Math. Phys., **2**, pp 124-148.

Kraichnan, R.H., 1966, Phys. Fluids, **9**, pp 1728-1752.

Kraichnan, R.H., 1976 J. Atmos. Sci., **33**, pp 1521-1536.

Leith, C.E., 1971, J. Atmos. Sci., **28**, pp 145-161.

Lele, S.K., 1988, in Proc. of the *Conference on the Physics of Compressible Turbulent Mixing*, Oct. 24-27, Princeton.

Leray, J., 1934, J. Acta. Math, **63**, pp 193-248.

Lesieur, M. and Schertzer, D., 1978, J. Mécanique, **17**, pp 609-646.

Lesieur, M., 1987, *Turbulence in Fluids*, Nijhoff Publishers, Dordrecht. Second edition, 1990, Kluwer Publishers.

Lesieur, M., Montmory, C. and Chollet, J.P., 1987, Phys. Fluids., **30**, pp 1278-1286.

Lesieur, M., Staquet, C., Le Roy, P. and Comte, P., 1988, J. Fluid Mech., **192**, pp 511-534.

Lesieur, M. and Rogallo, B., 1989, Phys. Fluids A, **1**, pp 718-722.

Lesieur, M., Métais and Rogallo, R., 1989, C.R. Acad. Sci., **308**, Ser. II, pp 1395-1400.

McLaughlin, Papanicolao and Pironneau, 1985 SIAM J. Appl. Math.

Millionshtchikov, M., 1941, Dokl. Akad. Nauk. SSSR, **32**, pp 615-618.

Moin, P. and and Kim, J., 1982, J. Fluid Mech., **118**, pp 341-378.

Marion, J.D. 1988, Thèse de l'Ecole Centrale de Lyon.

Métais, O. and Lesieur, M. 1992, J. Fluid Mech., **239**, pp 157-194.

Normand, X. and Lesieur, M., 1992, Theor. Comput. Fluid Dynamics, **3**, pp 231-252.

Ogura, Y., 1963, J. Fluid Mech., **16**, pp 33-40.

Orszag, S. A., 1970, J. Fluid Mech., **41**, pp 363-386.

Orszag, S.A. and Patterson, G.K., 1972, in *Statistical models and turbulence*, Springer-Verlag, **12**, pp 127-147.

Papamoschou, D. and Roshko, A., 1988, J. Fluid Mech., **197**, pp 453-477.

Sandham, N.D. and Reynolds, W.C., 1991, J. Fluid Mech., **224**, pp 133-158.

Silveira-Neto, A., Grand, D., Métais, O. and Lesieur, M., 1991, Phys. Rev. Letters, **66**, pp 2320-2323.

Soetrisno, M., Greenough, J.A., Eberhardt, S. and Riley, J., 1989, "Confined compressible mixing layers: part I. Three-dimensional instabilities", AIAA 20 Fluid Dynamics, Plasma Dynamics and Lasers Conference, Buffalo, pp th st Nat. Fluid Dynamics Congress, Cincinnati, AIAA paper 89-1810.

Warhaft, Z. and Lumley, J.L., 1978, J. Fluid Mech., **88**, pp 659-684.

Figure captions

Figure 1: domain Δ_k in the $[p, q]$ plane corresponding to triadic interactions.

Figure 2: schematic kinetic energy spectrum with the spectral eddy-viscosity and conductivity.

Figure 3: evolution of the kinetic energy and temperature spectra in a spectral large-eddy simulation involving 128^3 grid points.

Figure 4: spanwise coherent structures in a temporal LES mixing layer calculation, visualized by a numerical dye initially located at the interface of the mixing layer (in white). In red is shown the longitudinal vorticity equal to the initial spanwise vorticity U/δ_0 .

Figure 5: top view of the vortex-lattice structure in the direct-numerical simulation of a periodic mixing-layer (from Comte et al., 1992).

Figure 6: top view of longitudinal low-pressure tubes in the direct-numerical simulation of a periodic compressible mixing layer of convective Mach number 1 (from Fouillet, 1991).

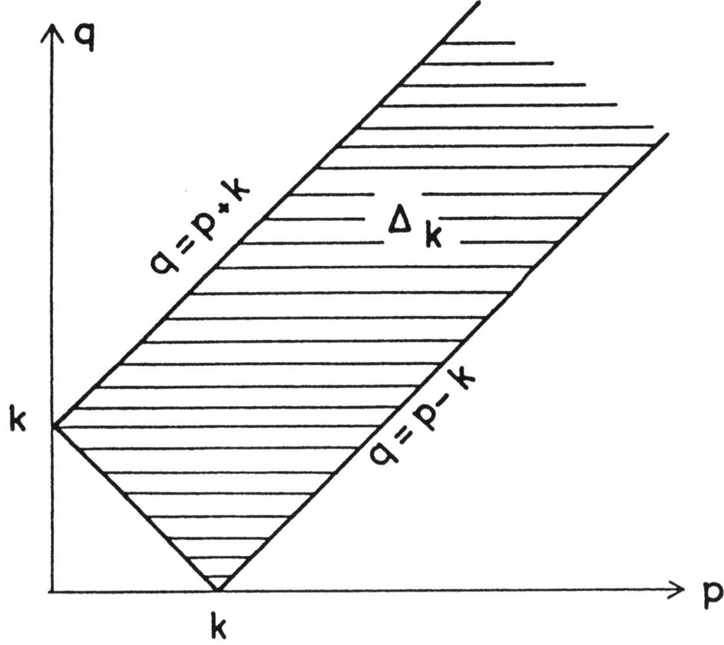

Figure 1.

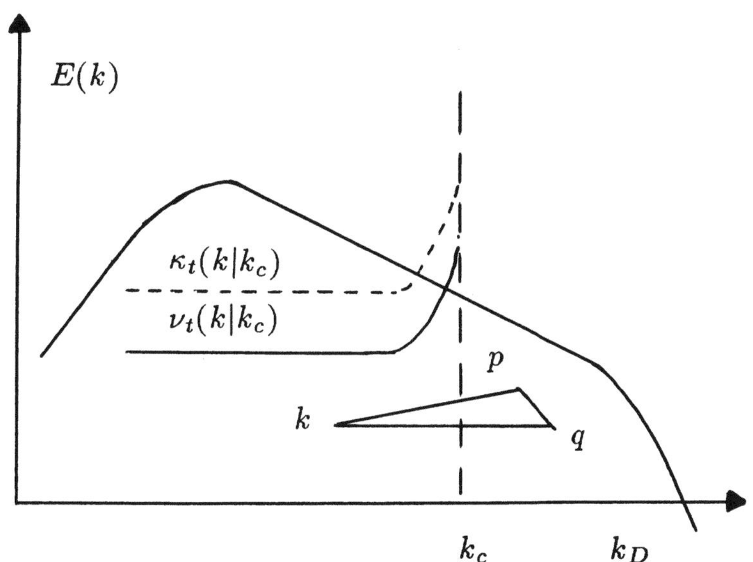

Figure 2.

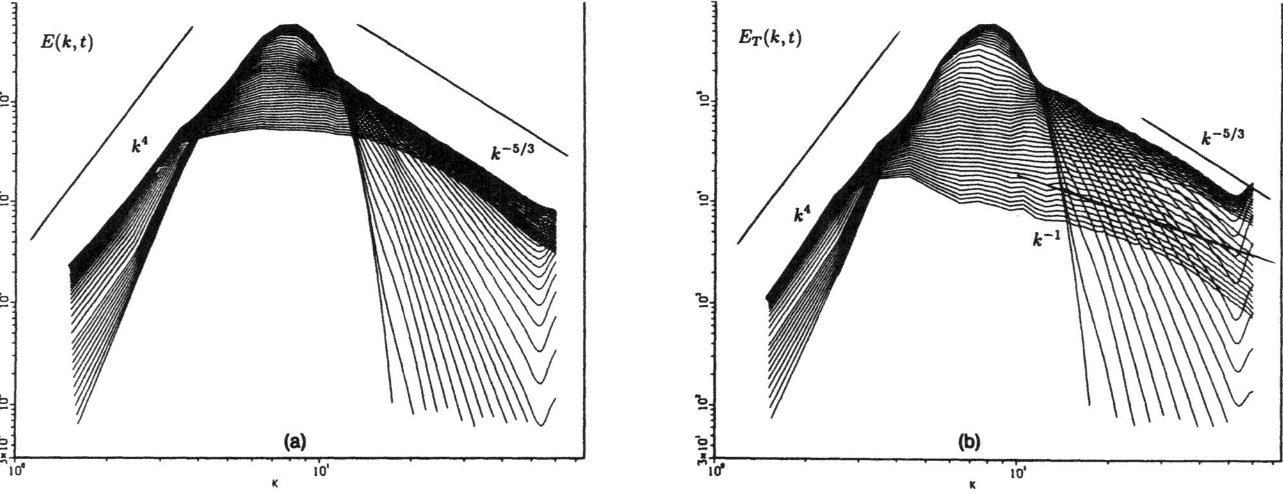

Figure 3.

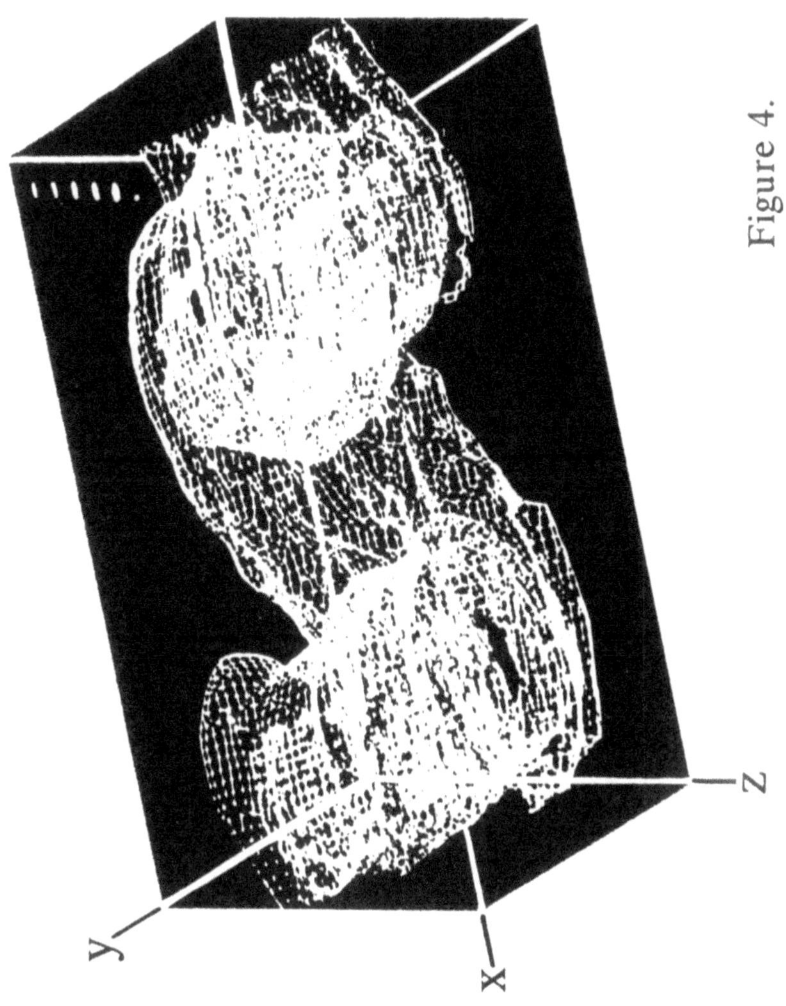

Figure 4.

Figure 5.

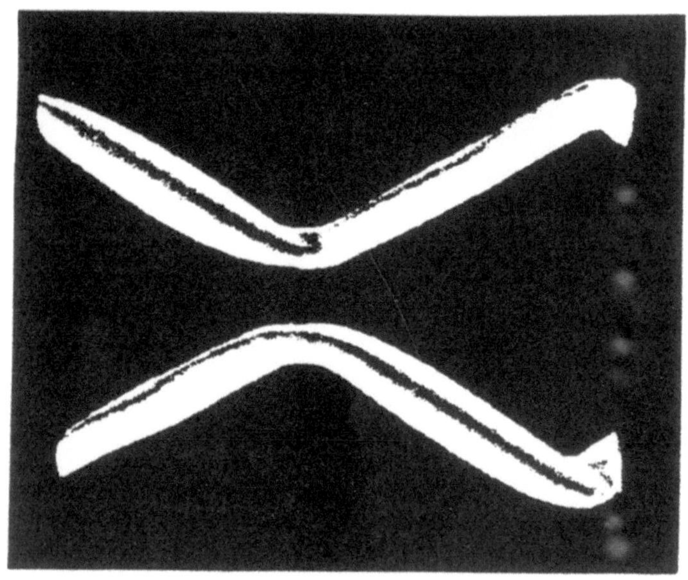

Figure 6.

Different Levels of Air Dissociation Chemistry and Its Coupling with Flow Models

J. Warnatz, U. Riedel, R. Schmidt

Institut für Technische Verbrennung, Universität Stuttgart
Pfaffenwaldring 12,
7000 Stuttgart, Germany

This work is supported in part by Deutsche Forschungsgemeinschaft, Bonn
SFB 259 "High Temperature Problems of Re-Usable Space Transportation Systems"

Abstract

Three selected topics relevant for hypersonic problems shall be considered in
this paper:

- **Thermal chemistry of air dissociation on different levels:** Thermodynamics and reaction kinetics of thermal air dissociation at high temperatures are now rather well known. Systematic sensitivity analysis leads to the conclusion that oxygen dissociation is the rate-limiting reaction.

- **Non-thermal chemistry:** Phenomena like vibrational relaxation lead to considerable complications. A mechanism is developed, taking into consideration vibrational–translational relaxation of nitrogen and oxygen molecules up to the dissociation limit. Results are presented and compared with that of a corresponding thermal mechanism.

- **Coupling of flow and chemistry:** Stiffness introduced by chemical reaction terms into the conservation equations demands for implicit solution procedures. The treatment of two–dimensional problems (Euler equations and Navier-Stokes equations) is described here to illustrate solution methods including detailed chemistry.

1 Introduction

At the elevated temperatures typical of hypersonic problems, the components of air N_2 and O_2 are dissociating within times shorter than the times typical of the processes considered [1–3], i. e. the residence time in the high temperature region after the shock front (see Figure 1). These data are presented for atmospheric pressure; for

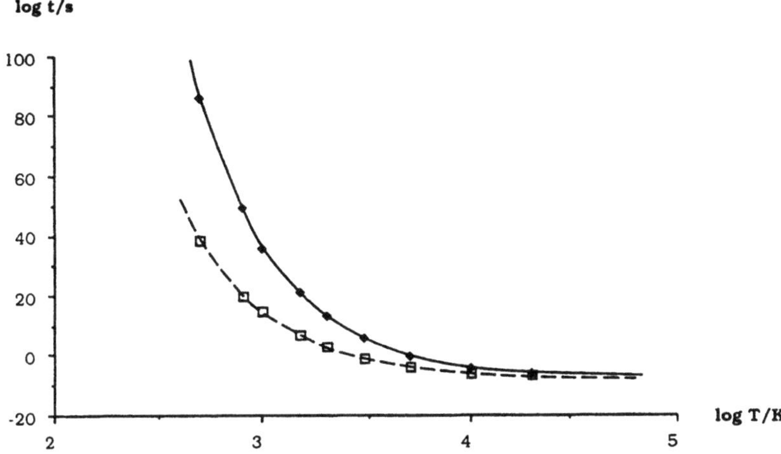

Figure 1: Mean decomposition times of O_2 (broken line) and N_2 (full line) at atmospheric pressure.

lower pressure the characteristic times for dissociation will move to smaller values. In the case of *thermal equilibrium* (i. e. establishment of Boltzmann distribution for translational, rotational, and vibrational degrees of freedom of the molecules considered, defining a common temperature) and neglecting ionization processes, the chemistry occurring can be described by the reaction mechanism of Table 1. M is an unspecified molecule removing or providing collision energy with an efficiency which will be specified later; the equal sign means that both forward and reverse reaction are considered. Thus, four independent additional conservation equations for the masses of the five species N, O, NO, N_2, and O_2 have to be added to that for total mass, momentum, and energy (one species mass is linearly dependent on the others because of total mass conservation). Though the number of dependent variables is not increased drastically, this expanded system is extremely cumbersome because of its "stiffness", a property first detected in combustion systems [4]. This stiffness is caused by characteristic times of the reaction rates of single reactions differing by

O_2	$+$	$M^{\cdot}$	$=$	O	$+$	O	$+ \quad M^{\cdot}$	$(R1), (R2)$
N_2	$+$	$M^{\cdot\cdot}$	$=$	N	$+$	N	$+ \quad M^{\cdot\cdot}$	$(R3), (R4)$
NO	$+$	$M^{\cdot\cdot\cdot}$	$=$	N	$+$	O	$+ \quad M^{\cdot\cdot\cdot}$	$(R5), (R6)$
O	$+$	N_2	$=$	NO	$+$	N		$(R7), (R8)$
NO	$+$	O	$=$	O_2	$+$	N		$(R9), (R10)$

Table 1: Reaction scheme.

many orders of magnitude. At least for time-dependent problems, explicit methods are not able to overcome this problem, so that time-consuming implicit integration methods (see [5,6]) have to be used. Some examples including detailed chemistry will be described in Chapter 4 below.

2 Thermal Chemistry of Air Dissociation

2.1 Equilibrium Chemistry and Thermodynamics

In many applications, in special at higher temperatures, the assumption of "thermodynamic" (or "chemical") equilibrium is a very useful approximation. Both forward and reverse reactions e. g. in Reactions (R1) – (R10) above are then fast enough to allow all concentrations to equilibrate in times short in comparison with the characteristic time of the problem considered. This means a drastic simplification: Instead by time-consuming solution of a differential equation system, the concentrations (or partial pressures) can be calculated from a simple non-linear equation system resulting from the principles of thermodynamics (minimization of free enthalpy at constant pressure). In the sense of Gibbs' phase law, the 5 species system mentioned above have 3 linearly independent chemical reactions (e. g. R1/R2, R3/R4, and R5/R6) and 2 components (e. g. N and O), thus leading to a non-linear equation system. From this, the partial pressures p_i can be calculated easily by standard methods (Newton iteration, minimization of Gibbs energy, etc.) if the total pressure and the composition r_{NO} (ratio of the atom numbers) are specified; the equilibrium constants can be taken from tables of thermodynamic values (for reference see e. g. [7–9]):

$$p_N + p_O + p_{NO} + p_{N_2} + p_{O_2} = p_{ges}$$

70

$$p_N + p_{NO} + 2p_{N_2} = r_{NO}(p_O + p_{NO} + 2p_{O_2})$$

$$\frac{(p_O)^2}{p_{O_2}} = K_{1,2}$$

$$\frac{(p_N)^2}{p_{N_2}} = K_{3,4}$$

$$\frac{p_N p_O}{p_{NO}} = K_{5,6}$$

Examples of results of these calculations are given in Figure 2 for air at pressure and temperature conditions corresponding to h = 71 km.

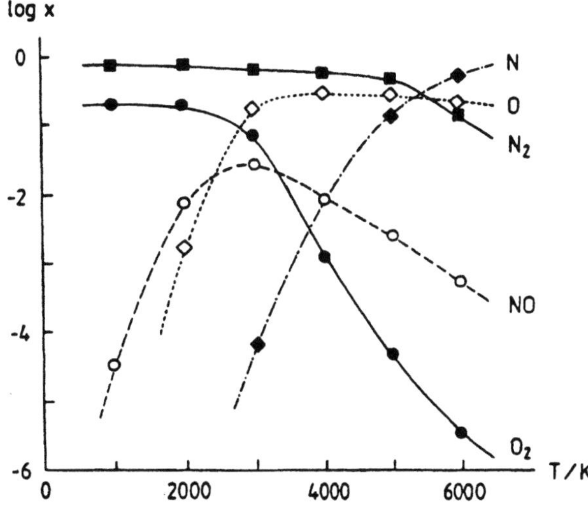

Figure 2: Equilibrium mole fractions in air at h = 71 km (p = $2.56.10^{-2}$bar).

2.2 Non–Equilibrium Chemistry

To write down a reaction mechanism describing air dissociation in a qualitative way is rather simple, as shown in Section 1 (Reactions (R1) - (R10)). Figure 3 demonstrates, how such a mechanism can describe the time dependence of transition from the frozen flow to the establishment of thermodynamic (or chemical) equilibrium. Of course, equilibrium values as calculated in Section 2.1 are the result of the calculations at sufficiently long times. The difficulties arise, if exact rate data are demanded. Here, the following problems have to be considered:

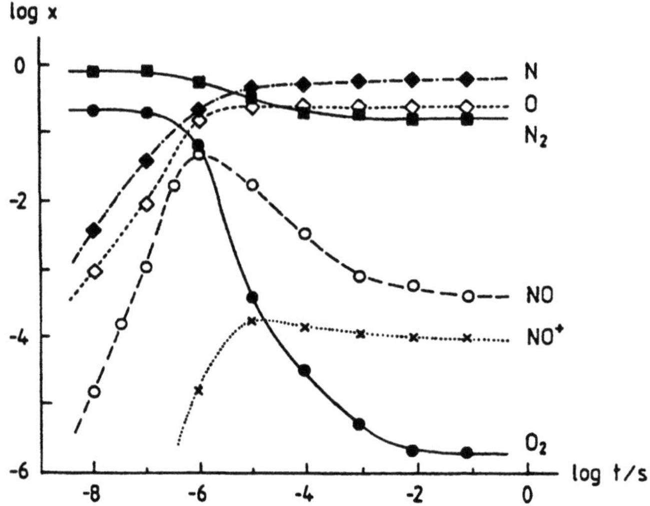

Figure 3: Temporal establishment of equilibrium in air at h = 71 km
(T_{eq} = 6002 K and p = $2.56.10^{-2}$ bar).

1. The temperature dependence of rate coefficients cannot be described by a simple Arrhenius form

$$k = A \, \exp\left(\frac{-E_a}{RT}\right)$$

as it was usually used up to the seventies (k = rate coefficient, A = pre-exponential factor, E_a = activation energy, R = gas constant, T = temperature). In any case, curved Arrhenius plots described in the simplest case by

$$k = A \, T^\beta \, \exp\left(\frac{-E_a}{RT}\right) \tag{1}$$

have to be considered, where the additional curvature caused by the temperature coefficient β of the pre-exponential (the measurement of which is very difficult due to the overwhelming influence of the exponential temperature dependence) makes the extrapolation from experimentally accessible to high temperatures rather difficult [1–3].

2. It is difficult to give good values for the collision efficiencies of the third bodies M in the termolecular reactions (R1) – (R6). In special, some problems arise if the unstable particles N and O are considered [1,2].

Reaction	A	β	E_a	$Ref.$
Dissociation				
$O_2 + M^{\cdot} \longrightarrow O + O + M^{\cdot}$	$2.70\ 10^{19}$	-1.00	494.00	*see text*
$N_2 + M^{\cdot\cdot} \longrightarrow N + N + M^{\cdot\cdot}$	$3.70\ 10^{21}$	-1.60	941.00	*Baulch* [15]
$NO + M^{\cdot\cdot\cdot} \longrightarrow N + O + M^{\cdot\cdot\cdot}$	$2.90\ 10^{15}$	0.00	621.00	*Roth* [16]
Exchange				
$O + N_2 \longrightarrow NO + N$	$1.82\ 10^{14}$	0.00	319.00	*Hanson* [3, 17]
$NO + O \longrightarrow O_2 + N$	$3.80\ 10^{09}$	1.00	173.10	*Hanson* [3, 17]

	O_2	N_2	O	N	NO	$Ref.$
$M^{\cdot}$	1.00	0.10	2.80	0.10	0.10	*see text*
$M^{\cdot\cdot}$	0.10	1.00	0.10	2.80	0.10	*like $O_2 + M$*
$M^{\cdot\cdot\cdot}$	0.05	0.05	1.00	1.00	1.00	*Wray* [18]

Table 2: Arrhenius parameters and collision efficiencies

(A in cm, mol, s; E_a in kJ/mol; $k = AT^{\beta} \exp(-\frac{E_a}{RT})$).

3. The reaction product NO is rather easily ionized and can initialize a complex chemistry of ionic species. Fortunately, rate coefficients of ionic species can be estimated much easier than that for neutral species' reactions [10].

4. Thermal non-equilibrium effects need a detailed chemistry of state-selected species and can expand the reaction mechanism by a factor of ten or more (see Section 3).

Mechanisms given earlier in the literature [11, 12] show serious deficiencies in the light of newer experimental results. A complete up-to-date literature research on Reactions (R1) – (R10) has been delivered recently (see Table 2 and [13, 14]). No effort was made here to give a complete review on the ionic reactions. Sensitivity analysis (see next section) shows the O_2 dissociation to be the rate-limiting step in air dissociation. For this reason, this reaction has once more been evaluated completely. Because of the overwhelming influence of the initial O_2 dissociation reaction and the reasons following below, selected reviewed values have been used for the other reactions involved:

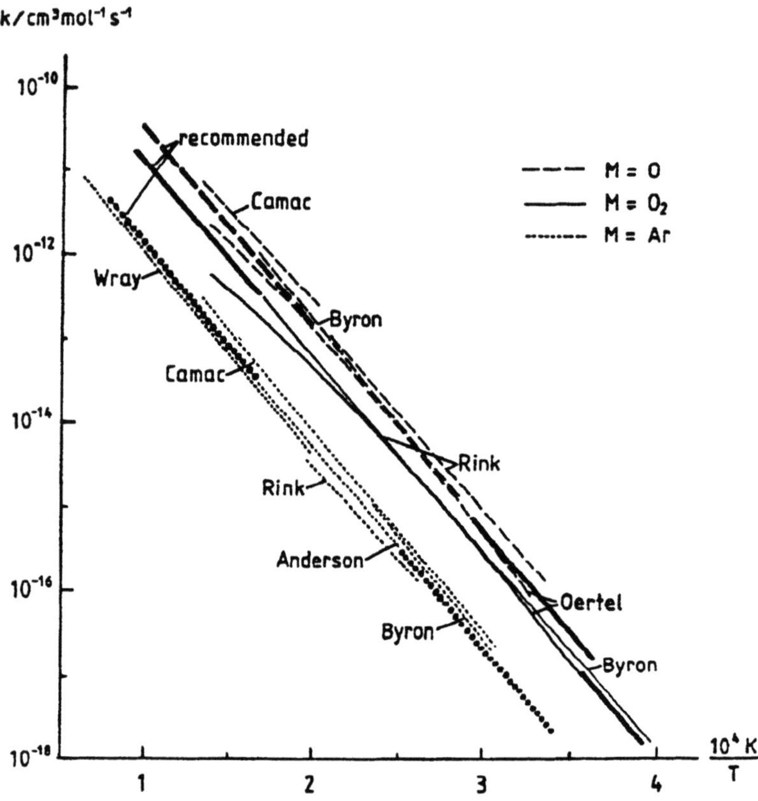

Figure 4: plot of the rate coefficient of the reaction $O_2 + M \longrightarrow O + O + M$.

1. N_2 decomposition is unimportant because of the large dissociation energy. A review value given by Baulch et al. [15] is used.

2. NO dissociation is not extremely rate-limiting. A new value of the rate coefficient determined by Roth et al. is used [16].

3. for the exchange reactions, a lot of experimental material is existing in the literature; careful reviews e. g. by Hanson et al., are available [3,17].

Figure 4 shows an Arrhenius plot of O_2 dissociation rate coefficients for different third molecules M = Ar, O_2 and O. Because of the scatter of the M = O data, the same temperature dependence has been chosen for this third body as for M = Ar, O_2. The recommendations are given as thick lines, giving the data listed in Table 2 and used in the calculations presented in the present work (see also [18]). It should be mentioned, that the mechanism shown in Table 2 in principle is not very different from that given recently by Park [19]. However, the weak point of this evaluation is the rather formal treatment of the third bodies' efficiencies which are the real problem in hypersonic applications and demand for further experimental study.

2.3 Sensitivity Analysis for Thermal Chemistry

Sensitivity analysis for 0D systems is extensively discussed in literature [21–24]; only a few results are presented here for completeness. The system considered can be written as $u(t) = f(t, u(t), k)$ with initial conditions $u(t = to) = u_o(k)$ where $u = (u_1, \ldots, u_S)$ is the vector of dependent variables and $k = (k_1, \ldots, k_R)$ is the set of parameters (e. g. rate coefficients). The $S * R$ matrix of sensitivities

$$S(t) = \frac{\partial u(t)}{\partial k}$$

is then determined by solution of the enlarged equation system given by

$$S'(t) - J(t)S(t) = \frac{\partial f(t, u(t), k)}{\partial k}$$

with i. c.'s. $S(t = t_o) = \frac{\partial u_o(k)}{\partial k}$ where $J(t) = \frac{\partial f(t, u(t), k)}{\partial u}$ is the Jacobian matrix which can be evaluated by numerical differentiation. Representative examples of sensitivity tests are given in Figure 5 showing the sensitivity for O atom concentration with respect to variation of all rate coefficients at eight different temperatures. Two important features are worth to be mentioned:

1. All reactions are nearly in the chemical equilibrium, except for reaction (R1/R2), which can be seen from the same absolute height of the bars for forward and reverse reaction. For this reason, differences of the forward and reverse reaction contributions have been calculated and plotted in Figure 6.

2. The sensitivity differences for reactions (R1/R2) are very large (see Figure 6); these reactions are not equilibrated, as can be seen from the different sensitivities for forward and reverse reaction shown in the diagram.

In summary these results mean that the oxygen dissociation is the dominating reaction.

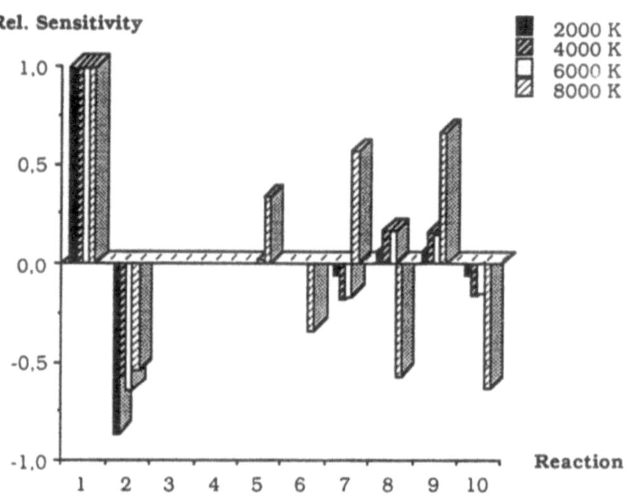

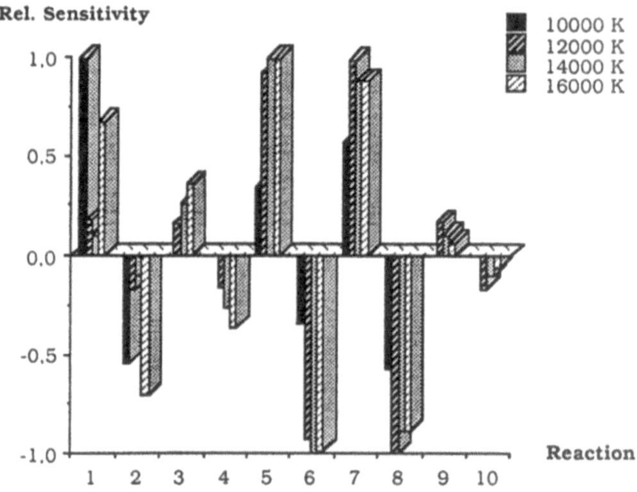

Figure 5: Sensitivity analysis for air dissociation at different temperatures (p = 2.56.10-2 bar).

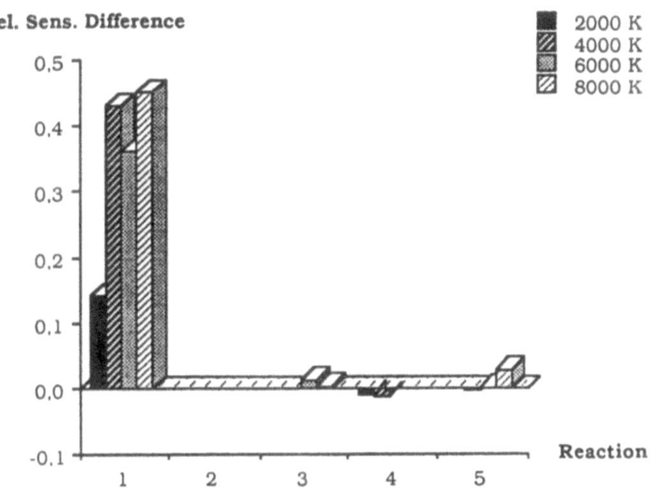

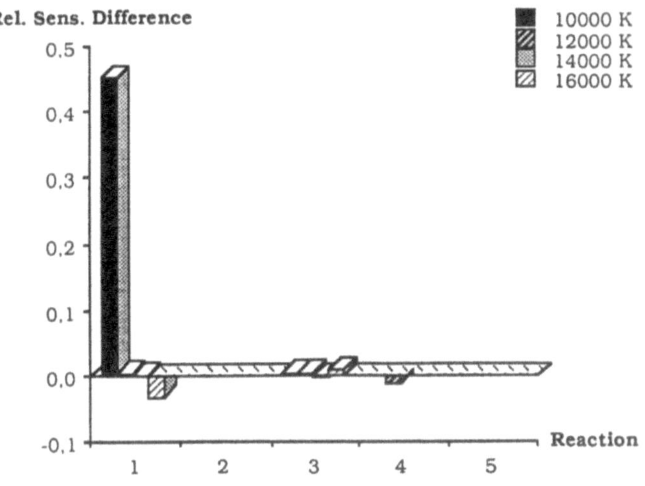

Figure 6: Sensitivity differences at different temperatures (p = 2.56.10-2 bar).

3 Non–Thermal Air Dissociation Chemistry

3.1 Thermal Non–Equilibrium

Under the extreme conditions encountered in reentry problems difficulties arise from the fact that thermal equilibrium is not established within times typical to the problem considered. Relaxation of translational and rotational degrees of freedom to thermal equilibrium takes, in general, only a few collisions. For vibrational degrees of freedom, however, this relaxation can be rather slow and control the establishment of thermal equilibrium. Therefore *thermal non–equilibrium* in addition to *chemical non–equilibrium* has to be included in a realistic modelling of hypersonic flows.

3.2 State Selected Reactions

In order to get a realistic description of the non–equilibrium situation elementary processes like vibrational–translational (VT), vibrational–rotational (VR) and vibrational–vibrational (VV) energy transfer have to be taken into account. Our method of attack is to include all internal states of the molecules in the reaction scheme. There is, however, the difficulty of providing all the detailed rate coefficients needed. Furthermore to handle such large mechanism in a 1D- or 2D--calculation leads to storage and computation time problems. Therefore the following simplifications are made: all vibrational states of N_2 and O_2 are treated as different species and their VT–transfer is modelled. The by–product NO is not vibrationally specified. Rotational states are not taken into account; VV–transfer is of relatively weak influence on our reaction scheme as compared with VT–transfer and is therefore neglected.

3.3 Rate Coefficients

3.3.1 Dissociation

Dissociation is described by an *anharmonic oscillator model* for the vibrational states of N_2 and O_2. The vibrational energy levels are calculated by

$$E(v) = hc \left\{ \nu_0(v + \frac{1}{2}) - \nu_0 x(v + \frac{1}{2})^2 \right\}$$

with $\nu_0=$ wave number of fundamental vibration, $\nu_0 x=$ anharmonicity correction, $h=$ Planck's constant, $c=$ speed of light. The detailed rate coefficients are assumed to fit in the Arrhenius expression

$$k(v,T) = A(v)\, T^\beta \exp\left(\frac{-E_a(v)}{RT}\right) \tag{2}$$

with

$$
\begin{aligned}
E_a(v) &= D - (E(v) - E(0))\,, \\
A(v) &= A(0)(v+1)\,.
\end{aligned}
$$

(D is the dissociation energy). The rate coefficients are related to the thermal rate coefficient k_0 by

$$k_0(T) = \sum_{v=0}^{\tilde{v}} f_0(v,T) k(v,T)\,. \tag{3}$$

f_0 is the thermal vibrational population and $\tilde{v}$ is the highest vibrational level. This normalization is necessary to guarantee the same long-time behaviour of thermal and non-thermal mechanism. The $A(v,T)$ resulting from the normalization procedure (3) are temperature dependent; they are approximated by a temperature-independent A which gives rise to a modification of β in the Arrhenius form of $k(v,T)$ (Equation (2)).

3.3.2 Exchange Reactions

Modelling of $k(v,T)$ for exchange reactions is quite similar to dissociation:

$$k(v,T) = A(v)\, T^\beta \exp\left(\frac{-E_a(v)}{RT}\right)$$

with thermal activation energy E_a and

$$
\begin{aligned}
E_a(v) &= E_a - (E(v) - E(0)) && \text{if} && E(v) - E(0) \le E_a \\
E_a(v) &= 0 && \text{if} && E(v) - E(0) > E_a\,.
\end{aligned}
$$

The same normalization procedure as for the dissociation reactions is applied.

3.3.3 Vibrational–Translational Energy Transfer

The basic quantity is the rate coefficient $k_{1,0}(T)$ for the transition from vibrational level $v=1$ to $v=0$ and is calculated by a semiempirical equation given by Lifshitz [26]

$$k_{1,0}(T) = A \exp\left(\frac{-B}{T^{\frac{1}{3}}}\right) \tag{4}$$

with empirical constants A and B including oscillator frequencies and species masses. The temperature dependence of $k_{1,0}$ corresponds with the theoretically derived Landau–Teller relation (see e. g. [27]). The Lifshitz formula above is related to the Millikan–White formula [28] which connects the vibrational relaxation time with temperature.

The deexcitation rate coefficients $k_{v,v-1}$ of vibrational levels with $v > 1$ are assessed by a scaling relation from Billing [29]:

$$k_{v,v-1}(T) = k_{1,0}(T) \left\{ \frac{v(1-x)}{1-vx} \exp\left[\omega x \left(\frac{4\pi}{3\alpha} \sqrt{\frac{\mu}{2kT}} \, (v-1) - \frac{\hbar}{kT} \right) \right] \right\} \quad (5)$$

with $\omega=$ fundamental vibrational frequency, $\omega x=$ anharmonicity correction, $\mu=$ reduced mass of collision partners and $\alpha=$ parameter of the exponential interaction potential. In the high temperature limit the $k_{v,v-1}$ approach a limiting expression via an effective cross section (Park [30]). All transitions are restricted to $\Delta v = \pm 1$.

It is obvious from the above considerations that in contrast to dissociation and exchange reactions it is not suitable to represent the VT–transfer by an Arrhenius form. We use five parmeters to calculate the rate coefficients as a function of temperature from (5)

$$k_{v,v-1}(T) = \frac{a_1 \, \exp\left(a_2 \, T^{-\frac{1}{2}} + a_3 \, T^{-\frac{1}{3}} + a_4 \, T^{-1} \right) \, a_5 \, T^{\frac{1}{2}}}{a_1 \, \exp\left(a_2 \, T^{-\frac{1}{2}} + a_3 \, T^{-\frac{1}{3}} + a_4 \, T^{-1} \right) + a_5 \, T^{\frac{1}{2}}} \quad (6)$$

collecting all fundamental constants of the molecules in the coefficients $a_i, (i = 1,\ldots,5)$ given in Appendix A (a_6 is an redundant parameter to check consistency of the scheme).

3.4   Comparison of Non–Thermal and Thermal Mechanism

For a homogeneous reaction system the profiles of mole fractions and temperature vs. time have been calculated for thermal and non–thermal mechanism. Initial conditions have been chosen typical to air behind a shock wave corresponding to a height of 71 kilometers. The initial temperature is choosen in such a way that both models arrive at the same equilibrium temperature.

Figure 7 shows a comparison of temperature profiles for thermal and non–thermal reaction scheme: The strong decrease of temperature for t=1...10 μs is due to endothermic dissociation processes. For t=0.1...1 ms thermal equilibrium is attained,

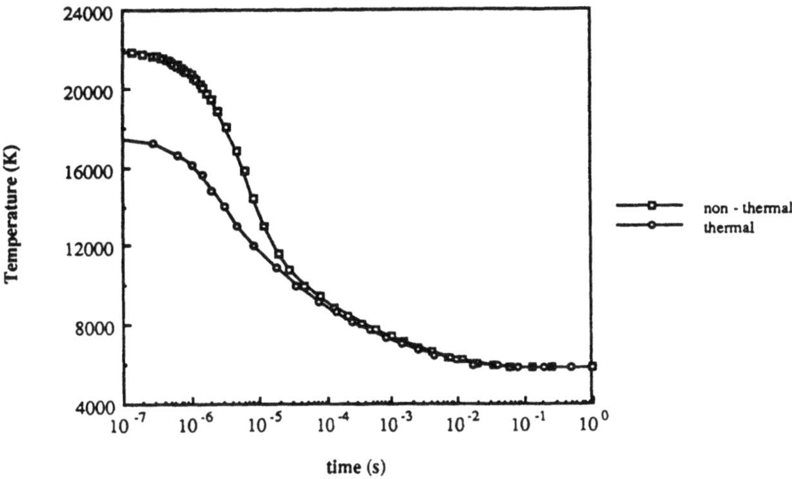

Figure 7: Thermal and non-thermal mechanism: comparison of temperatures.

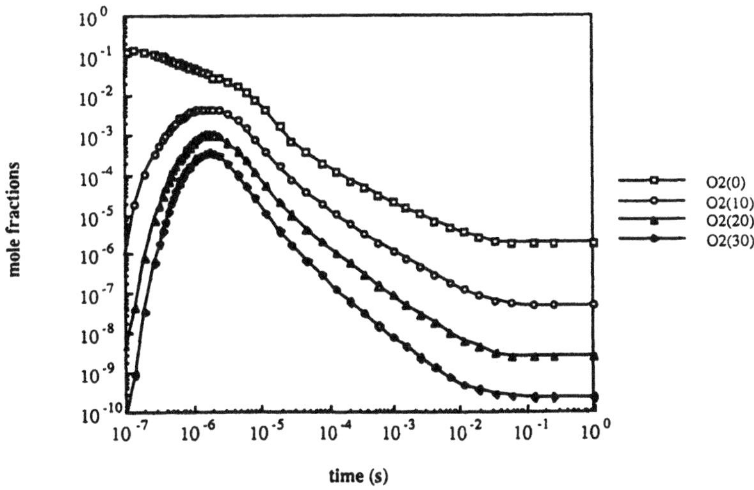

Figure 8: Population of selected vibrational levels of O_2 ($v = 0, 10, 20, 30$).

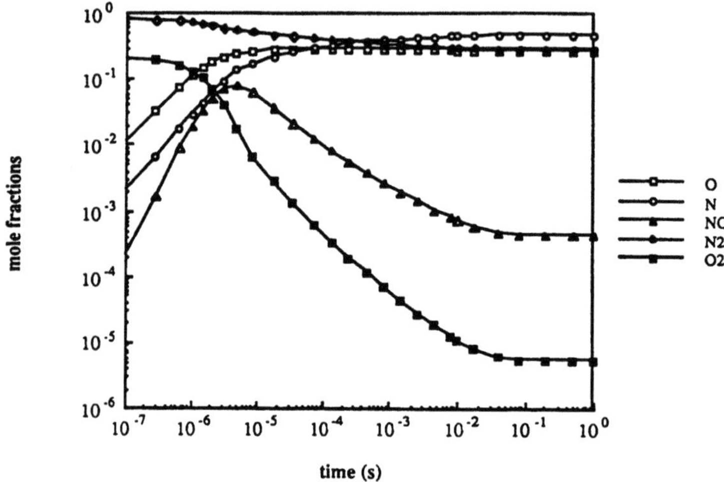

Figure 9: Mole fraction profiles for thermal mechanism, temperature range 17200 ... 5810 K.

both curves begin to coincide. The mole fractions of selected vibrational levels of $O_2(v)$ are shown in Figure 8 for an initial temperature of 22000 Kelvin. At t=0 s all O_2 molecules are in the ground state ($v = 0$) as prescribed by the initial conditions. For t>0 higher vibrational states are excited (maximum values at t=1... 10 μs) followed by relaxation to the equilibrium value T_{eq}=5810 K.

Figures 9 and 10 show calculated mole fraction profiles for thermal and non-thermal mechanism. For the non-thermal scheme the different vibrational states of $N_2(v)$ and $O_2(v)$ are summed up to N_2 and O_2. The dissociation reactions in the non-thermal case are delayed in comparison to their thermal counterparts. This can be explained by the time necessary to fully excite the vibrational states. The strong decrease of N_2 and O_2 for t = 1...10 μs is in accordance with the temperature profiles. The delayed dissociation in the non-thermal case gives rise to a later appearance of N and O atoms (and therefore of NO, too) if compared with the thermal mechanism.

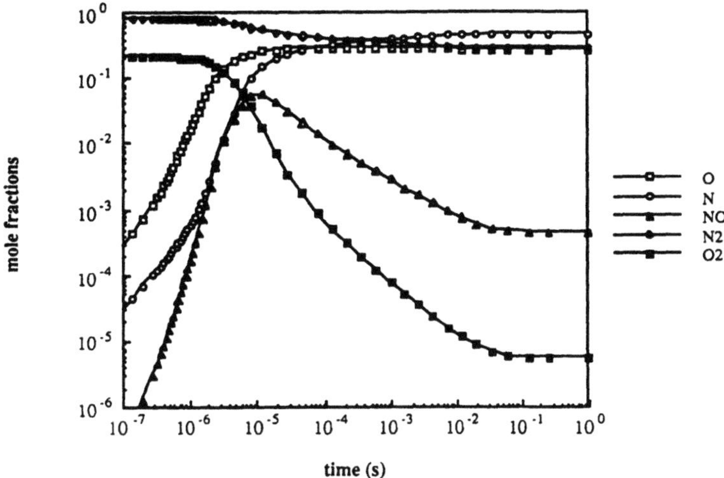

Figure 10: Mole fraction profiles for non–thermal mechanism, temperature range 22000 ... 5810 K.

4 Coupling of Flow and Chemistry

4.1 Euler Calculations Including Thermal Non–Equilibrium

Aim of the calculations is the incorporation of chemical reaction mechanisms as large as possible. For this purpose the code has to be as simple and fast as possible to allow for calculation times of hours at maximum if very large mechanisms are considered. Thus

1. a very simple geometry is taken to avoid costly gridding procedures. A half-sphere on the tip of a cylinder is used to simulate the reactive flow field in front of a reentry body. Only a few grid points are necessary in tangential direction, whereas about fifty equidistant points are used in radial direction.

2. A shooting method and shock fitting are used to attain simple integration. Disadvantage of the shooting method is that no meaningful boundary conditions can be formulated on the surface of the body.

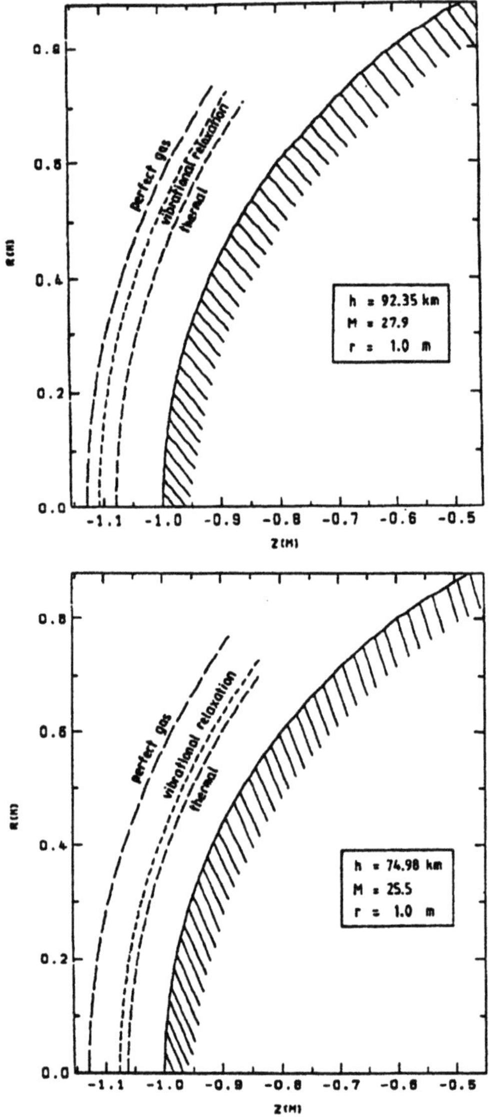

Figure 11: Influence of different levels of chemistry on the shock front position in front of a re-entry body for two cases (h = 92 km and h = 75 km).

Details can be found elsewhere [31]; at this place only some results are presented, demonstrating the very strong influence of different levels of chemistry on the flow field and in special on the position of the shock front. Figure 11 shows for two cases (h = 92 km and h = 75 km) the influence of different chemistry models on the position of the shock front. Equilibrium chemistry (see Section 2.1) predicts a shock front relatively near to the surface if compared with thermal chemistry (5 species, 10 reactions, see Table 2 in Section 2.2), demonstrating the necessity of finite rate chemistry. The importance of inclusion of non-thermal chemistry ($\sim$500 reactions of $\sim$80 species to describe vibrational excitation/relaxation) is distinctly shown, too.

4.2 Navier Stokes Calculations with Detailed Chemistry and Transport Model

4.2.1 Conservation Equations

The Navier Stokes equations for a chemically reacting flow in two space dimensions can be written in the following form

$$\frac{\partial \mathbf{U}}{\partial t} + \frac{\partial \mathbf{F}_1(\mathbf{U})}{\partial x} + \frac{\partial \mathbf{F}_2(\mathbf{U})}{\partial y} = \mathbf{S}(\mathbf{U}) \ . \tag{7}$$

$\mathbf{U} = (\rho, \rho u, \rho v, \rho y_1, \ldots, \rho y_n, e)^t$ is the vector of the conserved quantities and $\mathbf{S}(\mathbf{U}) = (0, 0, 0, M_1\dot{\omega}_1, \ldots, M_n\dot{\omega}_n, 0)^t$ denotes the source term due to chemical reactions. The flux vectors $\mathbf{F}_i$ consist of an inviscid and a viscous part:

$$\mathbf{F}_1(\mathbf{U}) = \mathbf{F}_1^{inv}(\mathbf{U}) + \mathbf{F}_1^{vis}(\mathbf{U}) = \begin{pmatrix} \rho u \\ \rho u u + p \\ \rho v u \\ \rho y_1 u \\ \vdots \\ \rho y_n u \\ (e+p)u \end{pmatrix} + \begin{pmatrix} 0 \\ \Pi_{xx} \\ \Pi_{xy} \\ j_{1,x} \\ \vdots \\ j_{n,x} \\ u\Pi_{xx} + v\Pi_{xy} + q_x \end{pmatrix}$$

$$\mathbf{F}_2(\mathbf{U}) = \mathbf{F}_2^{inv}(\mathbf{U}) + \mathbf{F}_2^{vis}(\mathbf{U}) = \begin{pmatrix} \rho v \\ \rho u v \\ \rho v v + p \\ \rho y_1 v \\ \vdots \\ \rho y_n v \\ (e+p)v \end{pmatrix} + \begin{pmatrix} 0 \\ \Pi_{xy} \\ \Pi_{yy} \\ j_{1,y} \\ \vdots \\ j_{n,y} \\ u\Pi_{xy} + v\Pi_{yy} + q_y \end{pmatrix} \tag{8}$$

Here ρ is the density, u and v are the velocity components, ρu and ρv are the x- and y–components of the momentum per unit volume, $e = \rho\left(\epsilon + \frac{1}{2}(u^2 + v^2)\right)$ is the total energy density, and ϵ is the specific internal energy. $\dot{\omega}_i$ is the molar scale rate of formation of species i ($1 \le i \le n$) and M_i the molar mass.

The components of the shear stress tensor $\overline{\overline{\Pi}}$ are [32]:

$$\begin{aligned} \Pi_{xx} &= -\frac{\mu}{3}\left(4u_x - 2v_y\right) \\ \Pi_{yy} &= -\frac{\mu}{3}\left(-2u_x + 4v_y\right) \\ \Pi_{xy} &= -\mu\left(u_y + v_x\right) \end{aligned}$$

where, for example, u_x is defined as $\frac{\partial u}{\partial x}$.

The molecular mass flux in a multicomponent system consists of two contributions [32, 33]: ordinary (concentration) diffusion $\vec{j}_i^{\,d}$ and thermal diffusion $\vec{j}_i^{\,T}$:

$$\begin{aligned} \vec{j}_i = (j_{i,x}, j_{i,y})^t &= \vec{j}_i^{\,d} + \vec{j}_i^{\,T} \\ &= -\rho\frac{y_i}{x_i}D_i^D \,\mathrm{grad}\, x_i - \frac{D_i^T}{T}\mathrm{grad}\, T \end{aligned}$$

with the mole fractions $x_i = y_i\frac{\overline{M}}{M_i}$ and mean molar mass $\overline{M}$. The total heat flux relative to the mass average velocity is

$$\begin{aligned} \vec{q} = (q_x, q_y)^t &= \vec{q}^{\,c} + \vec{q}^{\,d} \\ &= -\lambda\,\mathrm{grad}\, T + \sum_i h_i \vec{j}_i \; . \end{aligned}$$

Here $\vec{q}^{\,c}$ denotes the conductive energy flux and $\vec{q}^{\,d}$ the flux caused by inter-diffusion.

The pressure is determined from the equation of state

$$p(\rho, T) = \rho RT \sum_i \frac{y_i}{M_i} \; . \tag{9}$$

4.2.2 Transport Model and Thermodynamic Data

Reliable codes for the evaluation of transport coefficients of pure species [34] and mixtures [35] up to 2500K are now available from work in combustion processes. Extrapolation to higher temperatures is possible due to the fact that here the weak points of the theories used are becoming more and more unimportant. Furthermore, these weak points are mainly due to polar particles (like H_2O) which do not play a role in air dissociation chemistry. An extensive treatment of the theory of transport coefficients is given in [32, 33].

Thermodynamic data are needed to calculate enthalpies in the energy balance equation and for the computation of reverse reaction rate coefficients. For this purpose a database with polynominal fits of fifth order for each species molar heat capacity $c_{p,i}$ is used:

$$c_{p,i}(T) = a_1 + a_2 T + a_3 T^2 + a_4 T^3 + a_5 T^4 \; .$$

Two additional coefficients (integration constants) allow the computation of enthalpy and entropy. Mixture properties are given by

$$
\begin{aligned}
c_p &= \sum_i y_i \, c_{p,i} \\
h &= \sum_i y_i \, h_i \; .
\end{aligned}
$$

4.2.3 Numerical Implementation

In order to use a body fitted coordinate system the conservation equations have to be rewritten in curvilinear coordinates. The coordinate transformation to the new space variables ξ, η is given by

$$\xi = \xi(x,y) \; , \quad \eta = \eta(x,y)$$

and we can write for the partial derivatives

$$
\begin{pmatrix} \frac{\partial}{\partial x} \\ \frac{\partial}{\partial y} \end{pmatrix}
= \begin{pmatrix} \xi_x & \eta_x \\ \xi_y & \eta_y \end{pmatrix}
\begin{pmatrix} \frac{\partial}{\partial \xi} \\ \frac{\partial}{\partial \eta} \end{pmatrix} \; .
$$

Replacing $\frac{\partial}{\partial x}$ and $\frac{\partial}{\partial y}$ in equation (7) we get the *chain rule conservative law form* used in the computations:

$$\frac{\partial \mathbf{U}}{\partial t} + \xi_x \frac{\partial \mathbf{F}_1(\mathbf{U})}{\partial \xi} + \eta_x \frac{\partial \mathbf{F}_1(\mathbf{U})}{\partial \eta} + \xi_y \frac{\partial \mathbf{F}_2(\mathbf{U})}{\partial \xi} + \eta_y \frac{\partial \mathbf{F}_2(\mathbf{U})}{\partial \eta} = \mathbf{S}(\mathbf{U}) \; . \tag{10}$$

The metric coefficients are computed from the relations

$$\xi_x = \frac{y_\eta}{J} \,, \quad \xi_y = -\frac{x_\eta}{J} \,, \quad \eta_x = -\frac{y_\xi}{J} \,, \quad \eta_y = \frac{x_\xi}{J}$$

where $J = x_\xi y_\eta - x_\eta y_\xi$ denotes the Jacobian of the inverse transformation.

The viscous terms $\mathbf{F}_i^{vis}$ are approximated by central differences using five point stencils for first and nine point stencils for second derivatives. Flux splitting is applied to the inviscid fluxes. Rewriting (7) collecting only the inviscid fluxes gives:

$$\mathbf{S}(\mathbf{U}) = \frac{\partial \mathbf{U}}{\partial t} + \sum_{j=1}^{2} \left(\sum_{i=1}^{2} \frac{\partial \xi_j}{\partial x_i} \frac{\partial \mathbf{F}_i^{inv}(\mathbf{U})}{\partial \mathbf{U}} \right) \frac{\partial \mathbf{U}}{\partial \xi_j} \,, \tag{11}$$

and with recurrence to the Euler equations in primitive variables this equation can be transformed to give:

$$\mathbf{S}(\mathbf{U}) \;=\; \frac{\partial \mathbf{U}}{\partial t} + \sum_{j=1}^{2} \mathbf{T}\mathbf{A}_j\mathbf{T}^{-1}\frac{\partial \mathbf{U}}{\partial \xi_j} \tag{12}$$

with

$$A_j = \begin{pmatrix}
\tilde{u}_j & \rho\frac{\partial \xi_j}{\partial x_1} & \rho\frac{\partial \xi_j}{\partial x_2} & 0 & \cdots & \cdots & 0 \\
 & \tilde{u}_j & & & & & \frac{1}{\rho}\frac{\partial \xi_j}{\partial x_1} \\
 & & \tilde{u}_j & & 0 & & \frac{1}{\rho}\frac{\partial \xi_j}{\partial x_2} \\
 & & & \ddots & & & 0 \\
 & 0 & & & \ddots & & \vdots \\
 & & & & & \ddots & 0 \\
0 & c^2\rho\frac{\partial \xi_j}{\partial x_1} & c^2\rho\frac{\partial \xi_j}{\partial x_1} & 0 & \cdots & 0 & \tilde{u}_j
\end{pmatrix} .$$

$\tilde{u}_j$ is the flow velocity in the direction of the ξ_j coordinate:

$$\tilde{u}_j = u\frac{\partial \xi_j}{\partial x} + v\frac{\partial \xi_j}{\partial y} \,.$$

and c the *frozen speed of sound*. The matrics

$$T = \begin{pmatrix}
 & 1 & & & & & & \\
 & u & & \rho & & & & \\
 & v & & & \ddots & & 0 & \\
 & y_1 & & & & \ddots & & \\
 & \vdots & & 0 & & & \ddots & \\
 & y_n & & & & & & \rho \\
\frac{1}{2}(u^2+v^2)+h+\rho\frac{\partial h}{\partial \rho} & \rho u & \rho v & \rho\frac{\partial h}{\partial y_1} & \cdots & \rho\frac{\partial h}{\partial y_n} & \rho\frac{\partial h}{\partial p}-1
\end{pmatrix}$$

defines the transformation from primitive to conservative variables.

$\mathbf{A}_j$ is diagonalized to apply the flux vector splitting procedure:

$$\mathbf{A}_j = \mathbf{N}_j \mathbf{\Lambda}_j \mathbf{N}_j^{-1}$$

with the diagonal matrics of the eigenvalues

$$\mathbf{\Lambda}_j = diag(\tilde{u}_j - c, \tilde{u}_j, \ldots, \tilde{u}_j, \tilde{u}_j + c) \ .$$

The finite difference approximation of $\mathbf{TA}_j\mathbf{T}^{-1}\frac{\partial \mathbf{U}}{\partial \xi_j}$ is computed such that left sided (right sided) upwind differences are applied for flows coming from left (right). Let $\mathbf{\Lambda}_j^{\pm}$ be the diagonal matrices containing the positive and negative elements of $\mathbf{\Lambda}_j$, zero elsewhere, then (12) is approximated as:

$$\mathbf{S(U)} = \frac{\partial \mathbf{U}}{\partial t} + \sum_{j=1}^{2} \left\{ \mathbf{TN}_j \mathbf{\Lambda}_j^+ \mathbf{N}_j^{-1} \mathbf{T}^{-1} \delta_j^+ \mathbf{U} + \mathbf{TN}_j \mathbf{\Lambda}_j^- \mathbf{N}_j^{-1} \mathbf{T}^{-1} \delta_j^- \mathbf{U} \right\}$$

where $\delta_j^+ \mathbf{U}$ ($\delta_j^- \mathbf{U}$) is a left (right) sided difference, since positive (negative) eigenvalues correspond to right (left) going flows.

The structure of the spectrum of $\mathbf{\Lambda}_j$ implies that at most one eigenvalue can differ in sign from all the others. This is used to reduce the computational effort: $\mathbf{A}_j$ needs not to be decomposed at all if all the eigenvalues agree in sign. And even in the other case, e. g. if $\mathbf{\Lambda}_j^+$ has only a single nonzero element, only the left and right eigenvectors corresponding to the single positive eigenvalue of $\mathbf{A}_j$ are needed.

4.2.4 Temporal Integration

The system of ordinary differential-algebraic equations resulting from spatial discretization is solved using the extrapolation code LIMEX [36, 37]. This implicit method requires the evaluation of the Jacobian of the system which has, in our case, a block diagonal structure. The dimension of the Jacobian is given by $n_{eq} \cdot n_l \cdot n_k$ where n_{eq} is the dimension of the partial differential equation system and n_l, n_k are the number of grid points in the two space dimensions. The computation is performed numerically by difference approximation. To evaluate the Jacobian in a time saving way all entries having the same parameters i, l mod 5, k mod 5 are perturbed simultaneously (i is the index of the equation and l and k are the indices of the grid point system). Therefore the number of function evaluations necessary for the Jacobian does not depend on the number of grid points. The implicit time integration method requires the solution of a linear equation system $Ax = z$ with A having a block diagonal structure again. Solution is performed by a method based on incomplete LU–factorization. Details can be found in [38,39].

4.2.5 Adaptive Gridding

To improve the grid point density in regions with large gradients adaptive gridding is used. After each time step the grid point distribution for each spatial direction is determined by equipartitioning a mesh function and subsequent inverse interpolation. The mesh function for each coordinate ξ is obtained by a weighted summation of gradients and curvatures of n_v dependent variables:

$$F(\xi) = a_0 \frac{\xi}{\xi_{max}} + \sum_{l=1}^{n_l} \sum_{m=1}^{n_v} a_m \frac{\int_0^\xi \left| \frac{\partial u_{m,l}}{\partial \xi} \right| d\xi}{\int_0^{\xi_{max}} \left| \frac{\partial u_{m,l}}{\partial \xi} \right| d\xi} + b_m \frac{\int_0^\xi \left| \frac{\partial^2 u_{m,l}}{\partial \xi^2} \right| d\xi}{\int_0^{\xi_{max}} \left| \frac{\partial^2 u_{m,l}}{\partial \xi^2} \right| d\xi} \tag{13}$$

where $u_{m,l}$ is the value of variable m in the vector $\mathbf{U}$ of equation (7) at grid point l. The weighting factors a_m, b_m are chosen empirically.

In a next step the number of grid points to be inserted or deleted in each interval of the old mesh is computed and the final decision about regridding is made. If there are only slight changes the old grid point system is maintained and the integration continued. Old and new grid are assumed to differ sufficiently if either more than one new grid point has to be inserted in an interval of the old grid or if the total number of points to be added exceeds a certain limit depending on the overall number of grid points. If regridding has to be performed the solution of the last time step is interpolated onto the new mesh by piecewise monotonic cubic hermite interpolation [40,41].

4.2.6 Results

The flux splitting method described above has been used to calculate the non-equilibrium hypersonic flow around a half sphere with a radius of 1 meter. The detailed rate coefficients were computed from the parameters given in Table 2 according to the modified Arrhenius law (1). Freestream conditions are taken to approximately correspond to the *standard atmosphere* [42] at a geometric altitude of 75 kilometers ($p_\infty = 2.52$ Pa, $\rho_\infty = 0.43 \times 10^{-4}$ kg m^{-3}, $T_\infty = 205.30$ K). The freestream Mach number was $M_\infty = 25.0$.

Maximum temperature (see Figure 12) in the shock wave is 12300 Kelvin. The profiles of mass fractions indicate that dissociation occurs immediately behind the bow shock. All profiles have little variation downstream along the body. Oxygen molecules are completely dissociated; NO appears only in a small amount and thus O in constant amount (Figure 13). N_2 is less than 22% dissociated as shown in Figure 14. The stand off distance measured from the mach number plot is 0.086 m.

5 Conclusions

The examples discussed above demonstrate that the effects of chemical reaction and thermal non–equilibrium on the flow fields in hypersonic problems can be considerable. Nevertheless, they are tractable, and sufficient data can be provided to handle these reactive flows. However, chemistry cannot be treated by explicit integration methods. Much more costly partially or globally implicit methods have to be applied to handle these stiff systems. For stationary problems this means that operator splitting methods have to be used at least taking care of implicit treatment of the chemistry terms.

6 Acknowledgements

The authors are grateful to the "European Space Agency" for financial support of this work and thank Dr. U. Maas for valuable help.

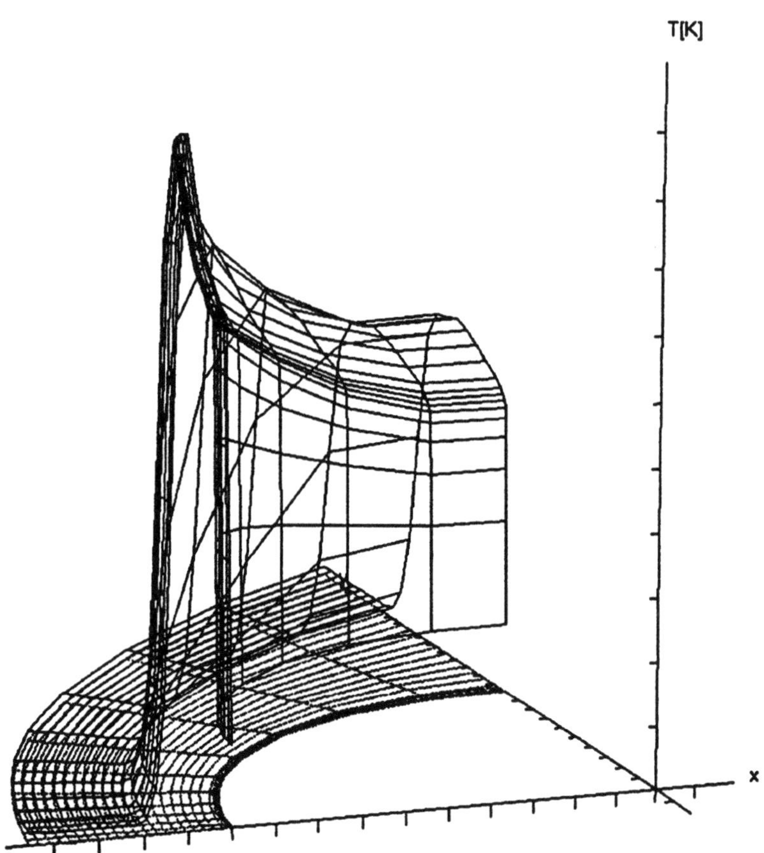

Figure 12: Temperature profile.

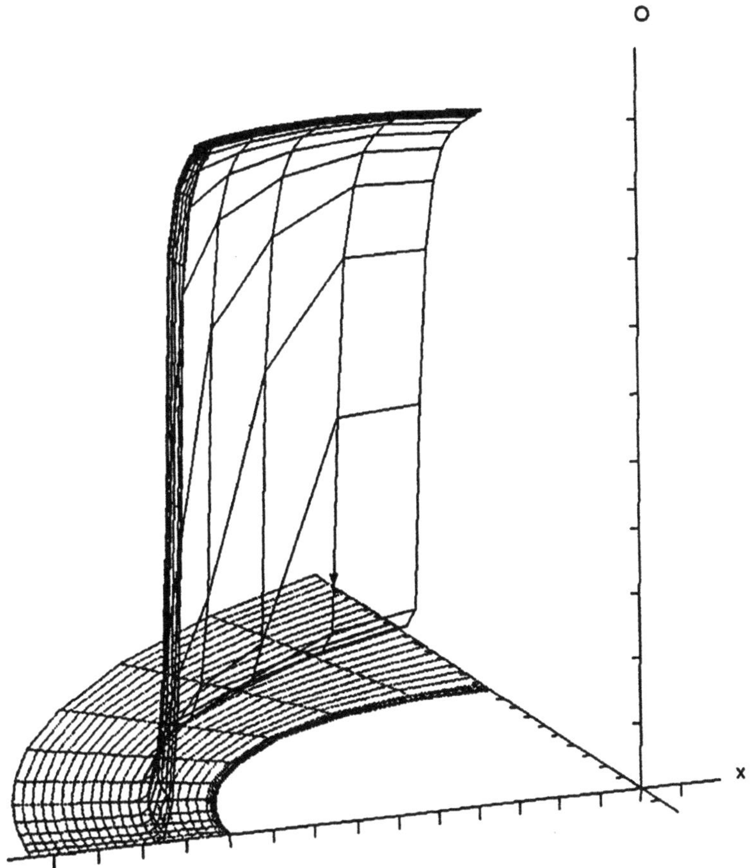

Figure 13: Profile of O atoms.

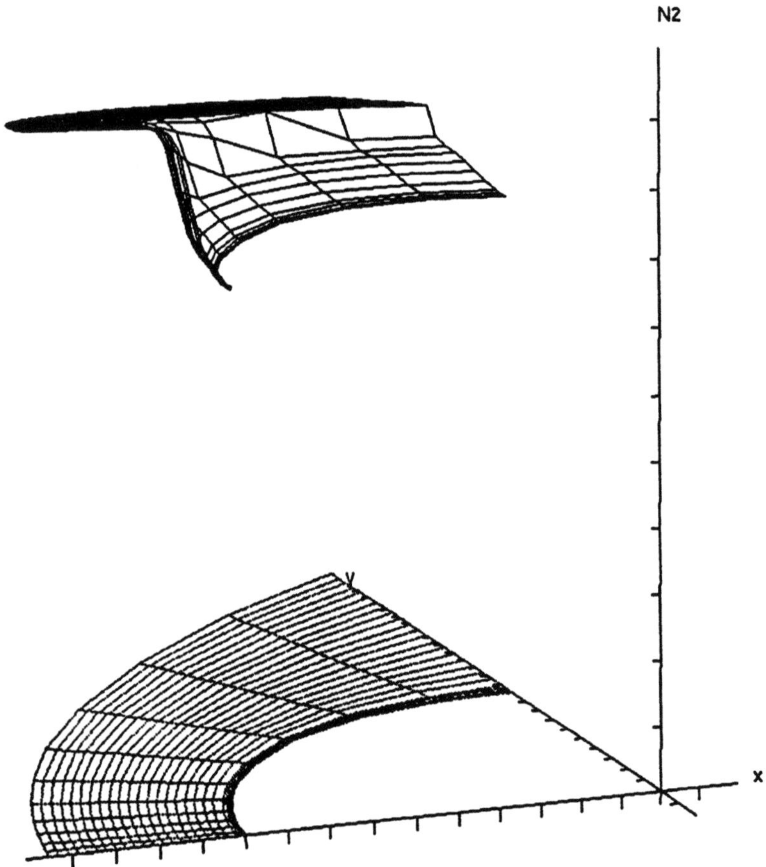

Figure 14: Profile of nitrogen molecules.

References

[1] J. Warnatz, Survey of Rate Coefficients in the C/H/O System, Sandia Report SAND83-8606. Sandia National Laboratories, Livermore (1983).

[2] J. Warnatz, *in:* W. C. Gardiner (ed.), *Combustion Chemistry.* Springer, New York (1984).

[3] R. K. Hanson, S. Salimian, *in:* W. C. Gardiner (ed.), *Combustion Chemistry.* Springer, New York (1984).

[4] J. O. Hirschfelder, 9th Symposium (International) on Combustion, p. 553. Academic Press, New York (1963).

[5] K. H. Ebert, P. Deuflhard, W. Jäger (eds.), *Modelling of Chemical Reaction Systems.* Springer, Heidelberg/New York (1981).

[6] J. Warnatz, W. Jäger (eds.), *Complex Chemical Reaction Systems: Mathematical Modelling and Simulation.* Springer, Heidelberg/New York (1987).

[7] D. R. Stull, H. Prophet, JANAF Thermochemical Tables. U. S. Department of Commerce, Washington D. C. (1971), and addenda.

[8] A. Burcat, *in:* W. C. Gardiner (ed.), *Combustion Chemistry.* Springer, New York (1984).

[9] B. J. McBride, to be published.

[10] J. Troe, *J. Chem. Phys.* **87**, 2773 (1987).

[11] T. R. A. Bussing, S. Eberhardt, Chemistry Associated with Hypersonic Vehicles. AIAA 22nd Thermophysics Conference, Honululu (1987).

[12] M. H. Bortner, A Review of Rate Constants of Selected Reactions of Interest in Re–Entry Flow Fields in the Atmosphere, NBS Technical Note 484. U. S. Department of Commerce, Washington D. C. (1969).

[13] ESA-HERMES Meeting on Reactive Flow. Kaiserslautern (1988).

[14] J. Warnatz, Air Dissociation Thermochemistry and Problems Resulting from the Coupling of Flow and Chemistry. *In* J. J. Bertin, R. ·Glowinski, J. Periaux (Eds.), *Hypersonics, Volume I – Defining the Hypersonic Environment.* Birkhäuser, Boston (1989).

[15] D. L. Baulch, D. D. Drysdale, D. G. Horne, Evaluated Kinetic Data for High Temperature Reactions, Vol. 2. Butterworths, London (1973)

[16] K. Thielen, P.Roth, 21st Symposium (International) on Combustion, p. 685. The Combustion Institute, Pittsburgh (1985).

[17] J. P. Monat, R. K. Hanson, C. H. Kruger, 17th Symposium (International) on Combustion, p. 543. The Combustion Institute, Pittsburgh (1979).

[18] K. L. Wray, 10th Symposium (International) on Combustion, p. 523. The Combustion Institute, Pittsburgh (1965).

[19] C. Park, AIAA Paper 85-0247, AIAA 23rd Aerospace Sciences Meeting, Reno, NV (1985).

[20] K. L. Wray, J. D. Teare, *J. Chem. Phys.* **36**, 2582 (1962).

[21] M. Caracotsios, W. E. Stewart, *Computers & Chemical Engineering* **4**, 359 (1983).

[22] J. R. Leis, M. A. Kramer, *Computers & Chemical Engineering* **9**, 93 (1985)).

[23] P. Glarbourg, J. A. Miller, R. J. Kee, *Combust. Flame* **65**, 177 (1986).

[24] U. Nowak, J. Warnatz, Proc. 11th International Colloquium on Dynamics of Explosions and Reactive Systems (1988), in press.

[25] P. Deuflhard, E. Hairer, J. Zugck, One-Step and Extrapolation Methods for Differential/ Algebraic Systems, Univ. Heidelberg, SFB 123: Tech. Rep. 318, (1985)

[26] A. Lifshitz, *J. Chem. Phys.* **61**, 2478 (1974).

[27] V. N. Kondratiev, E. E. Nikitin, *Gas Phase Reactions.* Springer, Berlin, Heidelberg, New York (1981).

[28] R. C. Millikan, D. R. White, *J. Chem. Phys.* **39**, 3209 (1963).

[29] G. D. Billing, Nonequilibrium Vibrational Kinetics. *In* M. Capitelli (Ed.), *Topics in Current Physics* **39**, Springer, Berlin, Heidelberg, New York (1986).

[30] C. Park, *Progr. in Astronautics and Aeronautics* **96**, 511 (1985).

[31] Y. Zhu, X. Wu, J. Warnatz, Computation of Non-Equilibrium Gas Flow Past Blunt Bodies. Report No. 488, SFB 123, Universität Heidelberg (1988). Submitted to *Computers and Fluids*.

[32] R. B. Bird, W. E. Stewart, E. N. Lightfoot, *Transport Phenomena*. Wiley, New York (1960).

[33] J. O. Hirschfelder, C. F. Curtiss, R. B. Bird, *Molecular Theory of Gases and Liquids*. Wiley, New York (1954).

[34] R. J. Kee, J. Warnatz, J. A. Miller, A Fortran Program Computer Code for the Evaluation of Gas–Phase Viscosities, Conductivities, and Diffusion Coefficients. SANDIA Report SAND83–8209 (1983).

[35] R. J. Kee, G. Dixon–Lewis, J. Warnatz, M. E. Coltrin, J. A. Miller, A Fortran Computer Code for the Evaluation of Gas–Phase Multicomponent Transport Properties. SANDIA Report SAND86–8246 (1986).

[36] P. Deuflhard, E. Hairer, J. Zugck, *Num. Math.* **51**, 501 (1987).

[37] P. Deuflhard, U. Nowak, Extrapolation Integrators for Quasilinear Implicit ODEs. *In* P. Deuflhard, B. Enquist (Eds.), *Large Scale Scientific Computing, Progress in Scientific Computing*, Vol. 7, p. 37. Birkhäuser (1987).

[38] U. Maas, J. Warnatz, *Impact of Computing in Science and Engineering* **1**, 394 – 420 (1989).

[39] U. Maas, J. Warnatz, Detatiled Numerical Simulation of H_2–O_2 Ignition in Two–Dimensional Geometries, Interdisziplinäres Zentrum für Wissenschaftliches Rechnen der Universität Heidelberg, Technical Report 1 (1990). To be presented at ICDERS 1991, Nagoja.

[40] F. N. Fritsch, R. E. Carlson, *SIAM J. Numer. Anal.* **17**, 238 (1980).

[41] F. N. Fritsch, J. Butland, *SIAM J. Sci. Stat. Comput.* **5**, 300 (1984).

[42] U. S. Standard Atmosphere (1976), U. S. Government Printing Office, Washington, D.C. 20402.

A   Non–Thermal Reaction Scheme

```
**************************************************
----   O2 Dissociation                                AV        BV     EAV
**************************************************
O2(V0)   +M'     =O       +O       +M'      2.001E+13   0.0   494.00
O2(V1)   +M'     =O       +O       +M'      4.001E+13   0.0   475.38
O2(V2)   +M'     =O       +O       +M'      6.002E+13   0.0   457.06
O2(V3)   +M'     =O       +O       +M'      8.002E+13   0.0   439.02
O2(V4)   +M'     =O       +O       +M'      1.000E+14   0.0   421.27
O2(V5)   +M'     =O       +O       +M'      1.200E+14   0.0   403.81
O2(V6)   +M'     =O       +O       +M'      1.400E+14   0.0   386.64
O2(V7)   +M'     =O       +O       +M'      1.600E+14   0.0   369.75
O2(V8)   +M'     =O       +O       +M'      1.800E+14   0.0   353.16
O2(V9)   +M'     =O       +O       +M'      2.001E+14   0.0   336.85
O2(V10)  +M'     =O       +O       +M'      2.201E+14   0.0   320.84
O2(V11)  +M'     =O       +O       +M'      2.401E+14   0.0   305.11
O2(V12)  +M'     =O       +O       +M'      2.601E+14   0.0   289.67
O2(V13)  +M'     =O       +O       +M'      2.801E+14   0.0   274.52
O2(V14)  +M'     =O       +O       +M'      3.001E+14   0.0   259.66
O2(V15)  +M'     =O       +O       +M'      3.201E+14   0.0   245.09
O2(V16)  +M'     =O       +O       +M'      3.401E+14   0.0   230.80
O2(V17)  +M'     =O       +O       +M'      3.601E+14   0.0   216.81
O2(V18)  +M'     =O       +O       +M'      3.801E+14   0.0   203.10
O2(V19)  +M'     =O       +O       +M'      4.001E+14   0.0   189.68
O2(V20)  +M'     =O       +O       +M'      4.201E+14   0.0   176.55
O2(V21)  +M'     =O       +O       +M'      4.401E+14   0.0   163.71
O2(V22)  +M'     =O       +O       +M'      4.601E+14   0.0   151.16
O2(V23)  +M'     =O       +O       +M'      4.801E+14   0.0   138.90
O2(V24)  +M'     =O       +O       +M'      5.001E+14   0.0   126.93
O2(V25)  +M'     =O       +O       +M'      5.201E+14   0.0   115.24
O2(V26)  +M'     =O       +O       +M'      5.401E+14   0.0   103.85
O2(V27)  +M'     =O       +O       +M'      5.601E+14   0.0    92.74
O2(V28)  +M'     =O       +O       +M'      5.801E+14   0.0    81.92
O2(V29)  +M'     =O       +O       +M'      6.002E+14   0.0    71.39
O2(V30)  +M'     =O       +O       +M'      6.202E+14   0.0    61.15
O2(V31)  +M'     =O       +O       +M'      6.402E+14   0.0    51.20
O2(V32)  +M'     =O       +O       +M'      6.602E+14   0.0    41.53
O2(V33)  +M'     =O       +O       +M'      6.802E+14   0.0    32.16
O2(V34)  +M'     =O       +O       +M'      7.002E+14   0.0    23.07
O2(V35)  +M'     =O       +O       +M'      7.202E+14   0.0    14.28
O2(V36)  +M'     =O       +O       +M'      7.402E+14   0.0     5.77

**************************************************
----   N2 Dissociation                                AV        BV     EAV
**************************************************
N2(V0)   +M''    =N       +N       +M''     1.143E+15  -0.6   941.00
N2(V1)   +M''    =N       +N       +M''     2.286E+15  -0.6   913.13
N2(V2)   +M''    =N       +N       +M''     3.429E+15  -0.6   885.60
N2(V3)   +M''    =N       +N       +M''     4.572E+15  -0.6   858.41
N2(V4)   +M''    =N       +N       +M''     5.715E+15  -0.6   831.55
N2(V5)   +M''    =N       +N       +M''     6.858E+15  -0.6   805.03
N2(V6)   +M''    =N       +N       +M''     8.001E+15  -0.6   778.84
N2(V7)   +M''    =N       +N       +M''     9.144E+15  -0.6   753.00
N2(V8)   +M''    =N       +N       +M''     1.029E+16  -0.6   727.48
N2(V9)   +M''    =N       +N       +M''     1.143E+16  -0.6   702.31
N2(V10)  +M''    =N       +N       +M''     1.257E+16  -0.6   677.47
N2(V11)  +M''    =N       +N       +M''     1.372E+16  -0.6   652.97
N2(V12)  +M''    =N       +N       +M''     1.486E+16  -0.6   628.80
N2(V13)  +M''    =N       +N       +M''     1.600E+16  -0.6   604.97
N2(V14)  +M''    =N       +N       +M''     1.714E+16  -0.6   581.47
N2(V15)  +M''    =N       +N       +M''     1.829E+16  -0.6   558.32
N2(V16)  +M''    =N       +N       +M''     1.943E+16  -0.6   535.50
N2(V17)  +M''    =N       +N       +M''     2.057E+16  -0.6   513.01
N2(V18)  +M''    =N       +N       +M''     2.172E+16  -0.6   490.86
N2(V19)  +M''    =N       +N       +M''     2.286E+16  -0.6   469.05
N2(V20)  +M''    =N       +N       +M''     2.400E+16  -0.6   447.58
N2(V21)  +M''    =N       +N       +M''     2.514E+16  -0.6   426.44
N2(V22)  +M''    =N       +N       +M''     2.629E+16  -0.6   405.63
N2(V23)  +M''    =N       +N       +M''     2.743E+16  -0.6   385.17
N2(V24)  +M''    =N       +N       +M''     2.857E+16  -0.6   365.04
N2(V25)  +M''    =N       +N       +M''     2.972E+16  -0.6   345.25
N2(V26)  +M''    =N       +N       +M''     3.086E+16  -0.6   325.79
N2(V27)  +M''    =N       +N       +M''     3.200E+16  -0.6   306.67
```

				AV	BV	EAV
N2(V28) +M''	=N	+N	+M''	3.315E+16	-0.6	287.88
N2(V29) +M''	=N	+N	+M''	3.429E+16	-0.6	269.44
N2(V30) +M''	=N	+N	+M''	3.543E+16	-0.6	251.32
N2(V31) +M''	=N	+N	+M''	3.657E+16	-0.6	233.55
N2(V32) +M''	=N	+N	+M''	3.772E+16	-0.6	216.11
N2(V33) +M''	=N	+N	+M''	3.886E+16	-0.6	199.01
N2(V34) +M''	=N	+N	+M''	4.000E+16	-0.6	182.24
N2(V35) +M''	=N	+N	+M''	4.115E+16	-0.6	165.81
N2(V36) +M''	=N	+N	+M''	4.229E+16	-0.6	149.72
N2(V37) +M''	=N	+N	+M''	4.343E+16	-0.6	133.96
N2(V38) +M''	=N	+N	+M''	4.458E+16	-0.6	118.54
N2(V39) +M''	=N	+N	+M''	4.572E+16	-0.6	103.46
N2(V40) +M''	=N	+N	+M''	4.686E+16	-0.6	88.71
N2(V41) +M''	=N	+N	+M''	4.800E+16	-0.6	74.30
N2(V42) +M''	=N	+N	+M''	4.915E+16	-0.6	60.23
N2(V43) +M''	=N	+N	+M''	5.029E+16	-0.6	46.49
N2(V44) +M''	=N	+N	+M''	5.143E+16	-0.6	33.09
N2(V45) +M''	=N	+N	+M''	5.258E+16	-0.6	20.02
N2(V46) +M''	=N	+N	+M''	5.372E+16	-0.6	7.29

```
****************************************************                  AV       BV      EAV
----   NO Dissociation
****************************************************
NO       +M'''       =N       +O       +M'''        2.900E+15      0.0    621.00
```

```
****************************************************                  AV       BV      EAV
----   O2 Exchange reaction
****************************************************
```

			AV	BV	EAV
O2(V0) +N	=NO	+O	1.151E+09	1.0	37.37
O2(V1) +N	=NO	+O	2.302E+09	1.0	18.75
O2(V2) +N	=NO	+O	3.453E+09	1.0	0.43
O2(V3) +N	=NO	+O	4.603E+09	1.0	0.00
O2(V4) +N	=NO	+O	5.754E+09	1.0	0.00
O2(V5) +N	=NO	+O	6.905E+09	1.0	0.00
O2(V6) +N	=NO	+O	8.056E+09	1.0	0.00
O2(V7) +N	=NO	+O	9.207E+09	1.0	0.00
O2(V8) +N	=NO	+O	1.036E+10	1.0	0.00
O2(V9) +N	=NO	+O	1.151E+10	1.0	0.00
O2(V10) +N	=NO	+O	1.266E+10	1.0	0.00
O2(V11) +N	=NO	+O	1.381E+10	1.0	0.00
O2(V12) +N	=NO	+O	1.496E+10	1.0	0.00
O2(V13) +N	=NO	+O	1.611E+10	1.0	0.00
O2(V14) +N	=NO	+O	1.726E+10	1.0	0.00
O2(V15) +N	=NO	+O	1.841E+10	1.0	0.00
O2(V16) +N	=NO	+O	1.956E+10	1.0	0.00
O2(V17) +N	=NO	+O	2.072E+10	1.0	0.00
O2(V18) +N	=NO	+O	2.187E+10	1.0	0.00
O2(V19) +N	=NO	+O	2.302E+10	1.0	0.00
O2(V20) +N	=NO	+O	2.417E+10	1.0	0.00
O2(V21) +N	=NO	+O	2.532E+10	1.0	0.00
O2(V22) +N	=NO	+O	2.647E+10	1.0	0.00
O2(V23) +N	=NO	+O	2.762E+10	1.0	0.00
O2(V24) +N	=NO	+O	2.877E+10	1.0	0.00
O2(V25) +N	=NO	+O	2.992E+10	1.0	0.00
O2(V26) +N	=NO	+O	3.107E+10	1.0	0.00
O2(V27) +N	=NO	+O	3.222E+10	1.0	0.00
O2(V28) +N	=NO	+O	3.337E+10	1.0	0.00
O2(V29) +N	=NO	+O	3.453E+10	1.0	0.00
O2(V30) +N	=NO	+O	3.568E+10	1.0	0.00
O2(V31) +N	=NO	+O	3.683E+10	1.0	0.00
O2(V32) +N	=NO	+O	3.798E+10	1.0	0.00
O2(V33) +N	=NO	+O	3.913E+10	1.0	0.00
O2(V34) +N	=NO	+O	4.028E+10	1.0	0.00
O2(V35) +N	=NO	+O	4.143E+10	1.0	0.00
O2(V36) +N	=NO	+O	4.258E+10	1.0	0.00

```
****************************************************                  AV       BV      EAV
----   N2 Exchange reaction
****************************************************
```

			AV	BV	EAV
N2(V0) +O	=NO	+N	4.168E+12	0.0	319.00
N2(V1) +O	=NO	+N	8.336E+12	0.0	291.13
N2(V2) +O	=NO	+N	1.250E+13	0.0	263.60
N2(V3) +O	=NO	+N	1.667E+13	0.0	236.41

N2(V4)	+O	=NO	+N	2.084E+13	0.0	209.55
N2(V5)	+O	=NO	+N	2.501E+13	0.0	183.03
N2(V6)	+O	=NO	+N	2.918E+13	0.0	156.84
N2(V7)	+O	=NO	+N	3.334E+13	0.0	131.00
N2(V8)	+O	=NO	+N	3.751E+13	0.0	105.48
N2(V9)	+O	=NO	+N	4.168E+13	0.0	80.31
N2(V10)	+O	=NO	+N	4.585E+13	0.0	55.47
N2(V11)	+O	=NO	+N	5.002E+13	0.0	30.97
N2(V12)	+O	=NO	+N	5.418E+13	0.0	6.80
N2(V13)	+O	=NO	+N	5.835E+13	0.0	0.00
N2(V14)	+O	=NO	+N	6.252E+13	0.0	0.00
N2(V15)	+O	=NO	+N	6.669E+13	0.0	0.00
N2(V16)	+O	=NO	+N	7.086E+13	0.0	0.00
N2(V17)	+O	=NO	+N	7.502E+13	0.0	0.00
N2(V18)	+O	=NO	+N	7.919E+13	0.0	0.00
N2(V19)	+O	=NO	+N	8.336E+13	0.0	0.00
N2(V20)	+O	=NO	+N	8.753E+13	0.0	0.00
N2(V21)	+O	=NO	+N	9.169E+13	0.0	0.00
N2(V22)	+O	=NO	+N	9.586E+13	0.0	0.00
N2(V23)	+O	=NO	+N	1.000E+14	0.0	0.00
N2(V24)	+O	=NO	+N	1.042E+14	0.0	0.00
N2(V25)	+O	=NO	+N	1.084E+14	0.0	0.00
N2(V26)	+O	=NO	+N	1.125E+14	0.0	0.00
N2(V27)	+O	=NO	+N	1.167E+14	0.0	0.00
N2(V28)	+O	=NO	+N	1.209E+14	0.0	0.00
N2(V29)	+O	=NO	+N	1.250E+14	0.0	0.00
N2(V30)	+O	=NO	+N	1.292E+14	0.0	0.00
N2(V31)	+O	=NO	+N	1.334E+14	0.0	0.00
N2(V32)	+O	=NO	+N	1.375E+14	0.0	0.00
N2(V33)	+O	=NO	+N	1.417E+14	0.0	0.00
N2(V34)	+O	=NO	+N	1.459E+14	0.0	0.00
N2(V35)	+O	=NO	+N	1.500E+14	0.0	0.00
N2(V36)	+O	=NO	+N	1.542E+14	0.0	0.00
N2(V37)	+O	=NO	+N	1.584E+14	0.0	0.00
N2(V38)	+O	=NO	+N	1.625E+14	0.0	0.00
N2(V39)	+O	=NO	+N	1.667E+14	0.0	0.00
N2(V40)	+O	=NO	+N	1.709E+14	0.0	0.00
N2(V41)	+O	=NO	+N	1.751E+14	0.0	0.00
N2(V42)	+O	=NO	+N	1.792E+14	0.0	0.00
N2(V43)	+O	=NO	+N	1.834E+14	0.0	0.00
N2(V44)	+O	=NO	+N	1.876E+14	0.0	0.00
N2(V45)	+O	=NO	+N	1.917E+14	0.0	0.00
N2(V46)	+O	=NO	+N	1.959E+14	0.0	0.00

```
********************************************
----  O2 V-T energy transfer
********************************************
```

O2(V1)	+M	=O2(V0)	+M	6.719E+17	0.000E+00
				-2.139E+02	-1.737E+01
				8.989E+12	1.862E+01
O2(V2)	+M	=O2(V1)	+M	1.354E+18	9.599E+00
				-2.139E+02	-1.737E+01
				8.989E+12	1.833E+01
O2(V3)	+M	=O2(V2)	+M	2.047E+18	1.920E+01
				-2.139E+02	-1.737E+01
				8.989E+12	1.804E+01
O2(V4)	+M	=O2(V3)	+M	2.751E+18	2.880E+01
				-2.139E+02	-1.737E+01
				8.989E+12	1.775E+01
O2(V5)	+M	=O2(V4)	+M	3.466E+18	3.839E+01
				-2.139E+02	-1.737E+01
				8.989E+12	1.746E+01
O2(V6)	+M	=O2(V5)	+M	4.193E+18	4.799E+01
				-2.139E+02	-1.737E+01
				8.989E+12	1.717E+01
O2(V7)	+M	=O2(V6)	+M	4.931E+18	5.759E+01
				-2.139E+02	-1.737E+01
				8.989E+12	1.688E+01
O2(V8)	+M	=O2(V7)	+M	5.681E+18	6.719E+01
				-2.139E+02	-1.737E+01
				8.989E+12	1.659E+01
O2(V9)	+M	=O2(V8)	+M	6.444E+18	7.679E+01
				-2.139E+02	-1.737E+01
				8.989E+12	1.631E+01

Reaction		7.219E+18	8.639E+01
O2(V10) +M	=O2(V9) +M	7.219E+18	8.639E+01
		-2.139E+02	-1.737E+01
		8.989E+12	1.602E+01
O2(V11) +M	=O2(V10) +M	8.007E+18	9.599E+01
		-2.139E+02	-1.737E+01
		8.989E+12	1.573E+01
O2(V12) +M	=O2(V11) +M	8.808E+18	1.056E+02
		-2.139E+02	-1.737E+01
		8.989E+12	1.544E+01
O2(V13) +M	=O2(V12) +M	9.623E+18	1.152E+02
		-2.139E+02	-1.737E+01
		8.989E+12	1.515E+01
O2(V14) +M	=O2(V13) +M	1.045E+19	1.248E+02
		-2.139E+02	-1.737E+01
		8.989E+12	1.486E+01
O2(V15) +M	=O2(V14) +M	1.130E+19	1.344E+02
		-2.139E+02	-1.737E+01
		8.989E+12	1.457E+01
O2(V16) +M	=O2(V15) +M	1.215E+19	1.440E+02
		-2.139E+02	-1.737E+01
		8.989E+12	1.428E+01
O2(V17) +M	=O2(V16) +M	1.303E+19	1.536E+02
		-2.139E+02	-1.737E+01
		8.989E+12	1.400E+01
O2(V18) +M	=O2(V17) +M	1.391E+19	1.632E+02
		-2.139E+02	-1.737E+01
		8.989E+12	1.371E+01
O2(V19) +M	=O2(V18) +M	1.482E+19	1.728E+02
		-2.139E+02	-1.737E+01
		8.989E+12	1.342E+01
O2(V20) +M	=O2(V19) +M	1.574E+19	1.824E+02
		-2.139E+02	-1.737E+01
		8.989E+12	1.313E+01
O2(V21) +M	=O2(V20) +M	1.668E+19	1.920E+02
		-2.139E+02	-1.737E+01
		8.989E+12	1.284E+01
O2(V22) +M	=O2(V21) +M	1.763E+19	2.016E+02
		-2.139E+02	-1.737E+01
		8.989E+12	1.255E+01
O2(V23) +M	=O2(V22) +M	1.860E+19	2.112E+02
		-2.139E+02	-1.737E+01
		8.989E+12	1.226E+01
O2(V24) +M	=O2(V23) +M	1.959E+19	2.208E+02
		-2.139E+02	-1.737E+01
		8.989E+12	1.197E+01
O2(V25) +M	=O2(V24) +M	2.060E+19	2.304E+02
		-2.139E+02	-1.737E+01
		8.989E+12	1.168E+01
O2(V26) +M	=O2(V25) +M	2.163E+19	2.400E+02
		-2.139E+02	-1.737E+01
		8.989E+12	1.140E+01
O2(V27) +M	=O2(V26) +M	2.268E+19	2.496E+02
		-2.139E+02	-1.737E+01
		8.989E+12	1.111E+01
O2(V28) +M	=O2(V27) +M	2.375E+19	2.592E+02
		-2.139E+02	-1.737E+01
		8.989E+12	1.082E+01
O2(V29) +M	=O2(V28) +M	2.484E+19	2.688E+02
		-2.139E+02	-1.737E+01
		8.989E+12	1.053E+01
O2(V30) +M	=O2(V29) +M	2.595E+19	2.784E+02
		-2.139E+02	-1.737E+01
		8.989E+12	1.024E+01
O2(V31) +M	=O2(V30) +M	2.708E+19	2.880E+02
		-2.139E+02	-1.737E+01
		8.989E+12	9.952E+00
O2(V32) +M	=O2(V31) +M	2.824E+19	2.976E+02
		-2.139E+02	-1.737E+01
		8.989E+12	9.663E+00
O2(V33) +M	=O2(V32) +M	2.942E+19	3.072E+02
		-2.139E+02	-1.737E+01
		8.989E+12	9.375E+00
O2(V34) +M	=O2(V33) +M	3.062E+19	3.168E+02
		-2.139E+02	-1.737E+01
		8.989E+12	9.086E+00

```
O2(V35) +M        =O2(V34) +M                    3.185E+19    3.264E+02
                                                 -2.139E+02   -1.737E+01
                                                  8.989E+12    8.797E+00
O2(V36) +M        =O2(V35) +M                    3.311E+19    3.360E+02
                                                 -2.139E+02   -1.737E+01
                                                  8.989E+12    8.508E+00


*********************************************
----  N2 V-T energy transfer
*********************************************
N2(V1)  +M        =N2(V0)  +M                    1.717E+18    0.000E+00
                                                 -2.761E+02   -2.023E+01
                                                  9.288E+12    2.787E+01
N2(V2)  +M        =N2(V1)  +M                    3.455E+18    8.116E+00
                                                 -2.761E+02   -2.023E+01
                                                  9.288E+12    2.753E+01
N2(V3)  +M        =N2(V2)  +M                    5.214E+18    1.623E+01
                                                 -2.761E+02   -2.023E+01
                                                  9.288E+12    2.719E+01
N2(V4)  +M        =N2(V3)  +M                    6.995E+18    2.435E+01
                                                 -2.761E+02   -2.023E+01
                                                  9.288E+12    2.686E+01
N2(V5)  +M        =N2(V4)  +M                    8.797E+18    3.246E+01
                                                 -2.761E+02   -2.023E+01
                                                  9.288E+12    2.652E+01
N2(V6)  +M        =N2(V5)  +M                    1.062E+19    4.058E+01
                                                 -2.761E+02   -2.023E+01
                                                  9.288E+12    2.618E+01
N2(V7)  +M        =N2(V6)  +M                    1.247E+19    4.870E+01
                                                 -2.761E+02   -2.023E+01
                                                  9.288E+12    2.585E+01
N2(V8)  +M        =N2(V7)  +M                    1.434E+19    5.681E+01
                                                 -2.761E+02   -2.023E+01
                                                  9.288E+12    2.551E+01
N2(V9)  +M        =N2(V8)  +M                    1.623E+19    6.493E+01
                                                 -2.761E+02   -2.023E+01
                                                  9.288E+12    2.518E+01
N2(V10) +M        =N2(V9)  +M                    1.815E+19    7.304E+01
                                                 -2.761E+02   -2.023E+01
                                                  9.288E+12    2.484E+01
N2(V11) +M        =N2(V10) +M                    2.009E+19    8.116E+01
                                                 -2.761E+02   -2.023E+01
                                                  9.288E+12    2.450E+01
N2(V12) +M        =N2(V11) +M                    2.206E+19    8.928E+01
                                                 -2.761E+02   -2.023E+01
                                                  9.288E+12    2.417E+01
N2(V13) +M        =N2(V12) +M                    2.406E+19    9.739E+01
                                                 -2.761E+02   -2.023E+01
                                                  9.288E+12    2.383E+01
N2(V14) +M        =N2(V13) +M                    2.607E+19    1.055E+02
                                                 -2.761E+02   -2.023E+01
                                                  9.288E+12    2.349E+01
N2(V15) +M        =N2(V14) +M                    2.812E+19    1.136E+02
                                                 -2.761E+02   -2.023E+01
                                                  9.288E+12    2.316E+01
N2(V16) +M        =N2(V15) +M                    3.019E+19    1.217E+02
                                                 -2.761E+02   -2.023E+01
                                                  9.288E+12    2.282E+01
N2(V17) +M        =N2(V16) +M                    3.229E+19    1.299E+02
                                                 -2.761E+02   -2.023E+01
                                                  9.288E+12    2.248E+01
N2(V18) +M        =N2(V17) +M                    3.442E+19    1.380E+02
                                                 -2.761E+02   -2.023E+01
                                                  9.288E+12    2.215E+01
N2(V19) +M        =N2(V18) +M                    3.658E+19    1.461E+02
                                                 -2.761E+02   -2.023E+01
                                                  9.288E+12    2.181E+01
N2(V20) +M        =N2(V19) +M                    3.876E+19    1.542E+02
                                                 -2.761E+02   -2.023E+01
                                                  9.288E+12    2.148E+01
N2(V21) +M        =N2(V20) +M                    4.098E+19    1.623E+02
                                                 -2.761E+02   -2.023E+01
                                                  9.288E+12    2.114E+01
N2(V22) +M        =N2(V21) +M                    4.322E+19    1.704E+02
                                                 -2.761E+02   -2.023E+01
```

```
                                                       9.288E+12    2.080E+01
N2(V23) +M          =N2(V22) +M                        4.550E+19    1.786E+02
                                                      -2.761E+02   -2.023E+01
                                                       9.288E+12    2.047E+01
N2(V24) +M          =N2(V23) +M                        4.781E+19    1.867E+02
                                                      -2.761E+02   -2.023E+01
                                                       9.288E+12    2.013E+01
N2(V25) +M          =N2(V24) +M                        5.015E+19    1.948E+02
                                                      -2.761E+02   -2.023E+01
                                                       9.288E+12    1.979E+01
N2(V26) +M          =N2(V25) +M                        5.252E+19    2.029E+02
                                                      -2.761E+02   -2.023E+01
                                                       9.288E+12    1.946E+01
N2(V27) +M          =N2(V26) +M                        5.493E+19    2.110E+02
                                                      -2.761E+02   -2.023E+01
                                                       9.288E+12    1.912E+01
N2(V28) +M          =N2(V27) +M                        5.738E+19    2.191E+02
                                                      -2.761E+02   -2.023E+01
                                                       9.288E+12    1.878E+01
N2(V29) +M          =N2(V28) +M                        5.985E+19    2.272E+02
                                                      -2.761E+02   -2.023E+01
                                                       9.288E+12    1.845E+01
N2(V30) +M          =N2(V29) +M                        6.237E+19    2.354E+02
                                                      -2.761E+02   -2.023E+01
                                                       9.288E+12    1.811E+01
N2(V31) +M          =N2(V30) +M                        6.492E+19    2.435E+02
                                                      -2.761E+02   -2.023E+01
                                                       9.288E+12    1.777E+01
N2(V32) +M          =N2(V31) +M                        6.750E+19    2.516E+02
                                                      -2.761E+02   -2.023E+01
                                                       9.288E+12    1.744E+01
N2(V33) +M          =N2(V32) +M                        7.013E+19    2.597E+02
                                                      -2.761E+02   -2.023E+01
                                                       9.288E+12    1.710E+01
N2(V34) +M          =N2(V33) +M                        7.280E+19    2.678E+02
                                                      -2.761E+02   -2.023E+01
                                                       9.288E+12    1.677E+01
N2(V35) +M          =N2(V34) +M                        7.550E+19    2.759E+02
                                                      -2.761E+02   -2.023E+01
                                                       9.288E+12    1.643E+01
N2(V36) +M          =N2(V35) +M                        7.825E+19    2.841E+02
                                                      -2.761E+02   -2.023E+01
                                                       9.288E+12    1.609E+01
N2(V37) +M          =N2(V36) +M                        8.104E+19    2.922E+02
                                                      -2.761E+02   -2.023E+01
                                                       9.288E+12    1.576E+01
N2(V38) +M          =N2(V37) +M                        8.387E+19    3.003E+02
                                                      -2.761E+02   -2.023E+01
                                                       9.288E+12    1.542E+01
N2(V39) +M          =N2(V38) +M                        8.675E+19    3.084E+02
                                                      -2.761E+02   -2.023E+01
                                                       9.288E+12    1.508E+01
N2(V40) +M          =N2(V39) +M                        8.967E+19    3.165E+02
                                                      -2.761E+02   -2.023E+01
                                                       9.288E+12    1.475E+01
N2(V41) +M          =N2(V40) +M                        9.264E+19    3.246E+02
                                                      -2.761E+02   -2.023E+01
                                                       9.288E+12    1.441E+01
N2(V42) +M          =N2(V41) +M                        9.565E+19    3.328E+02
                                                      -2.761E+02   -2.023E+01
                                                       9.288E+12    1.407E+01
N2(V43) +M          =N2(V42) +M                        9.871E+19    3.409E+02
                                                      -2.761E+02   -2.023E+01
                                                       9.288E+12    1.374E+01
N2(V44) +M          =N2(V43) +M                        1.018E+20    3.490E+02
                                                      -2.761E+02   -2.023E+01
                                                       9.288E+12    1.340E+01
N2(V45) +M          =N2(V44) +M                        1.050E+20    3.571E+02
                                                      -2.761E+02   -2.023E+01
                                                       9.288E+12    1.307E+01
N2(V46) +M          =N2(V45) +M                        1.082E+20    3.652E+02
                                                      -2.761E+02   -2.023E+01
                                                       9.288E+12    1.273E+01

*************************************************
```

MODELING OF HYPERSONIC REACTING FLOWS*

Chul Park†
NASA Ames Research Center, Moffett Field, CA 94035

Table of Contents

Nomenclature
Section 1. Importance of Thermochemical Nonequilibrium in Hypersonic Flow
Section 2. Breakdown of One-Temperature Kinetic Model
Section 3. Vibrational Relaxation
Section 4. Reaction Rates
Section 5. Conservation Equations for Chemical Variables
Section 6. Proof of Validity of Two-Temperature Model
Figures

Nomenclature

c	$= $ average molecular velocity ($\sqrt{8kT/\pi m}$)
C	$= $ reaction rate constant, $cm^3 mole^{-1} sec^{-1}$
C_p	$= $ heat capacity at constant pressure
C_v	$= $ heat capacity at constant volume
D	$= $ dissociation energy, erg
e	$= $ electronic charge
$\vec{E}$	$= $ electrical field
E_e	$= $ energy in electron translational and electronic excitation modes per unit volume ($\sum_i \epsilon_{ei} N_i + 1.5kT_e$), erg/cm^3
E_v	$= $ energy in vibrational mode per unit volume ($\sum_i \epsilon_{vi} N_i$), erg/cm^3
f	$= $ fraction of heat transfer by electron gas given to electron gas only
g	$= \tau_L/\tau_D$
I	$= $ ionization potential, erg
k	$= $ Boltzmann constant, 1.3805×10^{-16} erg/K
k_f	$= $ forward reaction rate coefficient
k_r	$= $ reverse reaction rate coefficient
$K(v, v')$	$= $ rate coefficient for collisional excitation from a vibrational state v to another state v', cm^3/sec
m_i	$= $ mass of a particle of species i
M	$= $ collisional transition moment, see Eq. (4)
n	$= $ pre-exponential power on temperature in the expression for reaction rate coefficient

† Head, Experimental Aerothermodynamics Section

N	$=$ number density, cm^{-3}
N_e	$=$ electron density, cm^{-3}
N_x	$=$ number density of colliding particles, cm^{-3}
p	$=$ pressure, atm
p_e	$=$ electron pressure $(N_e k T_e)$
q	$=$ heat transfer rate, $erg/(cm^2 sec)$
Q	$=$ partition function
r	$=$ the border between the low-lying and high-lying vibrational states
s	$=$ dissociation limit
t	$=$ time, sec
T	$=$ heavy particle translational-rotational temperature, K
T_a	$=$ average temperature $= \sqrt{T_v T}$, K
T_v	$=$ vibrational temperature, K
v, v'	$=$ vibrational quantum numbers in quantum mechanical description of a molecule, or vibrational energy indices in classical description of a molecule
$\vec{V_i}$	$=$ vectorial diffusion velocity of species i
$\vec{w}$	$=$ vectorial velocity (u, v, w)
ϵ_v^d	$=$ average vibrational energy per molecule of the dissociating molecule, erg
ϵ_e	$=$ average electronic excitation energy per particle, erg
ϵ_v	$=$ average vibrational excitation energy per particle, erg
ϵ_1	$=$ the energy level of the first excited vibrational state, erg
γ	$= C_p/C_v$
η	$=$ relative efficiency of collision in causing dissociation
κ	$=$ equivalent thermal conductivity
μ	$=$ equivalent mass of two colliding particles $(m_A m_B/(m_A+m_B)$
ν	$=$ collision frequency
ρ	$=$ normalized population (N/N_E)
σ	$=$ collision cross-section
τ_D	$=$ relaxation time in excitation of vibration by heavy particle collision given by diffusion theory
τ_e	$=$ relaxation time in excitation of vibration by electron collision
τ_L	$=$ relaxation time in excitation of vibration by heavy particle collision given by Landau-Teller theory
τ_v	$=$ relaxation time in excitation of vibration

Subscripts

c	$=$ continuum (dissociated) state
d	$=$ dissociation
e	$=$ electron
E	$=$ equilibrium
h	$=$ homogeneous solution
m	$=$ molecule
p	$=$ particular solution
r	$=$ chemico-kinetic phenomenon
v	$=$ vibration

x   = maximum value

Section 1. Importance of Thermochemical Nonequilibrium in Hypersonic Flows

The nonequilibrium thermochemical processes must be studied in order to better predict aerodynamic and heat transfer characteristics of a hypersonic vehicle (see Chapter 6 of Ref. 1). The effect of nonequilibrium chemistry on convective heat transfer rates is well known, and therefore will not be considered here. Chemical phenomena affect aerodynamic characteristics of the hypersonic vehicles, especially their pitching moments and trim angles of attack. Radiative heat transfer in gases are very sensitively affected by the thermochemical processes in the gases.

Figure 1 shows the experimental data and calculated results on the pitching moment characteristics of the raked-off blunted elliptic cone geometry[2] proposed for use with an aeroassisted orbital transfer vehicle[3]. As seen here, pitching moment is affected by the density ratio ρ_2/ρ_∞. The density ratio is dictated mostly by the $\gamma = C_p/C_v$, which is in turn determined by the thermochemical phenomena behind the shock wave. The intersect of a C_m curve with the horizontal axis, $C_m = 0$, is the trim angle of attack. As seen here, the trim angle is affected by the thermochemistry of the gas.

Figure 2 shows experimental data taken in a shock tube on the thickness of the shock layer over a 45° half-angle cone as a function of the distance from its apex, and compares them with the theoretical values based on the assumptions of a) frozen, b) equilibrium flow, and those calculated using a finite-rate chemistry model based on the conventional c) one-temperature model, and d) the recent multi-temperature model (see Ref. 4). The multi-temperature calculations have been carried out by Candler.[5] As seen here, only the multi-temperature model agrees with the experimental data. The conventional one-temperature model predicts the flow to be closer to equilibrium than actually is. Near the apex, the shock angle is large and is nearly equal to that of a frozen (that is, perfect-gas) flow value. At large distances away from the apex, the shock angle approaches that of an equilibrium flow. The wall pressure is dictated mostly by the local shock angle; it decreases with the distance. This decrease in wall pressure causes a positive (nose-up) pitching moment to a lifting body. As a result, the trim angle of attack and the slope of the pitching moment curve are both affected by this phenomenon.

In Figure 3, the measured shock stand-off distgances in a nitrogen flow over a circular cylinder are compared with the one-temperature model (see Ref. 6). The quantity Ω in the abscissa is the Damköhler number based on one-temperature model. Again, the one-temperature model predicts the flow to be closer to equilibrium than it actually is.

In Figure 4, an interferogram of the flow over a hemisphere-cylinder at an angle of attack is compared with the prediction made using the conventional one-temperature model with the rate coefficient arbitrarily divided by a factor of 10 (see Ref. 7 and Chapter 6 of Ref. 1). As seen here, theory and experiment are in fair agreement only when a slower rate coefficient is used.

In Figure 5, the shock stand-off distances over spheres measured in a ballistic range are shown (see Ref. 8). The theoretical predictions made using a one-temperature model were found to underpredict the shock stand-off distances.[5] This means that, again, the one-temperature model predicts the flow to be closer to equilibrium than it actually is.

In Figure 6, the existing experimental data on the chemical relaxation times are compared with the predictions made using the two-temperature model (see Ref. 9). Agreement is fairly good here. Although not shown, the one-temperature model also agrees with the experimental data at the low end of the shock velocity, i.e., at 4 km/sec. However, at higher velocities, the one-temperature model underestimates the reaction times.

Before the Apollo vehicles were flown, wind tunnel experiments were conducted to determine the trim angle of attack. During the entry flights of the Apollo vehicles, the trim angles were found to be different from what were predicted from the wind tunnel tests; a nose-up pitching moment developed in all cases (see Ref. 10 and Chapter 8 of Ref. 1). Figure 7 shows the difference between the trim angles observed during the flight and the predicted values.

A similar phenomenon occurred during the entry flights of the Space Shuttle vehicles: a nose-up pitching moment developed, that is, the center-of-pressure moved forward (see Ref. 11 and Chapter 8 of Ref. 1). In Figure 8, the difference between the measured center-of-pressure and that predicted by the wind tunnel tests are shown. The nose-up pitching moment seen in both Apollo and Space Shuttle flights is attributed commonly to the high-temperature real-gas effects.[11]

In Figure 9, the boundary-layer displacement phenomenon is illustrated for the perfect-gas (frozen), nonequilibrium, and equilibrium flows. The displacement thickness is thinner for reacting flows. Since the boundary-layer behaves as that of a perfect gas near the leading edge and as that of an equilibrium flow toward the trailing edge, the boundary-layer becomes comparatively thinner toward the trailing edge. This causes the shock wave to curve (convex) and the surface pressure to drop along the wall. In the flow over the windward side of a wing, this phenomenon causes a nose-up pitching moment.

These examples show that the conventional method of predicting chemical reactions always predict that the flow is closer to equilibrium than it actually is, and that such an incorrect assessment of chemical rate processes may lead to incorrect predictions of aerodynanamic characteristics of a vehicle. Because of this mistake, most people thought that the flight regime of the most hypersonic vehicles would be in the equilibrium regime, while, in reality, they would be in the nonequilibrium regime. The mistake is caused by using the one-temperature model. The purpose of this lecture is to learn how to deal with the problem correctly, namely, how to handle the chemical problems using the two-temperature model.

Section 2. Breakdown of One-Temperature Kinetic Model

Finite-rate chemical reactions have been studied for many decades mostly in connection with combustion or chemical manufacturing processes. In such applications, temperature was below about 3000 K. The rate coefficients for chemical reactions are known fairly accurately at such temperatures. Dissociation rate data for most gases exist to a temperature of up to about 8000 K, except for N_2 for which data exist to 14,000 K. The post-shock temperture behind the normal shock in the flight regime of the Space Shuttle orbiter is about 30,000 K. The use of the existing chemical model to this high temperature regime without modification risks the following errors:

(1) Pre-exponential Power - Existing experimental data on rate coefficients have been

fitted with an Arrhynius expression of the form

$$k_f = CT^n \exp(-D/kT) \tag{1}$$

Usually the range of temperatures over which the experimental data are taken is too small for the power n to be determined accurately. When an existing reaction rate data are extrapolated to high temperatures, one finds that the resulting k_f values imply a reaction cross-section which is unrealistically large.

(2) Influence of Vibrational Nonequilibrium[12] - It is intuitively clear that dissociation of a molecular species will occur preferentially from the high vibrational states. If the populations of varius vibrational states are characterized by a vibrational temperature T_v, then the rate coefficients must be a function not only of the translational-rotational temperature T but also of T_v. At relatively low temperatures (below 5000 K, say), the D/kT in Eq. (1) is so large that dissociation reactions do not occur very fast. However, such a temperaure is sufficiently large to overcome the vibrational energy gap. As a result, at such relatively low temperatures, vibrational relaxation reaches equilibrium prior to significant dissociation. Therefore, one is assured of T_v = T during the chemical reaction, and therefore there is only one temperature T controlling chemical reactions. However, at high temperatures, the exponential factor is sufficiently large for the reactions to occur simultaneously with the vibrational excitation. Hence, one can not assume $T = T_v$ at high temperatures.

(3) Breakdown of Landau-Teller Equation - The vibrational relaxation phenomenon has been traditionally described using the Landau-Teller equation for the harmonic oscillator of the form[13]

$$\frac{\partial \epsilon_v}{\partial t}\bigg|_r = \frac{\epsilon_{vE} - \epsilon_v}{\tau_L} \tag{2}$$

The relaxation time τ_L has been expressed by[13]

$$\tau_L = \frac{1}{p}\exp[A(T^{-1/3} - 0.015\mu^{1/4}) - 18.42] \quad \sec \tag{3}$$

whre p is pressure in atmospheres, and A is a parameter determined by the molecules involved. This expression breaks down at high temperatures for the following reasons: bimodal behavior of the vibration-vibration energy transfer, limiting cross-section phenomena, prefential removal of high vibrational states in dissociation, and diffusive nature of vibrational relaxation. These phenomena will be discussed in detail below.

Section 3. Vibrational Relaxation

3.1. Bimodal Behavior of Vibration-Vibration Energy Transfer

Figure 10 shows the so-called second moment of vibrational transitions in N_2 calculated by the Landau-Teller model and a more accurate model.[14] The second moment, $M(v)$, is defined by

$$M = \frac{1}{2}\int_{-\infty}^{\infty} K(v, v + \Delta v)(\Delta v)^2 d(\Delta v) \tag{4}$$

where $K(v, v + \Delta v)$ is the rate coefficients for vibrational transitions from the initial state v to the final state $v'=v+\Delta v$. As will be shown later, M controls vibrational relaxation behavior in a gas. In the Landau-Teller model, M increases monotonically with v. However, in a more accurate calculation, $M(v)$ dips deep in the mid-range of v. This is caused by the anharmonic nature of the molecular potential. In the relatively low temperature regime (T $\leq$ 5,000 K), only the first few vibrational levels are excited and so the dip does not matter. However, at higher temperatures, this dip significantly lowers the vibrational relaxation rates and dissociation/recombination rates. This causes a bimodal distribution of the vibrational states, as shown in Figure 11.

3.2. Limiting Cross-Section[15]

At very high temperatures, Eq. (3) implies unrealistically large reaction cross-sections. Denoting the cross-section at infinitely large temperure by σ_v, the relaxation time τ can not be smaller than the limiting, minimum time τ_x given by

$$\tau_x = \frac{1}{N_x \sigma_v c}$$

where c is the thermal speed and σ_v is a cross-section. This deficiency can be corrected for by

$$\tau'_L = \tau_L + \tau_x \tag{5}$$

By comparing the calculated radiation characteristics with experimental data, σ_v was deduced to be

$$\sigma_v = d(50000/T)^2 \quad cm^2$$

where d is between 10^{-17} and 3×10^{-17} cm^2 for most gases.[9].

3.3. Preferential High-Vibrational-State Removal by Dissociation[16,17]

Since dissociation occurs preferentially from the high vibrational states, the rate of change of vibrational energy per unit volume E_v is affected by the amount

$$\left(\frac{\partial E_v}{\partial t}\right)_r = \epsilon_v^d \left(\frac{\partial N_m}{\partial t}\right) \tag{6}$$

where ϵ_v^d is the average vibrational energy lost during dissociation. The exact value of ϵ_v^d is still unknown. However, it is estimated to be at least 30% of the dissociation energy of the molecule.[13,15,16]

3.4. Diffusive Nature of Vibrational Relaxation[18]

Imagine a molecular species AB undergoing dissociation: AB $\rightarrow$ A + B. The rate of change of the number density of the vibrational state v is described by the master equation

$$\frac{\partial N_v}{\partial t} = N_x \sum_0^s K(v',v)N_{v'} + K(c,v)N_A N_B N_x$$

$$- N_x \sum_0^s K(v,v')N_v - K(v,c)N_v N_x \tag{7}$$

We introduce the equilibrium number density of the v-th vibrational state N_{vE} in reference to the given number density of the molecule N_m

$$N_{vE} = N_m(Q_v/Q_m) \tag{8}$$

We define the normalized population ρ_v by

$$\rho_v = N_v/N_{vE} \tag{9}$$

We then divide both sides of Eq. (7) by $N_x N_{vE}$, use the detailed balance relationship

$$K(v,v')N_{vE} = K(v',v)N_{v'E} \tag{10}$$

Then, one obtains

$$\frac{1}{N_x}\frac{\partial \rho_v}{\partial t} = \sum_0^s K(v,v')(\rho_{v'} - \rho_v) + K(v,c)(\rho_A \rho_B - \rho_v) \tag{11}$$

For the high temperature region of concern where the kinetic energy kT is larger than the vibrational energy gap, collisional excitation and deexcitation of vibrational states occur almost in accordance with classical mechanics. According to the classical concept, vibrational levels are continuously distributed. The quantum numbers v and v' are now considered to represent vibrational energies. The summation in Equation (11) must be replaced by an integration

$$\frac{1}{N_x}\frac{\partial \rho_v}{\partial t} = \int_0^s K(v,v')(\rho_{v'} - \rho_v)dv' + K(v,c)(\rho_A \rho_B - \rho_v) \tag{12}$$

The integro-differential equation can be converted to a differential equation if M given by Eq. (4) is bounded. If it is bounded, one can show that the master Eq. (12) reduces to (see Chapter 3 of Ref. 1)

$$\frac{1}{N_x}\frac{\partial \rho(v)}{\partial t} = \frac{\partial}{\partial v}\left(M\frac{\partial \rho(v)}{\partial v}\right) + K(v,c)[\rho_A \rho_B - \rho(v)] \tag{13}$$

This is a diffusion equation in one dimension with distributed sources and sinks.

In Figure 11, the normalized vibrational state populations are shown behind a shock wave at different times.[14] The low vibrational states do indeed relax according to the diffusion description.

3.5. Vibrational Relaxation Time by Diffusion Theory

Vibrational energy is contained mostly in the low lying levels. To determine the behavior of the low vibrational states (see Chapter 3 of Ref. 1), the second term in the right-hand side of (13) can be neglected, so that

$$\frac{1}{N_x}\frac{\partial \rho}{\partial t} = \frac{\partial}{\partial v}\left(M\frac{\partial \rho}{\partial v}\right) \tag{14}$$

Let us consider the case where M is constant. Defining the effective diffusivity κ by $\kappa=N_z M$, the solution to Eq. (14) becomes an equation of heat conduction through a one-dimensional rod

$$\frac{\partial \rho}{\partial t} = \kappa \frac{\partial^2 \rho}{\partial v^2} \tag{15}$$

For this purpose, the rod can be considered semi-infinite.

In the beginning, let us assume that gas temperature T is low, and so all molecules are in their ground state $v = 0$. At time $t = 0$, the gas is suddenly brought to a high temperature T. We will assume that T remains constant throughout. The ρ values are all zero at $t = 0$ except at $v = 0$. For simplicity, the ρ value at $v = 0$ will be taken as 1. One can show for this case (Chapter 3 of Ref. 1)

$$\frac{\partial T_v}{\partial t} = \frac{\pi \kappa}{2} \frac{T_v}{T} \left(\frac{D/v_s}{T_v T}\right)^2 \frac{(T - T_v)^3}{k^2} \tag{16}$$

Eq. (16) can be expressed in the form

$$\frac{\partial T_v}{\partial t} = \frac{T - T_v}{\tau_D} = g \frac{T - T_v}{\tau_L} \tag{17}$$

where the relaxation time τ_D is related to the Landau-Teller relaxation time τ_D by

$$\frac{\tau_D}{\tau_L} = \frac{2}{\pi} \frac{k T_v}{\epsilon_1} \frac{T^2}{(T - T_v)^2} = \frac{1}{g} \tag{18}$$

3.6. Bridging Between Landau-Teller and Diffusion Theories

Lee[19] has derived an expression similar to Eq. (17) for the diffusive regime empirically through numerical integration of Eq. (14) for the case where colliding particles are electrons. His results can be written as

$$\frac{\partial \epsilon_v}{\partial t} = \frac{\epsilon_{vE} - \epsilon_v}{\tau_D} = \frac{\epsilon_{vE} - \epsilon_v}{\tau_L'} \left|\frac{T - T_v}{T_s - T_v}\right|^{s-1} \tag{19}$$

If $s = 1$, Eq. (19) becomes identical to the Landau-Teller equation. At high temperatures, the most appropriate value of s was found by Lee to be 3.5. For $s = 3.5$, Eq. (19) gives nearly the same relaxation rate values as Eq. (18). For an intermediate temperature, s must be between 1 and 3.5. Park[9,20] proposed a bridging formula

$$s = 3.5 \exp(-5000/T_s). \tag{20}$$

Refs. 9 and 20 show that this selection of s results in a fair agreement between the measured and calculated radiation intensities behind a shock wave in both nitrogen and air.

Eqs. (18) and (19) can be represented by a single equation

$$\frac{\partial \epsilon_v}{\partial t} = g \frac{\epsilon_{vE} - \epsilon_v}{\tau_L'} \tag{21}$$

wherein the correction factor g refers to either τ'_L/τ_D in Eq. (18) or that defined in Eq. (19), depending on the formula chosen.

Section 4. Reaction Rates

When the thermodynamic state of a gas is undergoing such a rapid change that the internal state $\rho(v)$ does not satisfy the quasi-steady-state condition, in general one observes the following features (see Chapter 3 of Ref. 1):

(1) The low-lying vibrational states establish a distribution which can be characterized by the slopes of the population with respect to the vibrational variables at $v=0$, $(\partial\rho/\partial v)_0$. Vibrational temperature T_v can be defined from this slope, and the function ρ for the low states can be considered to be a function of T_v.

(2) The high-lying vibrational and rotational states tend to establish a quasi-steady-state distribution.

The border between the low levels which are characterized by T_v and the high levels populated according to a quasi-steady-state condition will be denoted by the subscript r. Since the population distribution for the low levels $v < v_r$ is taken to be known, there is no need to solve for them using the master equation: the master equation needs to be solved only for the high states $v_r < v < v_s$. In solving the master equation for the high states, the low state ρ values can be transposed to the right-hand side as known quantities. Because of the quasi-steady-state condition for the high levels, the master equation becomes a steady-state equation of the form

$$\frac{\partial}{\partial v}[M\frac{\partial\rho}{\partial v}] - K(v,c)\rho(v) = -K(v,c)\rho_A\rho_B - \int_0^r K(v',v)\rho(v')dv' \tag{22}$$

Since ρ for the low states is a function of T_v and T, the right-hand side is a function also of T_v and T.

The complete solution of Eq. (22) can be written as (see Chapter 3 of Ref. 1)

$$\rho = Homogeneous\ Solution + Particular\ Solution$$

The particular solution can be written as a sum of two parts in the form

$$Particular\ Solution = \rho_q + \rho_p\rho_A\rho_B \tag{23}$$

Here the first term ρ_q is the particular solution of Equation (22) with $\rho_A\rho_B$ set to zero. The second term $\rho_p\rho_A\rho_B$ is the particular solution with finite $\rho_A\rho_B$ minus ρ_q. The ρ_q and ρ_p are functions of T_v and T, but ρ_p is a function also of $\rho_A\rho_B$. Thus, the complete solution of Eq. (22) can be written as

$$\rho = \rho_h(T) + \rho_q(T,T_v) + \rho_p(T,T_v,\rho_A\rho_B)\rho_A\rho_B \tag{24}$$

where ρ_h is the homogeneous solution.

Since the free-bound and bound-free transitions contributing to dissociation and recombination occur mostly to and from the high states, k_f and k_r can be expressed approximately as

$$k_f = \int_r^s K(v,c)\frac{Q_v}{Q_m}(\rho_h + \rho_q)dv = k_f(T,T_v) \tag{25}$$

$$k_r = \int_r^s K(v,c)\frac{N_m}{N_{AE}N_{BE}}(1 - \frac{Q_v}{Q_m}\rho_p)dv = k_r(T,T_v,\rho_A\rho_B) \qquad (26)$$

Note in Eqs. (25) and (26) that the rate coefficients are not the sums of the rates of transitions, but are those weighted by the factors containing the homogeneous and particular solutions of the master equation. Recombination is significant only when $\rho_A\rho_B$ is large. When $\rho_A\rho_B$ is large, the second term in the right-hand side of Equation (23) becomes smaller than the first term, and therefore can be neglected. The solution ρ_p in this case is a function only of T. Consequently,

$$k_r = k_r(T) \qquad (27)$$

Park[9,20] has shown that the existing experimental data on nitrogen and air can be reproduced computationally if the rate coefficients k_f and k_r are assumed to be functions only of the average temperature T_a defined by

$$T_a = \sqrt{T_v T} \qquad (28)$$

Section 5. Conservation Equations for Chemical Variables[21]

5.1. Vibrational Energy

Because of the strong coupling between the vibrational modes of different molecules in air (Chapter 2 of Ref. 1), it is appropriate to assume that there exists only one vibrational temperature T_v in air. The vibrational energy per unit volume E_v is affected by two mechanisms: heat conduction within the vibrational mode and chemico-kinetic processes (Chapter 4 of Ref. 1). Conservation of vibrational energy can thus be expressed as

$$\frac{\partial E_v}{\partial t} + \vec{\nabla} \cdot \vec{w} E_v = -\vec{\nabla} \cdot \vec{q}_v + \dot{E}_v \qquad (29)$$

where $\vec{q}_v$ is the vector representing the flow of latent heat of vibration flowing within the vibrational mode, and E_v is the rate of change of E_v by the kinetic processes. The latent heat of vibration flows through the fluid via diffusion of vibrationally excited molecules. Hence, $\vec{q}_v$ can be written as

$$\vec{q}_v = \sum_{i=m} \epsilon_{vi} N_i \vec{V}_i \ .$$

The kinetic term is given by

$$\dot{E}_v = \sum_{i=m} N_i g(\frac{\epsilon_{vE} - \epsilon_v}{\tau_L'})_i + N_{N_2}\frac{e_{vE}(T_e) - e_v}{\tau_e} + \sum_{i=m} \epsilon_{vi}^d(\frac{\partial N_i}{\partial t})$$
$$+ \sum_i \eta_{vi} D_i(\frac{\partial N_i}{\partial t}) \qquad erg \cdot cm^3 sec^{-1} \ . \qquad (30)$$

The first term in the right-hand side expresses the rate of vibrational excitation by collisions with heavy particles. The limit on the summation, $k = m$, symbolizes that the summation extends over all molecules. The coefficient g is the correction for the Landau-Teller equation due to the diffusive nature of the vibrational relaxation defined by Eq. (21). The second term is the rate of vibrational excitation by collisions with electrons. This process occurs only for N_2. The relaxation time τ_e is given by Lee[21]. The third term represents the loss of vibrational energy during dissociation due to the preferential removal of high vibrational states. The fourth term is the rate of removal of vibrational energy contained in the colliding molecules. The quantity η_{vk} is the efficiency of the vibrational mode of molecule k in causing dissociation, and needs to be calculated from the kinetic theory.

Note that Eqs. (29) and (30) are slightly different from those given by Lee[21].

5.2. Electron-Electronic Energy

The equation of conservation of momentum for electron gas can be written as[21]

$$m_e N_e \frac{D\vec{w}}{Dt} == -\vec{\nabla}p_e - \sum_{i \neq e} \frac{m_e}{m_i} N_i \nu_i (\vec{V_i} - \vec{V_e}) - eN_e\vec{E} \ .$$

Because m_e is very small, the left-hand-side can be set to zero. Under the assumption of no electrical current, the second term in the right-hand side is zero, and so the electron momentum equation becomes

$$eN_e\vec{E} = -\vec{\nabla}p_e \ . \tag{31}$$

It is known that the low-lying electronic states of both atoms and molecules tend to be populated according to Boltzmann distribution as dictated by electron translational temperature (Chapter 3 of Ref. 1). Since most electronic excitation energy in a species is contained in these low-lying states even in a nonequilibrium flow, it is appropriate to evaluate electronic excitation energy of a species using electron temperature T_e. It follows from this that the sum of the electronic excitation energy and the kinetic energy of the electrons, $E_e = \sum_i \epsilon_{ei} N_i + 1.5\ N_e k T_e$, is a quantity for which a conservation equation must be written.

The electron-electronic energy per unit volume, E_e, is affected by three mechanisms: conduction by viscous transport processes, the chemico-kinetic processes, and the work done to electrons by the electrical field (Chapter 4 of Ref. 1). Conservation of electron-electronic energy can be written as

$$\frac{\partial E_e}{\partial t} + \vec{\nabla} \cdot \vec{w}(E_e + p_e) = -\vec{\nabla} \cdot [f\kappa_e \vec{\nabla}T_e + \sum_{i=1}^{n} \epsilon_{ei}(T_e)N_i\vec{V_i}]$$
$$+ \dot{E}_e - eN_e\vec{E} \cdot \vec{w} \ . \tag{32}$$

The first term containing the gradient of T_e represents conduction of electron translational energy into electron gas. The quantities f and κ_e must be calculated from the kinetic theory (see Chapter 4 of Ref. 1). The second term containing $\vec{V_i}$ represents the rate of transfer of electronic excitation energy by diffusion. The last term represents gain of energy by electron gas due to electric field.

The third term is

$$\dot{E}_e = - \sum_{i=ion} I_i \left(\frac{\partial N_i}{\partial t} \right) + D_{N_2} \left(\frac{\partial N_{N_2}}{\partial t} \right) + 2 N_e \sum_{i=all} \nu_i \frac{m_e}{m_i} \frac{3}{2} k(T - T_e)$$
$$- N_{N_2} \frac{\epsilon_{vE}(T_e) - \epsilon_v}{\tau_e} + \sum_{i=all} \epsilon_{ei} \left(\frac{\partial N_i}{\partial t} \right) - q_R \ . \tag{33}$$

The first term in the right-hand side of Eq. (33) represents the rate of removal of electron energy during electron-impact ionization. The second term represents the rate of removal of electron energy during electron-impact dissociation of N_2. This process is significant only for N_2. The third term represents the rate of transfer of kinetic energy of the heavy particles into electron gas by elastic collisions. The fourth term represents the rate of transfer of electron translational energy into vibrational mode of N_2 molecules. The fifth term represents the rate of addition of electronic excitationenergy by newly-created particles. The sixth term represents the rate of energy loss by radiation.

Note that Eqs. (32) and (33) are slightly different from those given by Lee.[21]

5.3. Two-Temperature Model

In the regime where the degree of ionization is substantial, that is, of the order of 0.01% or higher, the coupling between the translational energy of the electrons and the vibrational energy of N_2 represented by the second term in the right-hand side of Eq. (30) and the fourth term in Eq. (33) is so strong that T_e becomes closely equal to T_v. In such a case, one can assume $T_e = T_v$, and consider the sum of E_e and E_v as the conserved variable characterized by the common temperature. This leads to the two-temperature model considered in Refs. 9 and 20. The conservation equation for the combined energies is obtained by adding Eqs. (29) and (32) in the form:

$$\frac{\partial (E_v + E_e)}{\partial t} + \vec{\nabla} \cdot \vec{w}(E_v + E_e + p_e) = - \vec{\nabla} \cdot [\vec{q_v} + f \kappa_e \vec{\nabla} T_e$$
$$+ \sum_i \epsilon_{ei} N_i \vec{V_i}] + \dot{E}_v + \dot{E}_e - e N_e \vec{E} \cdot \vec{w} \ . \tag{34}$$

Section 6. Proof of Validity of Two-Temperature Model

Direct proof of validity of the two-temperature model described above is found in two types of experimental data: radiation and shock shapes. Radiation emission behind a shock wave is sensitively affected by the population of the internal states. Therefore, correctly reproducing the radiation phenomena in a nonequilibrium environment is a severe test of the model. Refs. 9 and 20 (see also Chapter 8 of Ref. 1) show that only by the use of the two-temperature model, can one numerically reproduce the existing experimental data on radiation taken in shock tubes, ballistic ranges, and in flights.

The fact that only a two-temperature model can correctly reproduce the observed shock shape for a 45° half-angle cone has already been pointed out in Introduction (see

Fig. 2). The fact that the same is true for a sphere is mentioned in Introduction and shown in Ref. 5. Additionally, for a circular cylinder, Ref. 22 shows that only the two-temperature model can correctly reproduce the observed interferogram pattern. This result is reproduced in Fig. 12.

References

[1]Park, C., *Nonequilibrium Hypersonic Aerothermodynamics,* John Wiley and Sons, 1990.

[2]Scott, C. D., Ried, R. C., Maraia, R. J., Li, C. P., and Derry, M., "An AOTV Aeroheating and Thermal Protection Study," AIAA Paper 84-1710, June 1984.

[3]Wells, W. L., "Measured and Predicted Aerodynamic Coefficients and Shock Shapes for Aeroassist Flight Experiment (AFE) Configuration," NASA Technical Paper 2956, January, 1990.

[4]Spurk, J. H., "Experimental and Numerical Nonequilibrium Flow Studies," AIAA Journal, Vol. 8, No. 6, June 1970, pp. 1039-1045.

[5]Candler, G. V., "On the Computation of Shock Shapes in Nonequilibrium Hypersonic Flows," AIAA Paper 89–312, 1989.

[6]Hornung, H., "Non-Equilibrium Dissociating Nitrogen Flow Over Spheres and Circular Cylinders," Journal of Fluid Mechanics, Vol. 64, 1974, pp. 149-176.

[7] Macrossan, M. N., and Stalker, R. J., "Afterbody Flow of a Dissociating Gas Downstream of a Blunt Nose," AIAA Paper 87-0407, January 1987.

[8]Lobb, R. K., "Experimental Measurement of Shock Detachment Distance on Spheres Fired in Air at Hypervelocities," The High Temperature Aspects of Hypersonic Flow, AGARDograph 68, edited by W. C. Nelson, Pergamon Press, 1964, pp. 519-527.

[9]Park, C., "Assessment of Two-Temperature Kinetic Model for Ionizing Air," Journal of Thermophysics and Heat Transfer, Vol. 3, 1989, pp. 233-244. 87-1574, 1987.

[10]Hillje, E. R., and Savage, R., "Status of Aerodynamic Characteristics of the Apollo Entry Configuration," AIAA Paper 68-1143, 1968.

[11]Romere, P. O., and Whitnah, A. M., "Space Shuttle Entry Longitudinal Aerodynamic Comparisons of Flights 1-4 with Preflight Predictions," Shuttle Performance: Lessons Learned," NASA CP 2283, compiled by J. P. Arrington and J. J. Jones, 1983, pp. 283-307.

[12]Hammerling, P., Teare, J. D., and Kivel, B., "Theory of Radiation from Luminous Shock Wave sin Nitrogen," Physics of Fluids, Vol. 2, 1959, pp. 422-426.

[13] Millikan, R. C., and White, D. R., "Systematics of Vibrational Relaxation," Journal of Chemical Physics, Vol. 139, 1963, pp. 3209-3213.

[14]Sharma, S. P., Huo, W. M., and Park, C., "The Rate Parameters for Coupled Vibration-Dissociation in a Generalized SSH Approximation," AIAA Paper 88-2714.

[15]Park, C., "Problem of Rate Chemistry in the Flight Regimes of Aeroassited Orbital Transfer Vehilcles," Thermal Design of Aeroassisted Orbital Transfer Vehicles: Progress in Astronautics and Aeronautics, Vol. 96, edited by H. F. Nelson, pp. 511-537.

[16]Treanor, C. E., and Marrone, P. V., "Effect of Dissociation on the Rate of Vibrational Relaxation," Physics of Fluids, Vol. 5, 1962, pp. 1022-1-26.

[17]Marrone, P. V., and Treanor, C. E., "Chemical Relaxation with Preferential Dissociation from Excited Vibrational Levels," Physics of Fluids, Vol. 6, 1963, pp. 1215-1221.

[18]Keck, J. C., "Diffusion Theory of Nonequilibrium Dissociation and Recombination," Journal of Chemical Physics, Vol. 43, 1965, pp. 2284-2298.

[19]Lee, J. H., "Electron-Impact Vibrational Excitation Rates in the Flowfield of Aeroassisted Orbital Transfer Vehicles," Thermophysical Aspects of Reentry Flows: Progress in Astronautics and Aeronautics, Vol. 102, edited by J. N. Moss and C. D. Scott, 1980, pp. 152-196.

[20]Park, C., "Assessment of Two-Temperature Kinetic Model for Dissociating and Weakly-Ionizing Nitrogen," Journal of Thermophysics and Heat Transfer, Vol. 2, 1988, pp. 8-16.

[21]Lee, J. H., "Basic Governing Equations for the Flight Regimes of Aeroassisted Orbital Transfer Vehicles," Thermal Design of Aeroassisted Orbital Transfer Vehilces: Progress in Astronautics and Aeronautics, Vol. 96, edited by H. F. Nelson, AIAA, pp. 3-53.

[22]Park, C., and Yoon, S., "A Fully-Coupled Implicit Method for Thermo-Chemical Nonequilibrium Air at Sub-Orbital Flight Speeds," Journal of Spacecraft and Rockets, Vol. 28, 1991, pp. 31-39.

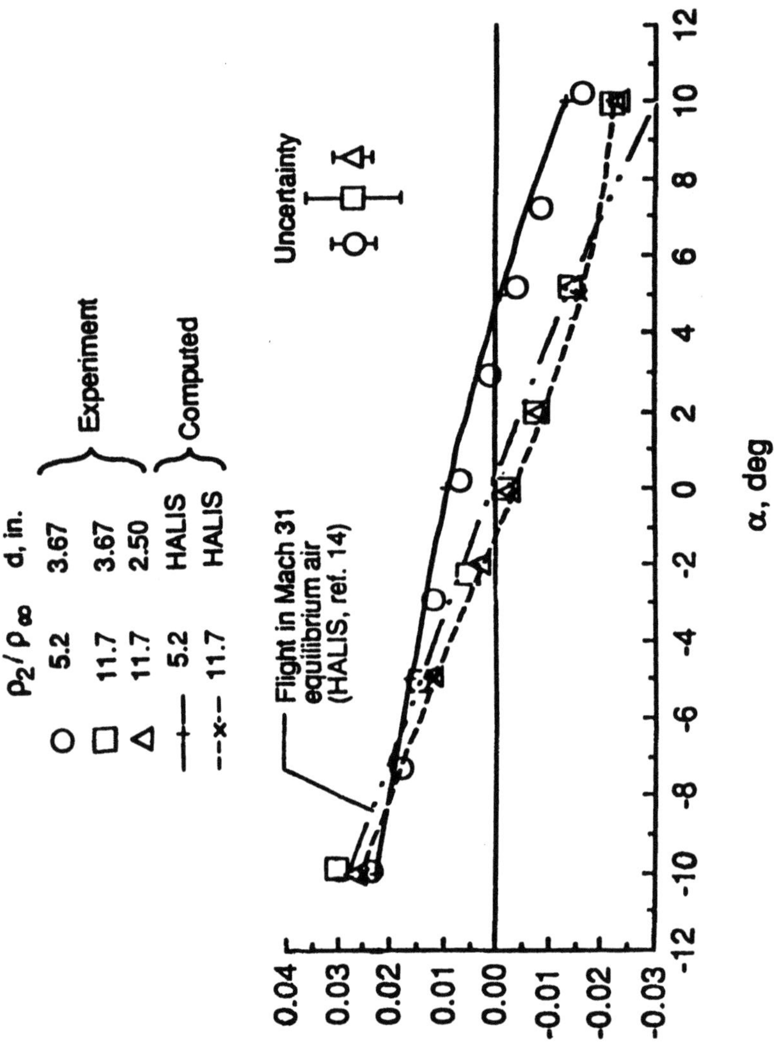

Figure 1. Pitching moment characteristics of the Aeroassist Flight Experiment (AFE) models tested in wind tunnels compared with theories (Ref. 3).

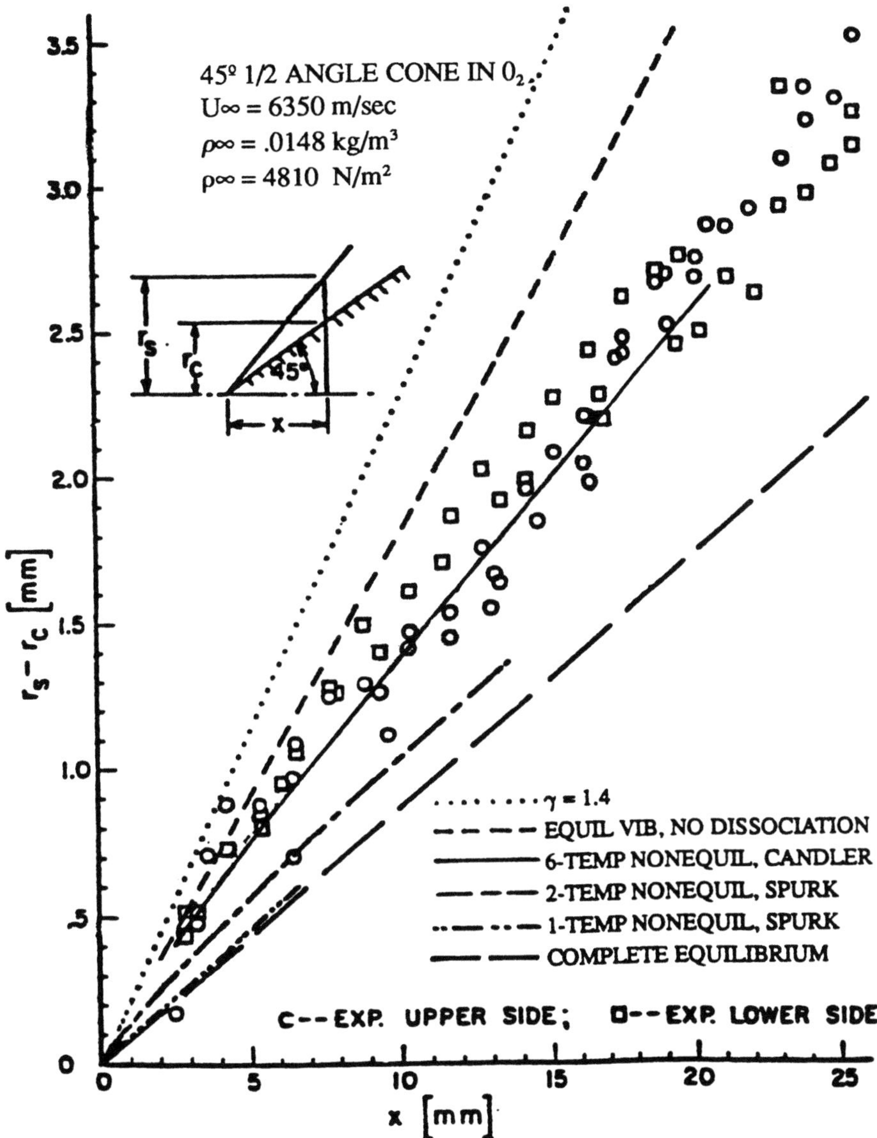

Figure 2. Experimental and theoretical shock layer thicknesses over a 45° cone at zero angle of attack (Refs. 4 and 5).

119

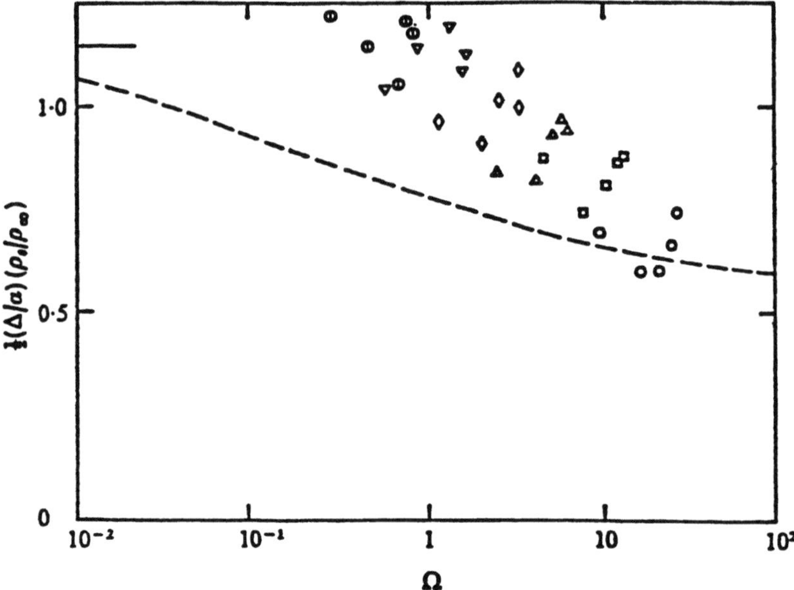

Figure 3. Shock stand-off distances over a circular cylinder in nonequilibrium nitrogen flow in a shock tunnel compared with those calculated using a one-tempeurature model, from Ref. 6. The abscissa are the flow Damköhler numbers. The ordinates are the normalized shock stand-off distances.

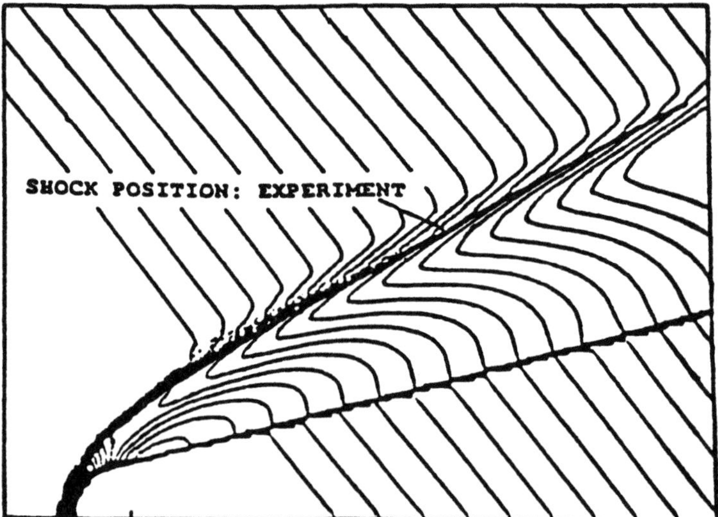

Figure 4. Interferogram of the flow over a sphere-cylinder inclined at an angle of attack in nitrogen flow compared with that calculated using a one-temperature model with the rate coefficient divided by 10, from Ref. 7.

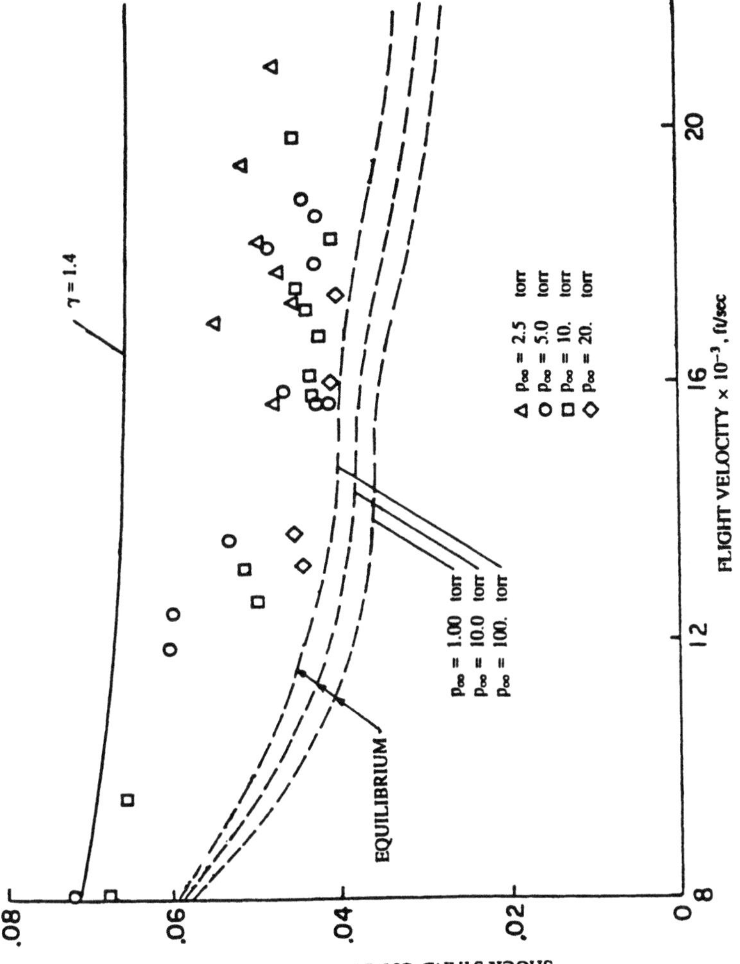

Figure 5. Shock stand-off distances over a sphere measured in a ballistic range, compared with the theoretical predictions for the perfect gas and the equilibrium gas, from Ref. 8.

121

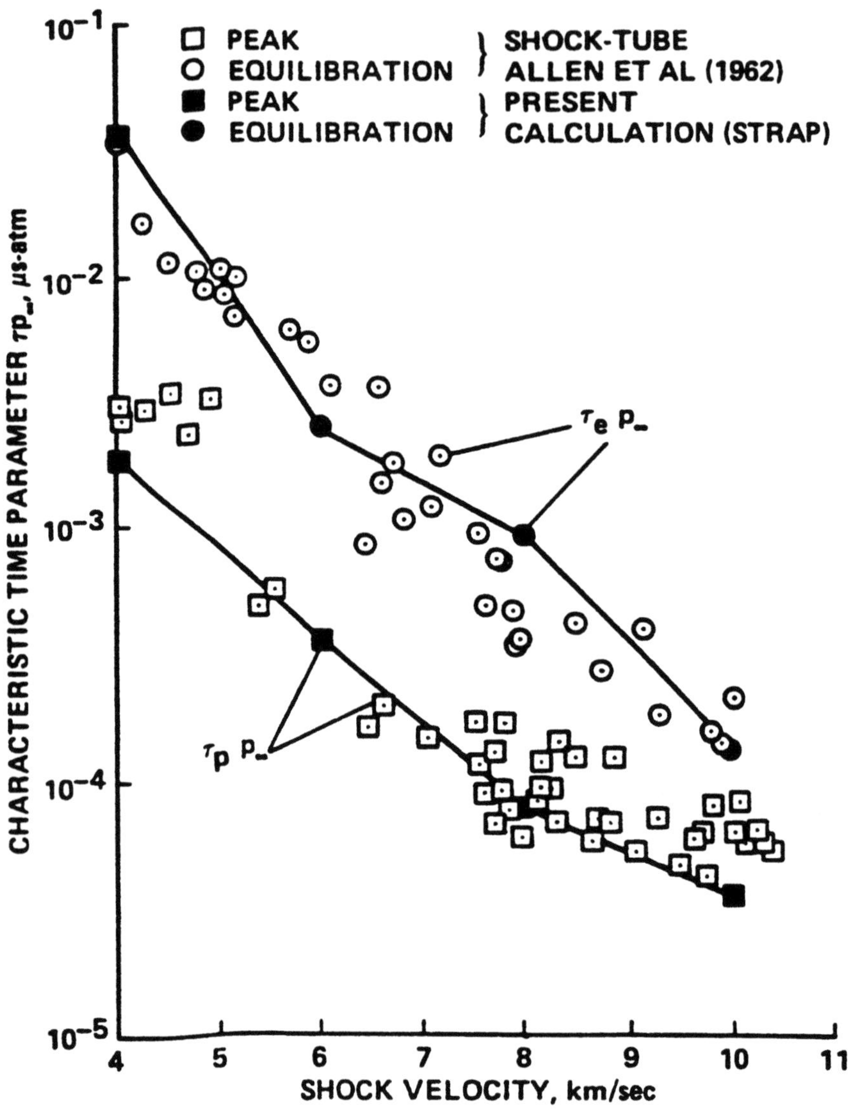

Figure 6. The experimental data on the characteristic relaxation times for chemical reactions in air compared with the theoretical predictions made using a two-temperature model, from Ref. 9.

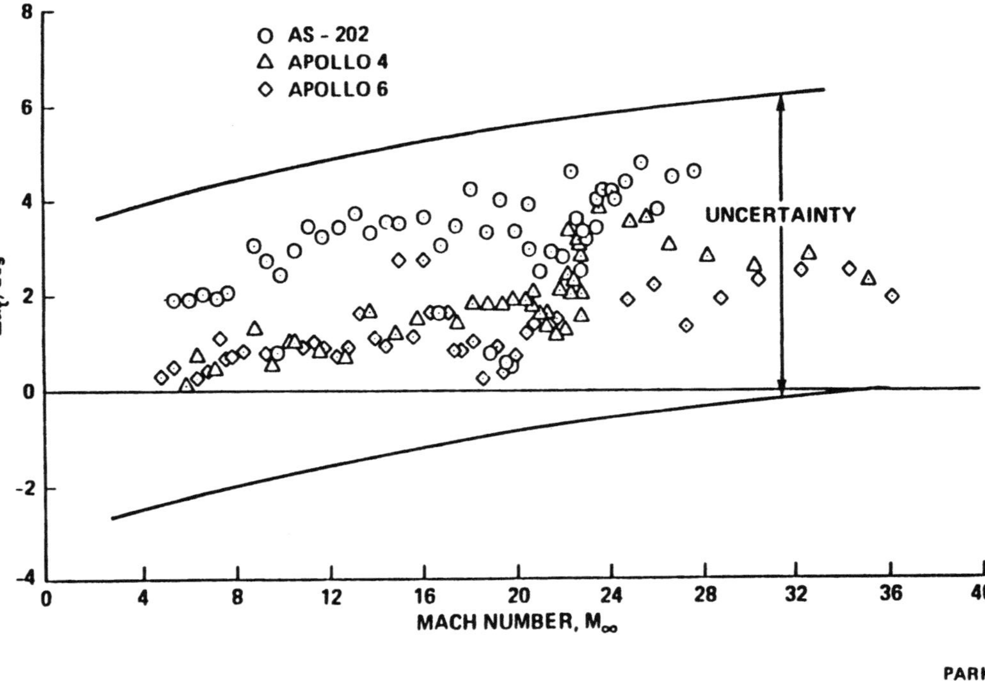

Figure 7. The difference between the trim angles of attack of the first three Apollo entry vehicles and those predicted from the wind tunnel tests, from Ref. 10.

123

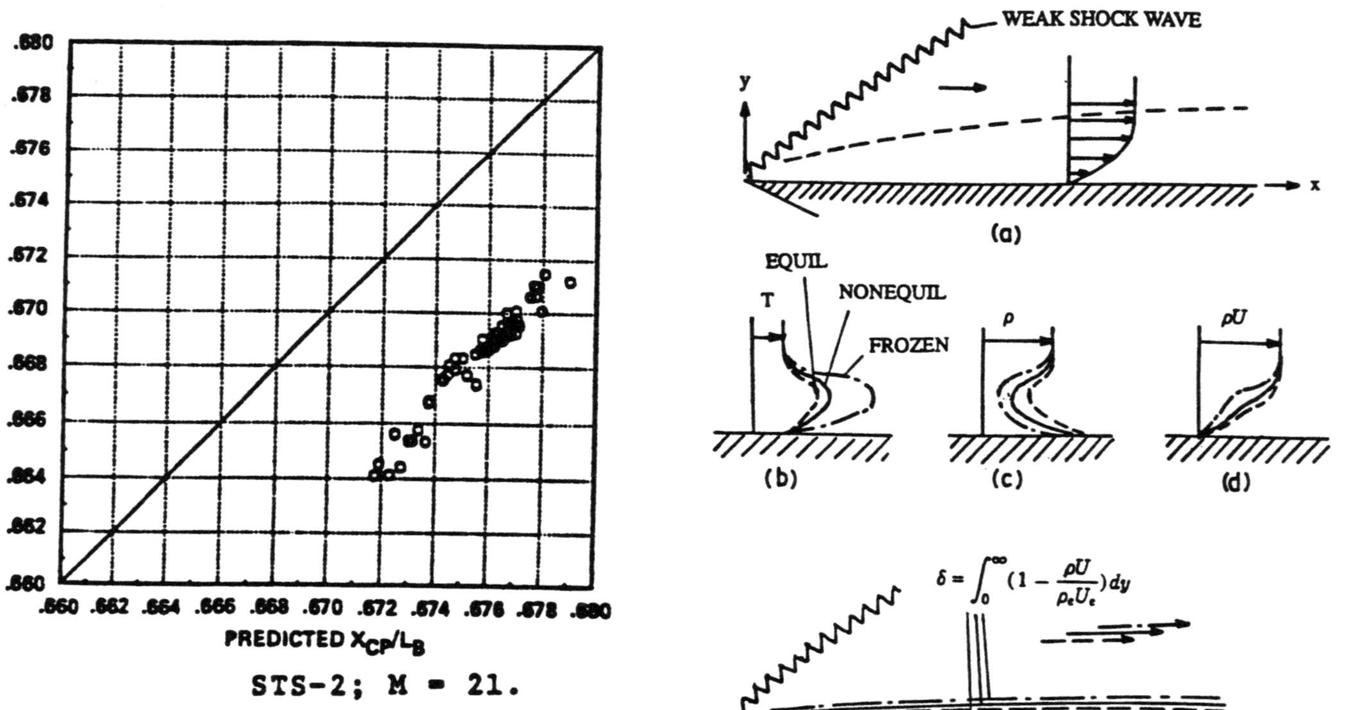

Figure 8. The difference between the center of pressure for the Space Shuttle vehicle SST-2 and that predicted by the wind tunnel tests, from Ref. 11.

Figure 9. The schematic of the boundary layer displacement phenomenon in a reacting flow.

124

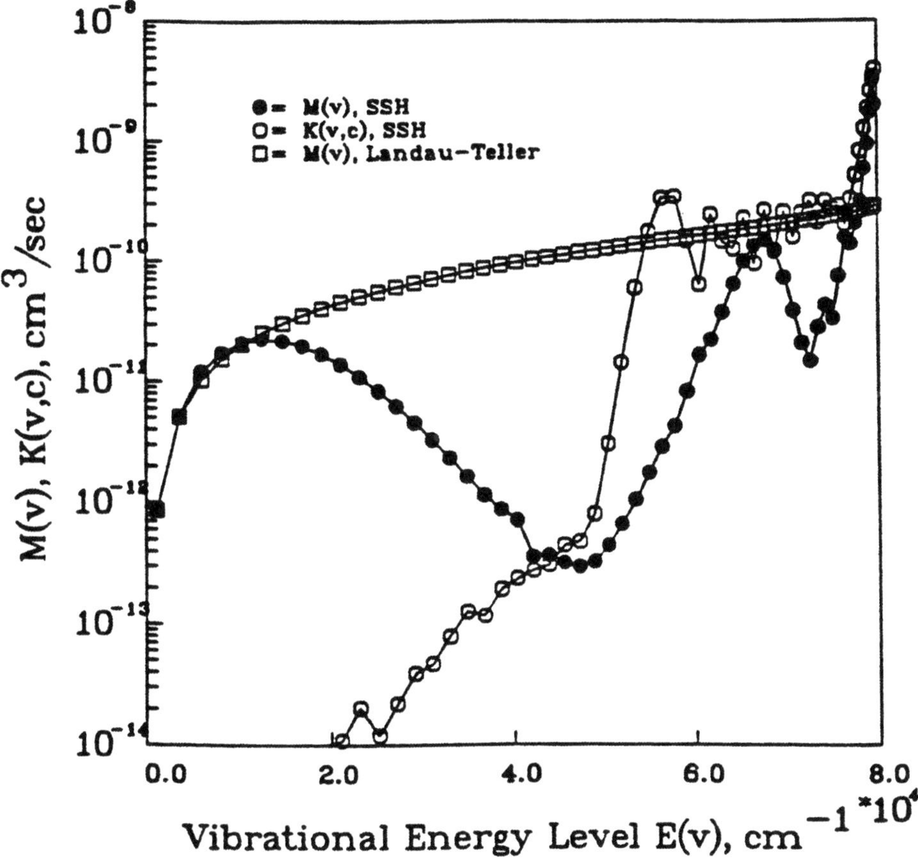

Figure 10. The second moment of the vibrational excitation rate coefficients, M(v), and the bound-free transition rates K(v,c) for N_2 calculated by the SSH (Schwartz, Slawsky, and Herzfeld) theory compared with the M(v) calculated by the Landau-Teller theory, from Ref. 14.

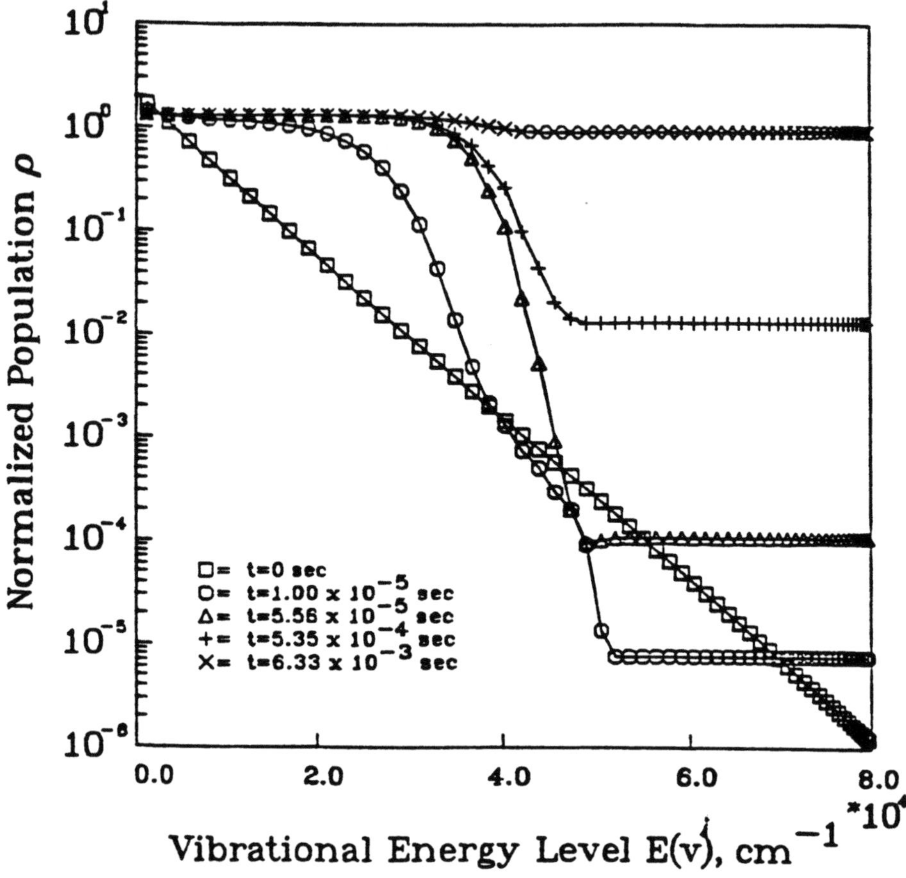

Figure 11. Normalized vibrational population in N_2 behind a shock wave as a function of time, from Ref. 14.

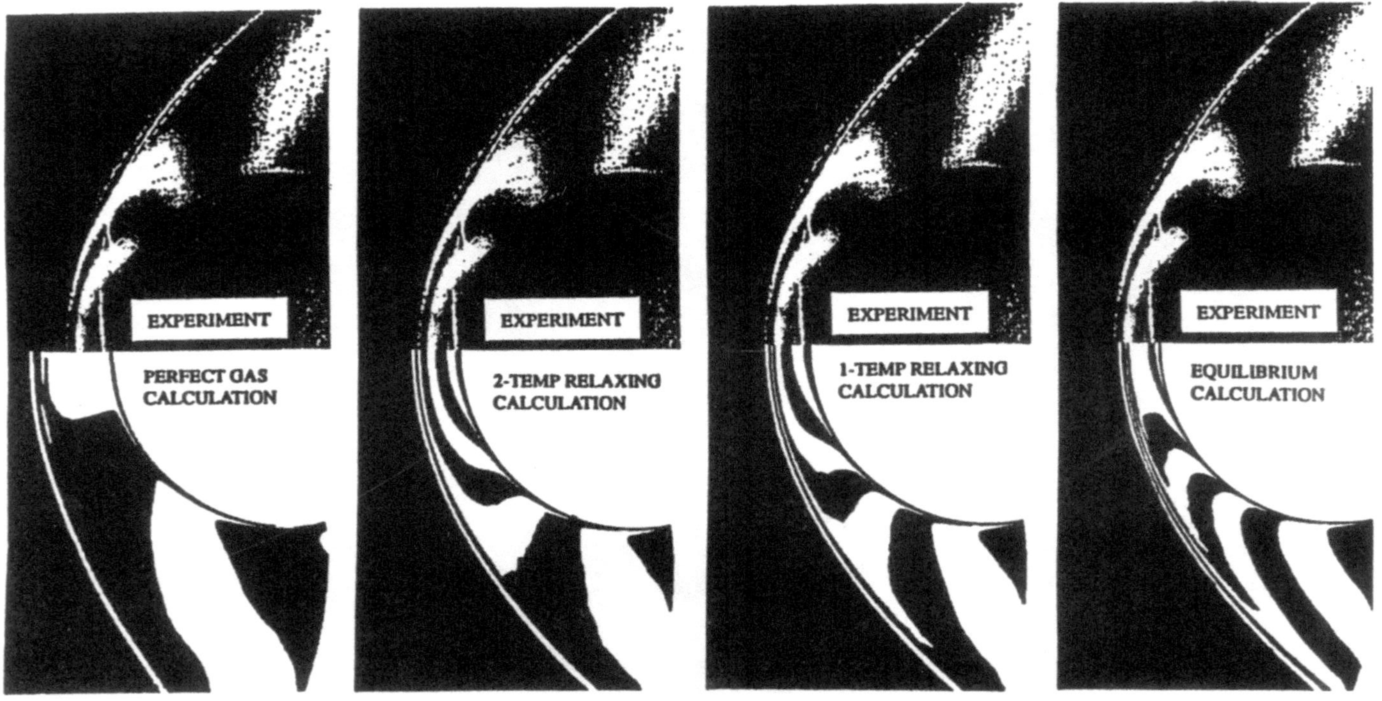

Figure 12. Comparison between experimental and computed interferograms for a flow around a circular cylinder: gas = nitrogen, Mach number = 6.1, model diameter = 5 cm, from Ref. 22.

Modeling of Hypersonic Non Equilibrium Flows

F. Grasso and V. Bellucci

Department of Mechanics and Aeronautics
University of Rome "La Sapienza"
Via Eudossiana, 18 – 00184 – Rome, Italy

ABSTRACT

In the present work the influence of hypersonic flow models in the presence of finite rate chemistry and gas-surface interaction has been critically evaluated, and a solution methodology based on a finite volume approach has been developed. The technique relies on a second order accurate total variation diminishing formulation and it uses an efficient point implicit algorithm coupled with a multistage Runge Kutta technique. Applications of the model to compute viscous hypersonic flows over a wedge have shown the overwhelming influence of wall catalyticity.

INTRODUCTION

The analysis of hypersonic flows differs substantially from that of supersonic flows, and the development of advanced space transportation systems requires a detailed simulation of the high temperatures effects (usually referred to as "real gas" effects).

At ambient temperature air can be represented as a perfect gas having only the translational and rotational modes (fully) excited.

High speed flows are characterized by a high kinetic energy content that is converted into translational energy (for example past bow shocks and within the boundary layer due to viscous dissipation), thus increasing the brownian motion of the particles, and, as a consequence, the (translational) temperature of the gas. The high temperature is then responsible for the vibrational and electronic excitation, dissociation of diatomic molecules, and ionization. Therefore, under typical hypersonic conditions air must be considered as a mixture of reacting gases whose thermodynamic state is characterized by: *i)* a translational temperature (identifying the translational and rotational energy modes); *ii)* a vibrational temperature for each of the poliatomic species (identifying the vibrational energy contribution); *iii)* and an electronic temperature (for the electronic energy contribution of the heavy particles and for the translational contribution of the free electrons).

Assuming that the energy is rapidly transfered to all vibrational modes [1], then air can be described via a three-temperature model: the translational (T), the vibrational (T_v), and the electronic (T_e) temperatures. A two-temperature model (T, T_v) is defined by assuming a rapid exchange of energy between the vibrational and electronic degrees of freedom [2,3,4]. A one temperature model is obtained under the assumption of an infinitely fast energy exchange between translation and vibration.

Depending upon the characteristic scales of the thermal and chemical exchange processes, different situations may arise. The flow is said to be in equilibrium if both thermal and chemical time scales are small compared to the fluid-dynamic time scale. A frozen situation arises if thermal and chemical time scales are large compared to the fluid-dynamic time scale. Finite rate processes must be taken into account when all time scales are of the same order.

Two parameters can be defined to characterize the thermal (vibrational) and chemical relaxation processes: the ratio of the fluid-dynamic time to the vibrational relaxation time (Ψ_v), and the ratio of the former to the chemical time (Ψ_c).

The vibrational relaxation time scale (τ_v) can be estimated from the Millikan and White relation

$$\frac{1}{\tau_v} = \frac{p}{C} \, \exp\left(-\frac{A}{T^{1/3}}\right)$$

Then, in the shock layer, Ψ_v is given by

$$\Psi_v = \frac{L}{U_\infty} \frac{R}{C} \, \rho_{shk} \, T_{shk} \, \exp\left(-\frac{A}{T_{shk}^{1/3}}\right)$$

where ρ_{shk} and T_{shk} are the post-shock density and temperature values, and R is the gas constant. Then, using the hypersonic limit for an estimation of ρ_{shk}/ρ_∞ one obtains:

$$\Psi_v = K_v \, \exp\left(-\frac{A}{T_{shk}^{1/3}}\right) \frac{\rho_\infty L}{U_\infty} \, T_{shk} \tag{1}$$

where K_v is a constant.

For dissociating air, assuming that diatomic oxygen dissociates due to collisions with diatomic oxygen and nitrogen, the chemical time scale (τ_c) can be approximately estimated as follows:

$$\frac{1}{\tau_c} = C \, T^{-1} \, \exp\left(-\frac{B}{T}\right) \frac{Y_{O_2}}{\overline{W}} \frac{\rho^2}{\rho_\infty}$$

where $\overline{W}$ is the average molecular weight.

Then, in the shock layer, Ψ_c is given by

$$\begin{aligned}
\Psi_c &= \frac{L}{U_\infty} \, C \, T_{shk}^{-1} \, \exp\left(-\frac{B}{T_{shk}}\right) \frac{Y_{O_2}}{\overline{W}} \frac{\rho_{shk}^2}{\rho_\infty} \\
&= K_c \, \exp\left(-\frac{B}{T_{shk}}\right) \frac{\rho_\infty L}{U_\infty} \, T_{shk}^{-1}
\end{aligned} \tag{2}$$

where K_c is a constant.

From Eqns.(1),(2) we observe that both parameters Ψ_v and Ψ_c are most strongly influenced by the post shock translational temperature, and that non equilibrium effects depend upon the flight conditions. Limit situations arise depending upon the values of Ψ_v and Ψ_c. In particular, Eqns.(1),(2) show that the flow can reach thermal equilibrium ($\Psi_v \gg 1$) and be in chemical non equilibrium ($\Psi_c \sim 1$).

From the above considerations one can say that the modeling of hypersonic flows has different degrees of difficulties due to: *i)* uncertainties in the description of transport and chemical kinetics mechanisms; *ii)* uncertainties in the thermodynamic relations; *iii)* lack of high temperature data; *iv)* equilibrium vs. non equilibrium effects, etc. The question then arises as to which model describes adequately hypersonic flows and how to estimate the range of applicability of the model(s).

In the present paper the influence of the different physical submodels is assessed in the presence of finite rate chemistry, thermal equilibrium and gas-surface interaction. Moreover, an upwind biased total variation diminishing formulation for "real gases" is developed. Finally, applications of the model to compute non equilibrium flows over a wedge are presented.

GOVERNING EQUATIONS

Two approaches are in general possible in formulating the governing equations for a multicomponent mixture. The first one solves the continuity equation and $N-1$ species conservation equations (where N is the total number of species) [3,5–9]. The q-th species partial density is then obtained by subtracting the (computed) $N-1$ species densities from the total density. In the second approach all species conservation equations are solved [4,10–12]. It is obvious that from a physical point of view the two approaches are equivalent. However, from the numerical point of view the computed solution may be affected by the selected formulation. Indeed, in the first approach any numerical error is concentrated in the omitted species, thus leading to inaccuracies and even instabilities, particularly during the first stage of the computation.

In the present work the governing equations are formulated accounting for all species conservation equations. The equations are the conservation equations for a mixture of gases in thermal equilibrium and chemical non equilibrium under the continuum assumption (i.e. the phenomena associated with large values of the

Knudsen number are neglected). The model assumes: *i)* a single translational temperature (T) characterizing the translational modes of all species; *ii)* fully excited rotational modes; *iii)* vibrational and electronic modes in equilibrium with the translational ones; *iv)* absence of ionization. In vector form the two-dimensional conservation equations are

$$\frac{\partial}{\partial t} \int_S \mathbf{W}\, dS + \oint_{\partial S} \left[(\mathbf{F}_E - \mathbf{F}_V)\, n_x + (\mathbf{G}_E - \mathbf{G}_V)\, n_y \right] ds = \int_S \mathbf{H}\, dS \qquad (3)$$

The vector unknown $\mathbf{W}$, the x- $(y$-$)$ components of the inviscid $\mathbf{F}_E$ $(\mathbf{G}_E)$ and viscous fluxes $\mathbf{F}_V$ $(\mathbf{G}_V)$, and the source term $\mathbf{H}$ are defined as follows

$$\mathbf{W} = \left[\rho_q,\, \rho u,\, \rho v,\, \rho E \right]^T$$

$$\mathbf{F}_E = \left[\rho_q u,\, \rho u^2 + p,\, \rho u v,\, \rho u H \right]^T$$

$$\mathbf{G}_E = \left[\rho_q v,\, \rho u v,\, \rho v^2 + p,\, \rho v H \right]^T$$

$$\mathbf{F}_V = \left[-\rho_q u_q,\, \sigma_{xx},\, \sigma_{xy},\, u\,\sigma_{xx} + v\,\sigma_{xy} - Q_x \right]^T$$

$$\mathbf{G}_V = \left[-\rho_q v_q,\, \sigma_{xy},\, \sigma_{yy},\, u\,\sigma_{xy} + v\,\sigma_{yy} - Q_y \right]^T$$

$$\mathbf{H} = \left[\dot{w}_q,\, 0,\, 0,\, 0 \right]^T$$

Furthermore

$$\sigma = \mu \left[\nabla\mathbf{u} + (\nabla\mathbf{u})^T \right] - \frac{2}{3}\,\mu\,(\nabla \cdot \mathbf{u})\,\mathbf{I}$$

$$Q = -\eta\,\nabla T + \sum_q h_q\,\rho_q\,\mathbf{u}_q$$

$$E = \sum_q Y_q\,e_q + \frac{u^2 + v^2}{2}$$

$$H = E + \frac{p}{\rho}$$

$$\rho = \sum_q \rho_q$$

where $\mathbf{I}$, $\mathbf{u}_q$, Y_q, e_q and h_q are, respectively, the unit tensor, the diffusion velocity, the mass fraction, the internal energy and the enthalpy of the q-th species.

Thermodynamic Relations

In general the internal energy and the enthalpy of species q are functions of the translational, vibrational and electronic temperatures. However, under the assumption of thermal equilibrium they are only functions of the translational temperature. Assuming that the gas is a multicomponent mixture of perfect gases, two different approaches for determining thermodynamic properties are possible.

Model TR1

The first model assumes that the internal energy of each species is the sum of the translational, rotational, vibrational and electronic contributions (according to the factorization property of the partition function) [13].

For atomic species one has

$$e_q = \frac{3}{2} R_q T + e_{el,q} + \Delta h_q^o \tag{4}$$

For diatomic species, assuming that the rotational modes are fully excited, one obtains

$$e_q = \frac{5}{2} R_q T + e_{v,q} + e_{el,q} + \Delta h_q^o \tag{5}$$

where $e_{v,q}$, $e_{el,q}$ and Δh_q^o are, respectively, the vibrational and electronic energy contributions, and the enthalpy of formation (see Table 1).

The expressions for the energy contributions due to the excited vibrational and electronic states are obtained assuming Boltzmann distributions and a harmonic oscillator behaviour of the diatomic molecules, thus yielding

$$e_{v,q} = R_q \theta_q^v \frac{1}{\exp\left(\theta_q^v/T\right) - 1} \tag{6}$$

$$e_{el,q} = R_q \frac{\sum\limits_{i=1,N_q^e} g_{q,i}\, \theta_{q,i}^e \, \exp\left(-\theta_{q,i}^e/T\right)}{\sum\limits_{i=1,N_q^e} g_{q,i}\, \exp\left(-\theta_{q,i}^e/T\right)} \tag{7}$$

where θ_q^v, $\theta_{q,i}^e$ and $g_{q,i}$ are, respectively, the characteristic vibrational temperature, the electronic characteristic temperature of state i and its degeneracy (see Tables 1–2).

Observe that in the presence of dissociation, due to the coupling between vibration and dissociation, the expression of the vibrational energy is approximated (even under the assumption of thermal equilibrium). Indeed, on account of the fact that dissociation occurs at the higher vibrational levels, only a finite number of such levels should be accounted for. The number of vibrational levels before the occurrence of dissociation is given by

$$N_q^v = \text{int}\left(\frac{D_q}{\kappa\,\theta_q^v}\right) \tag{8}$$

where D_q is the dissociation energy, and κ is the Boltzmann constant.

For nitrogen oxide, molecular oxygen and molecular nitrogen one obtains $N_{NO}^v = 27$, $N_{O_2}^v = 26$, $N_{N_2}^v = 33$. Let ϵ_q^v be the percentage error defined as

$$\epsilon_q^v = \frac{e_{v,q} - \tilde{e}_{v,q}}{\tilde{e}_{v,q}} \tag{9}$$

where $e_{v,q}$ and $\tilde{e}_{v,q}$ are the vibrational energies computed assuming respectively an infinite and a finite number of vibrational levels.

For temperatures ≤ 10000 K it is found that the maximum value of the percentage error is $O(1\%)$ as shown in Fig. 1 (where the value of the forward activation energy given in Table 6 has been used to estimate D_q). Therefore, the use of the equilibrium expression for vibrational energy is justified even in the presence of chemical nonequilibrium.

In the present work ionization has been neglected, and only the following five major species have been accounted for (O, N, NO, O_2, N_2). For these species the number of electronic states is $N_q^e = 19, 22, 15, 19, 11$, as given in Ref.[14]. However, spectroscopic data for the higher electronic states are uncertain. Moreover, if the temperature is not greater than 10000 K, a reduced number of electronic states can be accounted for. Park [6] argues that the only excited electronic states are those that have an energy exceeding that of the ground state by a factor less than $2\,eV$, thus obtaining $N_q^e = 2, 2, 0, 3, 0$. Likewise, Palmer [3] neglects the electronic excitation of nitrogen oxide and diatomic nitrogen, and uses $N_q^e = 3, 3, 0, 3, 0$. Candler [15] accounts for the first two levels of all species.

In the present model the number of electronic states has been determined by imposing that the percentage error of the internal energy of each species computed with a reduced number of states is less than 1% of the value obtained by accounting for all electronic states. This gives $N_q^e = 2, 3, 2, 7, 2$. Fig. 2 shows indeed that even at higher temperatures the internal energy computed with the reduced model is in reasonable agreement with that obtained with the complete model.

Model TR2

The second model uses a polynomial curve fit as a function of temperature in the range 300 K $\leq T \leq$ 30000 K, as proposed by Ref.[16]:

$$e_q = R_q \sum_{k=1}^{5} \frac{A_{q,k}}{k} T^k - R_q T + \Delta h_q^o \tag{10}$$

where the polynomial coefficients $A_{q,k}$ are given in Table 3.

Transport Coefficients

The transport properties (viscosity, thermal conductivity and diffusion coefficients) need to be modeled to account for the effects of high temperature and variable composition on the transport mechanism in a dissociating gas mixture. In general the transport coefficients are evaluated by using either polynomial curve-fit of temperature, or expressions obtained from kinetic theory.

Model TC1

In the first model the transport coefficients are based on the Chapman-Enskog theory and on an extension of Yos' formula [1,2,16]. Chapman-Enskog theory amounts to solve Boltzmann's equation for the singlet-velocity distribution function (that coincides with the maxwellian one when the gas is in equilibrium), and strictly holds for monoatomic gases. Yos has extended the approach to account for the effects of momentum and energy transfer between different species by collisions. Based on the relations developed by Yos the mixture viscosity is defined as

$$\mu = \sum_q \frac{m_q\,\gamma_q}{\sum_r \gamma_r\,\Delta_{q,r}^{(2)}\,(T)} \tag{11}$$

where m_q, $\Delta_{qr}^{(2)}$ and γ_q are, respectively, the mass of the species q, the modified collision integral, and the molar concentration, the latter being defined as

$$\gamma_q = \frac{\rho_q}{\rho\,W_q} \tag{12}$$

where W_q is the molecular weight of species q.

For a mixture of gases the thermal conductivity can be split in two contributions that account for the transport of translational energy and the transport of energy due to the internal structure. Thus one has

$$\eta = \eta_{tr} + \eta_{int} \tag{13}$$

$$\eta_{tr} = \frac{15}{4}\,\kappa\,\sum_q \frac{\gamma_q}{\sum_r a_{q,r}\,\gamma_r\,\Delta_{q,r}^{(2)}\,(T)} \tag{14}$$

$$a_{q,r} = 1 + \frac{[1 - (m_q/m_r)]\,[0.45 - 2.54\,(m_q/m_r)]}{[1 + (m_q/m_r)]^2} \tag{15}$$

$$\eta_{int} = \kappa\,\sum_q \frac{\gamma_q\,(C_{v,q}/R_q - 3/2)}{\sum_r \gamma_r\,\Delta_{q,r}^{(1)}\,(T)} \tag{16}$$

where $C_{v,q}$ is the specific heat coefficient at constant volume (defined according to the selected thermodynamic relations).

The diffusion coefficient of species q in the mixture and the diffusion flux are given by

$$D_q = \frac{\kappa\,T}{p}\,\frac{\gamma_t^2\,W_q\,(1 - W_q\,\gamma_q)}{\sum_{r \neq q} \gamma_r\,\Delta_{q,r}^{(1)}\,(T)} \tag{17}$$

$$\rho_q\,\mathbf{u}_q = -\rho\,D_q\,\nabla\,X_q$$

where p is the pressure, $\gamma_t = \sum_q \gamma_q$ and X_q is the molar fraction of species q.

The modified collision integrals $\Delta_{q,r}^{(k)}$ are defined as [16]

$$\Delta_{q,r}^{(1)}(T) \;=\; \frac{8}{3}\left[\frac{2\,W_{qr}}{\pi\,R\,T}\right]^{1/2}\pi\,\overline{\Omega}_{q,r}^{(1,1)} \tag{18}$$

$$\Delta_{q,r}^{(2)}(T) \;=\; \frac{16}{5}\left[\frac{2\,W_{qr}}{\pi\,R\,T}\right]^{1/2}\pi\,\overline{\Omega}_{q,r}^{(2,2)} \tag{19}$$

where W_{qr} is the "reduced" molecular weight of the colliding species

$$\frac{1}{W_{qr}} = \frac{1}{W_q} + \frac{1}{W_r}$$

The collision integrals $\overline{\Omega}_{q,r}^{(k,k)}$ depend upon the dynamics of the collisions between particles of type q and r, and on the energy potential. In particular the same linear variation of the $\log(\overline{\Omega}_{q,r}^{(k,k)})$ with $\ln(T)$, as proposed in Ref.[2], has been used, i.e.

$$\log_{10}\left(\pi\,\overline{\Omega}_{q,r}^{(k,k)}\right) \;=\; \log_{10}\left(\pi\,\overline{\Omega}_{q,r}^{(k,k)}\right)_{2000} +$$

$$+\;\frac{\log_{10}\left(\pi\,\overline{\Omega}_{q,r}^{(k,k)}\right)_{4000} - \log_{10}\left(\pi\,\overline{\Omega}_{q,r}^{(k,k)}\right)_{2000}}{\ln(4000) - \ln(2000)} \cdot \left[\ln(T) - \ln(2000)\right]$$

$$(k = 1, 2)$$

In Table 4 the collision integrals of each species at $T = 2000$ K and $T = 4000$ K are given.

Model TC2

In the second model the transport coefficients are evaluated by a polynomial curve fit of temperature and Eucken approximation.

If the temperature is greater than 1000 K the viscosity of the species q is calculated according to the following formula [16]

$$\mu_q = 0.1\left(\exp C_{\mu_q}\right)T^{[A_{\mu_q}\ln T + B_{\mu_q}]} \tag{20}$$

where the viscosity coefficients A_μ's, B_μ's and C_μ's are given in Table 5.

The viscosity of the mixture is obtained by using the semi-empirical formula of Wilke

$$\mu = \sum_q \frac{X_q \, \mu_q}{\sum_r X_r \, \phi_{qr}} \tag{21}$$

where ϕ_{qr} is

$$\phi_{qr} = \frac{1}{\sqrt{8}} \left(1 + \frac{W_q}{W_r} \right)^{-1/2} \left[1 + \left(\frac{\mu_q}{\mu_r} \right)^{1/2} \left(\frac{W_r}{W_q} \right)^{1/4} \right]^2 \tag{22}$$

From kinetic theory [13] it can be shown that the transport of energy is proportional to that of momentum. Consequently, the coefficient of heat conduction is proportional to the viscosity:

$$\eta_q = \beta \, \mu_q \, C_{v,q} \tag{23}$$

where β is a proportionality constant.

For a monoatomic species the specific heat coefficient at constant volume accounts only for the translational contribution. Assuming that the transport of the translational energy correlates with the velocity, one obtains a value of $\beta = 5/2$.

For poliatomic gases the contributions due to rotation and vibration can be lumped all in one term. The latter involves no correlation with the velocity, and $\beta = 1$. Hence, one obtains

$$\eta_q = \frac{5}{2} \, \mu_q \, C_{v,q}^t + \mu_q \, C_{v,q}^{int} \tag{24}$$

where $C_{v,q}^t$ and $C_{v,q}^{int}$ are, respectively, the translational and "internal" specific heat coefficients at constant volume, defined as

$$C_{v,q}^t = \frac{3}{2} \, R_q$$

$$C_{v,q}^{int} = C_{v,q} - C_{v,q}^t$$

Thus one obtains

$$\eta_q = \mu_q \left(C_{v,q} + \frac{9}{4} \, R_q \right) \tag{25}$$

The thermal conductivity of the mixture is obtained by means of Wilke's rule

$$\eta = \sum_q \frac{X_q \, \eta_q}{\sum_r X_r \, \phi_{qr}} \tag{26}$$

The diffusion coefficient of species q in the mixture is computed according to the definition of the Lewis number (Le)

$$D_q = \frac{\eta \, Le}{\rho \, C_p} \tag{27}$$

where C_p is the frozen specific heat coefficient of the mixture, and the Lewis number is assumed to be constant $(Le = 1.4)$. The diffusion flux is given by

$$\rho_q \, \mathbf{u}_q = - \rho \, D_q \, \nabla Y_q$$

If the temperature is less than 1000 K, then Sutherland's Law is used. Depending on the value of the freestream temperature (T_∞) the following formulas are used.

If $T_\infty < 120$ K

$$\mu(T) \;=\; \begin{cases} \mu_\infty \, \tilde{T} & \text{if} \quad T \leq 120 \text{ K} \\[2mm] \mu_\infty \, \left(s_1 \, \tilde{T}\right)^{1/2} \dfrac{1 + s_2/s_1}{1 + s_2 \big/ \tilde{T}} & \text{if} \quad T > 120 \text{ K} \end{cases}$$

If $T_\infty \geq 120$ K

$$\mu(T) \;=\; \begin{cases} \mu_\infty \, s_1^{1/2} \, \tilde{T} \, \dfrac{1 + s_2}{s_1 + s_2} & \text{if} \quad T < 120 \text{ K} \\[2mm] \mu_\infty \, \tilde{T}^{1/2} \, \dfrac{1 + s_2}{1 + s_2 \big/ \tilde{T}} & \text{if} \quad T \geq 120 \text{ K} \end{cases}$$

where $s_1 = 120/T_\infty$, $s_2 = 110/T_\infty$, and $\tilde{T} = T/T_\infty$.

The thermal conductivity is then obtained from the Prandtl number definition (Pr)

$$\eta = \frac{\mu \, C_p}{Pr} \tag{28}$$

where a value of $Pr = .71$ is assumed.

The diffusion coefficients are computed according to Eqn.(27).

Chemistry Model

If air is the reacting gas and if one neglects ionization phenomena, then only five major species can be assumed (O, N, NO, O_2, N_2). Following Penner notation the generic reaction (r) can be written as

$$\sum_s \nu'_{s,r} [M_s] \;\leftrightarrows\; \sum_s \nu''_{s,r} [M_s]$$

where $\nu'_{s,r}$, $\nu''_{s,r}$, and $[M_s]$ are, respectively, the stoichiometric coefficients of the reactants and the products, and the molar concentration of species s.

The rate of production of species s is given by

$$\frac{d}{dt}[M_s] \;=\; \sum_r \left(\nu''_{s,r} - \nu'_{s,r}\right) R_r$$

$$R_r \;=\; k_{f,r} \prod_q [M_q]^{\nu'_{q,r}} - k_{b,r} \prod_q [M_q]^{\nu''_{q,r}} \tag{29}$$

where R_r is the (molar) species production of reaction r.

The forward and backward reaction rate constants are defined as follows:

$$k_{f,r} \;=\; C_{f,r}\, T^{\eta_{f,r}} \exp\left(-E_{f,r}/\kappa T\right) \tag{30}$$

$$k_{b,r} \;=\; C_{b,r}\, T^{\eta_{b,r}} \exp\left(-E_{b,r}/\kappa T\right) \tag{31}$$

The constants C's, η's and E's are, respectively, the Arrhenius constants, the pre-exponential factors, and the activation energies, and they are determined experimentally. In the absence of ionization the chemical reaction mechanism is [14]:

$$\begin{aligned}
O_2 + M &\;\leftrightarrows\; O + O + M \\
NO + M &\;\leftrightarrows\; N + O + M \\
N_2 + M &\;\leftrightarrows\; N + N + M \\
N_2 + O &\;\leftrightarrows\; NO + N \\
NO + O &\;\leftrightarrows\; O_2 + N
\end{aligned} \tag{32}$$

where M is anyone of the five species.

The first three reactions in Eqn.(32) are heavy-particle impact dissociation reactions, the fourth and fifth reactions are exchange reactions.

For each reaction the species production is defined as follows

$$R_1 = \sum_m \left(k_{f,1m} \frac{\rho_{O_2}}{W_{O_2}} \frac{\rho_m}{W_m} - k_{b,1m} \frac{\rho_O}{W_O} \frac{\rho_O}{W_O} \frac{\rho_m}{W_m} \right)$$

$$R_2 = \sum_m \left(k_{f,2m} \frac{\rho_{NO}}{W_{NO}} \frac{\rho_m}{W_m} - k_{b,2m} \frac{\rho_N}{W_N} \frac{\rho_O}{W_O} \frac{\rho_m}{W_m} \right)$$

$$R_3 = \sum_m \left(k_{f,3m} \frac{\rho_{N_2}}{W_{N_2}} \frac{\rho_m}{W_m} - k_{b,3m} \frac{\rho_N}{W_N} \frac{\rho_N}{W_N} \frac{\rho_m}{W_m} \right)$$

$$R_4 = k_{f,4} \frac{\rho_{N_2}}{W_{N_2}} \frac{\rho_O}{W_O} - k_{b,4} \frac{\rho_{NO}}{W_{NO}} \frac{\rho_N}{W_N}$$

$$R_5 = k_{f,5} \frac{\rho_{NO}}{W_{NO}} \frac{\rho_O}{W_O} - k_{b,5} \frac{\rho_{O_2}}{W_{O_2}} \frac{\rho_N}{W_N}$$

Thus, the source terms for the species are

$$\dot{w}_O = W_O \ (2\,R_1 + R_2 - R_4 - R_5)$$

$$\dot{w}_N = W_N \ (R_2 + 2\,R_3 + R_4 + R_5)$$

$$\dot{w}_{NO} = W_{NO} \ (-R_2 + R_4 - R_5)$$

$$\dot{w}_{O_2} = W_{O_2} \ (-R_1 + R_5)$$

$$\dot{w}_{N_2} = W_{N_2} \ (-R_3 - R_4)$$

Model CK1

In this model the forward reaction rate constants are given by Eqn.(30), where the coefficients are those of Table 6 [14]. The backward reaction rate constants are not evaluated via Arrhenius-type expressions as given by Eqn.(31), but rather via the equilibrium constant

$$k_{b,r} = \frac{k_{f,r}}{k_{eq,r}} \tag{33}$$

where $k_{eq,r}$ represents the equilibrium constant for the r-th reaction, defined as

$$k_{eq,r} = \exp\left(B_{1,r} + B_{2,r}\ln z + B_{3,r}\, z + B_{4,r}\, z^2 + B_{5,r}\, z^3\right) \tag{34}$$

where $z = 10000/T$, and the values of the parameters $B_{i,r}$ are given in Table 7 [17].

Model CK2

The seventeen reaction mechanism of the full Park model can be reduced to a seven reaction mechanism by neglecting the thermal dissociation reactions of N_2 and NO, under the assumption that they are slower than the two exchange reactions [6].

Let $\tau_{f,r}^q$ identify the characteristic (forward reaction) time of the generic reaction r

$$\tau_{f,r}^q = \rho_q \left(k_{f,r} \prod_s \rho_s^{\nu'_{s,r}}\right)^{-1} = Y_q \left(k_{f,r}\, \rho^{\nu'_r - 1} \prod_s Y_s^{\nu'_{s,r}}\right)^{-1} \tag{35}$$

where $\nu'_r = \sum_s \nu'_{s,r}$.

Eqn.(35) indicates that, in general, reactions characterized by smaller values of the forward rate constant are the slowest (thus taking longer time to reach equilibrium). Then, comparison of the reaction rate constants of the full Park model indicates that exchange reactions are faster than the reactions of thermal dissociation of NO and N_2, as shown in Figs. 3–4. As a consequence, the latter reactions can be neglected, thus yielding the following reduced mechanism:

$$O_2 + M \;\rightleftharpoons\; O + O + M$$

$$N_2 + O \;\rightleftharpoons\; NO + N$$

$$NO + O \;\rightleftharpoons\; O_2 + N$$

The rate constants for the above mechanism are evaluated as previously indicated.

Model CK3

This model differs from Park's full model in the evaluation of the rate constants, that are determined according to the Arrhenius-type law (Eqns.(30),(31)), whose parameters are given in Table 8 [16].

NUMERICAL SOLUTION

The solution of the governing equations for high speed flows requires the use of robust and accurate schemes. The flux-difference splitting of Roe [18], and the flux-vector splitting of Steger and Warming [19] and Van Leer [20] have been widely used for (perfect gas) high speed flow computations. Glaister [21] has extended Roe's approximate Riemann solver to (real gas) hypersonic flows in equilibrium. Flux-vector splitting for real gases has been developed by Liou and Van Leer [22], and Grossman and Cinnella [4]. Montagné *et al.* [23] have implemented a second order symmetric total variation diminishing scheme for inviscid flows in chemical equilibrium. A more general methodology for the solution of hypersonic flows in non equilibrium has been presented by Liu and Vinokur [10].

In the present work, following the approach of Ref.[10], we have extended a second order total variation diminishing method [24] to include the effects of non equilibrium chemistry.

Space and time discretizations are separated by using the method of lines, and a system of ordinary differential equations is obtained for every computational cell. A cell centered finite volume formulation is employed. By approximating surface and boundary integrals by means of the mean value theorem and mid-point rule, the governing equations (Eqn.(3)) are cast in the following discretized form:

$$S_{i,j}\, \frac{d\mathbf{W}_{i,j}}{dt} + \sum_{\beta=1}^{4} \left(\mathbf{F}_{num} \cdot \mathbf{n}\Delta s\right)_{\beta} = S_{i,j}\, \mathbf{H}_{i,j} \tag{36}$$

where β stands for the generic cell face, $\mathbf{n}$ is the positive unit normal to cell face whose length is Δs, and $S_{i,j}$ is the cell area.

The numerical flux vector is:

$$\mathbf{F}_{num} \cdot \mathbf{n} = \left(\mathbf{F}_{E,num} - \mathbf{F}_{V,num}\right) n_x + \left(\mathbf{G}_{E,num} - \mathbf{G}_{V,num}\right) n_y \tag{37}$$

Numerical inviscid flux discretization

An upwind biased second order total variation diminishing (TVD) scheme has been used to evaluate the inviscid flux contribution. The scheme is based on the modified Harten-Yee method generalized for multispecies in chemical non equilibrium [24]. The scheme has good properties of monotonicity and conservativity in the presence of discontinuities, and it yields second order accuracy and oscillation free solutions. By enforcing consistency at cell face $(i + 1/2, j)$ one obtains:

$$
\left(\mathbf{F}_{E,num}\, n_x + \mathbf{G}_{E,num}\, n_y \right)_{i+1/2,j} = \frac{1}{2} \left[\left(\mathbf{F}_{E_{i,j}} + \mathbf{F}_{E_{i+1,j}} \right) n_x + \left(\mathbf{G}_{E_{i,j}} + \mathbf{G}_{E_{i+1,j}} \right) n_y \right.
$$

$$
\left. + \mathbf{R}_{i+1/2,j}\, \boldsymbol{\Phi}_{i+1/2,j} \right] \tag{38}
$$

The term $\boldsymbol{\Phi}_{i+1/2,j}$ represents the numerical antidiffusive flux contribution, that modifies the inviscid flux to make the scheme upwind biased TVD and second order accurate. The expression of its elements $(\varphi^{\ell}_{i+1/2,j})$ is obtained by characteristic decomposition in the direction normal to cell face, thus obtaining:

$$
\varphi^{\ell}_{i+1/2,j} = \frac{1}{2}\, \psi \left(a^{\ell}_{i+1/2,j} \right) \left(g^{\ell}_{i+1,j} + g^{\ell}_{i,j} \right) - \psi \left(a^{\ell}_{i+1/2,j} + \gamma^{\ell}_{i+1/2,j} \right) \alpha^{\ell}_{i+1/2,j}
$$

$$
\gamma^{\ell}_{i+1/2,j} = \frac{1}{2}\, \psi \left(a^{\ell}_{i+1/2,j} \right) \begin{cases} \left(g^{\ell}_{i+1,j} - g^{\ell}_{i,j} \right) / \alpha^{\ell}_{i+1/2,j} & \alpha^{\ell}_{i+1/2,j} \neq 0 \\[2mm] 0 & \alpha^{\ell}_{i+1/2,j} = 0 \end{cases}
$$

$$
\alpha_{i+1/2,j} = \mathbf{R}^{-1}_{i+1/2,j} \left(\mathbf{W}_{i+1,j} - \mathbf{W}_{i,j} \right)
$$

where

$$
\psi\left(z \right) = \begin{cases} |z| & |z| \geq \delta_1 \\[2mm] \left(z^2 + \delta_1^2 \right) / 2\, \delta_1 & |z| < \delta_1 \end{cases}
$$

is an entropy correction to $|z|$, and $a^{\ell}_{i+1/2,j}$ are the eigenvalues of $\mathbf{A}(\mathbf{W}_{i+1/2,j})$. The parameter δ_1 enforces the entropy condition for vanishing eigenvalues, and its value depends on the type of flow to be computed.

A *minmod* limiter is selected for its better computational efficiency and speed of convergence:

$$
g^{\ell}_{i,j} = minmod \left(\alpha^{\ell}_{i+1/2,j}\, , \; \alpha^{\ell}_{i-1/2,j} \right)
$$

where

$$
minmod\left(x, y \right) = sgn\left(x \right) \cdot \max \left\{ 0, \min[|x|, y \cdot sgn\left(x \right)] \right\}
$$

For second order accuracy, boundary conditions on g are required. At all boundaries the normal derivative of g has been set equal to zero. Conservation is enforced by imposing that the contribution due to the modified flux $\mathbf{R}\,\boldsymbol{\Phi}$ is zero for all cells along the walls.

To construct linearly independent eigenvectors basis vectors are chosen in such a way as to be orthogonal to the cell face normal. Consequently, the right eigenvector matrix $(\mathbf{R})$ of the normal inviscid flux jacobian is defined as

$$
\mathbf{R} = \begin{bmatrix}
\delta_{qr} & 0 & Y_q & Y_q \\[2mm]
u & -c\,n_y & u + c\,n_x & u - c\,n_x \\[2mm]
v & c\,n_x & v + c\,n_y & v - c\,n_y \\[2mm]
\dfrac{\mathbf{u}\cdot\mathbf{u}}{2} - \dfrac{\chi_r}{K} & c\,(\mathbf{u}\cdot\mathbf{b}) & H + c\,u_n & H - c\,u_n
\end{bmatrix}
\tag{39}
$$

where $\mathbf{b}$ is the span basis vector orthogonal to $\mathbf{n}$, and K and χ_q are the pressure derivatives defined as

$$
K = \left(\frac{\partial p}{\partial \rho\, e} \right)_{\rho_q} = \frac{\sum \rho_q\, R_q}{\sum \rho_q\, C_{v,q}}
\tag{40}
$$

$$
\chi_q = \left(\frac{\partial p}{\partial \rho_q} \right)_{\rho e} = R_q\, T - K\, e_q
\tag{41}
$$

and

$$
c = \left[(K+1)\, \frac{p}{\rho} \right]^{1/2}
\tag{42}
$$

is the frozen speed of sound.

The left eigenvector matrix $\mathbf{R}^{-1}$ is

$$
\mathbf{R}^{-1} = \frac{1}{c^2} \begin{bmatrix}
\delta_{qr}\, c^2 - Y_q \left(\dfrac{K}{2}\, \mathbf{u}\cdot\mathbf{u} + \chi_r \right) & Y_q\, K\, \mathbf{u} & -Y_q\, K \\[3mm]
-c\,(\mathbf{u}\cdot\mathbf{b}) & c\,\mathbf{b} & 0 \\[3mm]
\dfrac{1}{2}\left[K\, \dfrac{\mathbf{u}\cdot\mathbf{u}}{2} + \chi_r - c\,(\mathbf{u}\cdot\mathbf{n}) \right] & \dfrac{1}{2}\,(c\,\mathbf{n} - K\,\mathbf{u}) & \dfrac{1}{2}\, K \\[3mm]
\dfrac{1}{2}\left[K\, \dfrac{\mathbf{u}\cdot\mathbf{u}}{2} + \chi_r + c\,(\mathbf{u}\cdot\mathbf{n}) \right] & -\dfrac{1}{2}\,(c\,\mathbf{n} + K\,\mathbf{u}) & \dfrac{1}{2}\, K
\end{bmatrix}
\tag{43}
$$

The difference of characteristic variables (α) is given by

$$\alpha_{i+1/2,j} = \begin{bmatrix} \Delta\rho_q - \dfrac{\overline{Y}_q}{\overline{c}^2}\,\Delta p \\[2ex] \dfrac{\overline{\rho}}{\overline{c}}\,\Delta\mathbf{u}\cdot\mathbf{b} \\[2ex] \dfrac{1}{2}\left(\dfrac{\Delta p}{\overline{c}^2} + \overline{\rho}\,\dfrac{\Delta\mathbf{u}\cdot\mathbf{n}}{\overline{c}}\right) \\[2ex] \dfrac{1}{2}\left(\dfrac{\Delta p}{\overline{c}^2} - \overline{\rho}\,\dfrac{\Delta\mathbf{u}\cdot\mathbf{n}}{\overline{c}}\right) \end{bmatrix}$$

where

$$\overline{(\cdot)} = (\cdot)_{i+1/2,j}$$
$$\Delta(\cdot) = (\cdot)_{i+1,j} - (\cdot)_{i,j}$$

The values at the interfaces are calculated by using a generalization of Roe's averaging to account for chemical non equilibrium [10]

$$\Delta\left(\mathbf{F}_E\,n_x + \mathbf{G}_E\,n_y\right) = \overline{\mathbf{A}}\left(\overline{\mathbf{W}}^{*},\,\overline{\chi}_q,\,\overline{K}\right)\Delta\mathbf{W} \tag{44}$$

where

$$\overline{\mathbf{W}}^{*} = b_{i,j}\,\mathbf{W}'_{i,j} + b_{i+1,j}\,\mathbf{W}'_{i+1,j}$$

$$\mathbf{W}' = \begin{bmatrix} \rho_q \\ \rho\,u \\ \rho\,v \\ \rho\,E + p \end{bmatrix}$$

$$b_{i+1,j} = \frac{1}{\rho_{i+1,j} + \tilde{\rho}}$$

$$b_{i,j} = \frac{1}{\rho_{i,j} + \tilde{\rho}}$$

$$\tilde{\rho} = \sqrt{\rho_{i,j}\,\rho_{i+1,j}}$$

The derivatives of pressure are given by the discretized equation of state

$$\Delta p = \sum_q \overline{\chi}_q \, \Delta \rho_q + \overline{K} \, \Delta \, (\rho \, e) \tag{45}$$

The solution of Eqn.(45) can be formulated geometrically by projecting the point $(\hat{\chi}_q, \hat{K})$ onto the hyperplane defined by the discretized equation of state, thus yielding

$$\overline{K} \;=\; \frac{D \, \hat{K}}{D - \Delta p \, \delta p} \tag{46}$$

$$\overline{\chi}_q \;=\; \frac{D \, \hat{\chi}_q + \hat{\sigma}^2 \, \Delta \rho_q \, \delta p}{D - \Delta p \, \delta p} \tag{47}$$

where

$$\hat{\chi}_q \;=\; \frac{(\chi_q)_{i,j} + (\chi_q)_{i+1,j}}{2}$$

$$\hat{K} \;=\; \frac{K_{i,j} + K_{i+1,j}}{2}$$

$$\hat{\sigma} \;=\; \frac{c^2_{i,j} + c^2_{i+1,j}}{2}$$

$$D \;=\; (\Delta p)^2 + \sum_q (\hat{\sigma} \, \Delta \rho_q)^2$$

$$\delta p \;=\; \Delta p - \sum_q \hat{\chi}_q \, \Delta \rho_q - \hat{K} \, \Delta(\rho e)$$

Numerical viscous flux discretization

According to the constitutive equations, the viscous fluxes depend upon the gradients of the primitive variables $(\mathbf{u}, T, Y_q)$. The numerical counterpart is obtained by applying Gauss theorem to a computational cell whose vertices are the two grid nodes (I, J) and $(I, J-1)$ and the centers of the two adjacent cells (i, j) and $(i+1, j)$. For an arbitrary function φ the numerical derivatives at $(i+1/2, j)$ are evaluated according to the following formulas:

$$\left(\frac{\partial \varphi}{\partial x} \right)_{i+1/2, j} \;=\; \frac{\Delta_i \varphi \, \Delta_j y - \Delta_j \varphi \Delta_i y}{\Delta_i x \, \Delta_j y - \Delta_j x \, \Delta_i y}$$

$$\left(\frac{\partial \varphi}{\partial y}\right)_{i+1/2,j} = -\frac{\Delta_i \varphi \, \Delta_j x - \Delta_j \varphi \, \Delta_i x}{\Delta_i x \, \Delta_j y - \Delta_j x \, \Delta_i y}$$

where

$$\Delta_i \left(\cdot \right) = \left(\cdot \right)_{i+1,j} - \left(\cdot \right)_{i,j} \quad ; \quad \Delta_j \left(\cdot \right) = \left(\cdot \right)_{I,J} - \left(\cdot \right)_{I,J-1}$$

The grid values $\varphi_{I,J}$, $\varphi_{I,J-1}$ are obtained by bilinear interpolation of cell center values. Hence, the discretized viscous flux contribution at cell face $(i+1/2,j)$ is an algebraic function of grid and cell center values, i.e.

$$(\mathbf{F}_{V,num})_{i+1/2,j} = f \left(\mathbf{W}_{i,j} ; \, \mathbf{W}_{i+1,j} ; \, \mathbf{W}_{I,J} ; \, \mathbf{W}_{I,J-1} \right)$$

<u>Time integration</u>

In the presence of chemical non equilibrium the system of ordinary differential equations is stiff due to the presence of the production term $\mathbf{H}$. Stiffness arises for the disparity between the characteristic chemical and fluid-dynamic times. Hence, the time integration would require an extremely small time step. However, for steady flows stiffness can be reduced by introducing a precondition matrix P, and the system of ordinary differential equations is modified accordingly

$$\mathbf{P}_{i,j} \frac{d\mathbf{W}_{i,j}}{dt} + \frac{1}{S_{i,j}} \sum_{\beta=1}^{4} (\mathbf{F}_{num} \cdot \mathbf{n} \, \Delta s)_\beta = \mathbf{H}_{i,j} \tag{48}$$

The integration in time of Eqn.(48) is performed by a three-stage Runge-Kutta point implicit algorithm [7]:

$$\mathbf{W}_{i,j}^{(0)} = \mathbf{W}_{i,j}^{n}$$

$$\mathbf{P}_{i,j}^{(k-1)} \left(\mathbf{W}_{i,j}^{(k)} - \mathbf{W}_{i,j}^{(0)} \right) = \alpha_k \, \Delta t \left\{ -\frac{1}{S_{i,j}} \sum_{\beta=1}^{4} \left((\mathbf{F}_{E,num}^{(k-1)} - \mathbf{F}_{V,num}^{(k-1)}) \cdot \mathbf{n} \, \Delta s \right)_\beta + \mathbf{H}_{i,j}^{(k-1)} \right\}$$

$$\mathbf{W}_{i,j}^{n+1} = \mathbf{W}_{i,j}^{(3)}$$

The precondition matrix scales all the characteristics times to the same order, and it is defined as

$$\mathbf{P}^{(k-1)} = \mathbf{I} - \alpha_k \, \Delta t \left(\frac{\partial \mathbf{H}}{\partial \mathbf{W}} \right)^{(k-1)} \tag{49}$$

For computational efficiency, a partial jacobian of the source term $(\partial \mathbf{H}/\partial \mathbf{W})$ is used by neglecting the dependency of $\mathbf{H}$ on ρu, ρv, ρE, without affecting the accuracy of the steady-state solution.

Boundary Conditions

The numerical solution of the governing equations requires boundary conditions to be imposed along the boundaries. Referring to Fig. 5 typical boundary conditions are: *i)* inflow; *ii)* outflow; *iii)* freestream; *iv)* symmetry conditions; *v)* solid wall.

Inflow conditions

At the inflow boundary (Γ_1) freestream values are imposed

$$p = p_\infty \quad ; \quad T = T_\infty \quad ; \quad Y_q = Y_q^{eq}(T_\infty) \quad ; \quad u = c_\infty \, \mathrm{M}_\infty \quad ; \quad v = 0$$

Outflow conditions

Along the outflow boundary Γ_3 first order extrapolation conditions are imposed on all variables

$$\frac{\partial \mathbf{W}}{\partial x} = 0$$

Freestream conditions

Freestream conditions are imposed along the boundary Γ_2. However, if the boundary is an outflow boundary, then all variables are extrapolated from the interior.

Symmetry conditions

In the presence of a symmetry boundary (Γ_5) the following conditions are imposed

$$\frac{\partial p}{\partial n} = 0 \quad ; \quad \frac{\partial \rho E}{\partial n} = 0 \quad ; \quad \frac{\partial Y_q}{\partial n} = 0 \quad ; \quad \frac{\partial u}{\partial n} = 0 \quad ; \quad v = 0$$

Solid wall

On a solid wall (Γ_4) continuum-type boundary conditions are set and the no-slip condition is enforced on the velocity: $u = v = 0$.

The pressure is obtained by assuming a zero normal pressure gradient, and fixed wall temperature or adiabatic conditions are set.

If the wall is non catalytic the normal (species) diffusion fluxes is zero at the wall, thus yielding

$$\frac{\partial Y_q}{\partial n} = 0$$

However, the surfaces of most hypersonic vehicles are made up of either a metal, a metal oxide, or carbon materials, and gas-surface reactions should be accounted for. In the present work the effects of the gas surface interaction have been accounted for by assuming that the characteristic times of surface reactions are smaller than the residence time (i.e. $\Psi_c \gg 1$). Consequently, equilibrium reactions are assumed to occur at the wall, thus yielding

$$Y_q = Y_q^{eq}\,(T_w)$$

where T_w is the wall temperature.

RESULTS

The model has been applied to compute the flow over a 10° wedge at an altitude of 61 km. The flow conditions are given in Table 9.

This test case has been investigated by other authors [5,25], and shows a small degree of non equilibrium in the boundary layer. The test case is a "simple" one considering the simplicity of the geometry. However, it contains all the relevant features of complex hypersonic flows, and has been selected to analyze the different aspects of modeling of hypersonic flows with and without gas-surface interaction phenomena.

All computations have been performed on a 176×48 grid with (non dimensional) normal mesh spacing ranging from $1.5\ 10^{-3}$ to $6.9\ 10^{-2}$, and cell aspect ratio varying between .5 and 13. Referring to Fig. 5, the upstream boundary (Γ_1) has been positioned at $x = 0$, the outflow boundary (Γ_3) has been set at $x = 4$ m, and the freestream boundary (Γ_2) has been set at $y = 1.35$ m. Along the wall the effects of catalyticity have been taken into account. In particular, for the same test case fully catalytic and non catalytic wall boundary conditions are imposed on (Γ_4). For this test case a total of 10 computations have been performed to analyze the influence of thermodynamic properties, transport and chemical kinetics mechanisms, as well as wall catalyticity. The conditions for the different test cases are given in Table 10.

Figs. 6–7 show the effects of molecular transport and wall catalyticity, and the results are compared versus the Ref.[5] test case solution. The use of the two different models for the molecular transport yields some differences on the temperature field and on the level of dissociation. The model based on Eucken's approximation yields a higher peak temperature in the viscous layer and, consequently, greater dissociation, as shown by the oxygen and nitrogen oxide mass fraction distributions. If the wall is non catalytic, the primary cause of oxygen and nitrogen oxide at the wall is by diffusion. When the wall is fully catalytic, a dramatic change in the chemical boundary layer is observed, as shown in Fig. 7. In particular, note that there exists a "critical" layer (approximately at $y/L = .02$ where the peak temperature occurs) from which O and NO are diffused toward both the wall and the edge of the boundary layer. Then, due to the low wall temperature, recombination occurs and O and NO disappear at the wall. However, the influence of molecular transport is negligible if the wall is fully catalytic.

Figs. 8–9 show the effects of chemical kinetics coupled with wall catalyticity. For the non catalytic case (see Fig. 8) the use of the reduced mechanism (CK2) yields a temperature profile that does not differ much from that obtained with the full Park model (CK1). Some differences are observed in the mass fraction distributions near the wall. Observe that the (smaller) oxygen dissociation of CK2 lowers the temperature peak, thus reducing the production of nitrogen oxide. Greater differences are observed with the use of Blottner's model that yields lower oxygen dissociation and greater production of nitrogen oxide. A fully catalytic wall (see Fig. 9) yields small effects on the temperature field. The concentration boundary layers are affected the most, mainly within the "critical" layer (approximately at $y/L = .02$).

The effects of thermodynamic relations and wall catalyticity are shown in Figs.

10–11. Small differences are observed in the value of the peak temperature. For the non catalytic case a (slightly) smaller value is predicted with the curve-fits model (TR2), which indicates (slightly) greater dissociation. If the wall is fully catalytic the effects are even smaller.

The effects on the velocity field are shown in Figs. 12–13 where the velocity component parallel to the wedge computed with the different models is plotted vs. the distance normal to the wall. The two figures seem to indicate that the velocity field is not affected by the physical models and wall catalyticity.

CONCLUSIONS

In the present work the crucial question that arises as to which model describes adequately viscous hypersonic flows in chemical non equilibrium has been addressed, and the influence of the models has been analyzed. In particular, emphasis has been put on the definition of thermodynamic relations, the description of transport processes and of chemical kinetics mechanisms in the presence of gas-surface interaction.

The solution methodology has been developed within a finite volume approach based on a second order accurate total variation diminishing formulation that accounts for real gas effects. In general, hypersonic flows in chemical non equilibrium are intrinsically stiff due to the disparity between the characteristic chemical and fluid dynamic times. A precondition matrix, related to the partial jacobian of the source terms, is introduced to scale all characteristic times to the same order, and to reduce the stiffness. For computational efficiency and robustness a point implicit algorithm is employed.

Applications of the model to two-dimensional flows have shown that for a non catalytic wall the definition and/or selection of the models are a critical issue that affect the solution prediction. In the presence of gas-surface interactions the issue is not as critical, at least for the fully catalytic wall behaviour that has been simulated in the applications.

ACKNOWLEDGEMENTS

This work was partially supported by the European Space Agency, through Dassault Aviation and Centro Italiano Ricerche Aerospaziali.

REFERENCES

[1] Lee J.H., "Basic Governing Equations for the Flight Regimes of Aeroassisted Orbital Transfer Vehicles", *Thermal Design of Aeroassisted Orbital Transfer Vehicles*, Nelson H.F. ed., Volume 96 in Progress in Astronautics and Aeronautics, American Inst. of Aeronautics and Astronautics, Inc., 1985.

[2] Gnoffo P.A., Gupta R.N., Shinn J.L., "Conservation Equations and Physical Models for Hypersonic Air Flows in Thermal and Chemical Nonequilibrium", NASA Technical Paper 2867, 1989.

[3] Palmer G., "The Development of an Explicit Thermochemical Nonequilibrium Algorithm and Its Applications to Compute Three Dimensional AFE Flowfields", AIAA-89-1701.

[4] Grossman B., Cinnella P., Garrett J., "A Survey of Upwind Methods for Flows with Equilibrium and Non-Equilibrium Chemistry and Thermodinamics", AIAA-89-1653.

[5] Prabhu D.K., Tannehill J.C., Marvin J.G., "A New PNS Code for Chemical Nonequilibrium Flows", AIAA J., Vol. 26, 1988.

[6] Park C., Yoon S., "Calculation of Real-Gas Effects on Blunt-Body Trim Angles", AIAA-89-0685.

[7] Bussing T.R.A., Murman E.M., "Finite-Volume Method for the Calculation of Compressible Chemically Reacting Flows", AIAA J., Vol. 26, 1988.

[8] Desideri J.A., Glinsky N., Hettena E., "Hypersonic Reactive Flow Computation", Comp. and Fluids, Vol. 18, n. 2, pp. 151–182, 1990.

[9] Shuen J.S., Yoon S., "Numerical Study of Chemically Reacting Flows Using a Lower-Upper Symmetric Sucessive Overrelaxation Scheme", AIAA J., Vol. 27, 1989.

[10] Liu Y., Vinokur M., "Upwind Algorithms for General Thermo-Chemical Nonequilibrium Flows", AIAA–89–0201.

[11] Hollanders H., Marraffa L., Montagné J.L., Morice Ph., Viviand H., "Computational Methods for Hypersonic Flows Special Techniques and Real Gas Effects", ONERA.

[12] Yee H.C., Shinn J.L., "Semi-Implicit and Fully Implicit Shock Capturing Methods for Nonequilibrium Flows", AIAA J., Vol. 27, 1989.

[13] Vincenti W.G., Kruger C.H. Jr., *Introduction to Physical Gas Dynamics*, John Wiley and Sons, Inc., New York, 1965.

[14] Park C., *Nonequilibrium Hypersonic Aerothermodynamics*, John Wiley and Sons, Inc., New York, 1990.

[15] Deiwert G., Candler G., "Three-Dimensional Supersonic and Hypersonic Flows Including Separation", AGARD Rep. No. 764, 1989.

[16] Gupta R.N., Yos J.M., Thompson R.A., Lee K.P., "A Review of Reaction Rates and Thermodynamic and Transport Properties for an 11-Species Air Model for Chemical and Thermal Nonequilibrium Calculations to 30 000 K", NASA Reference Publication 1232, 1990.

[17] Park C., "Convergence of Computation of Chemical Reacting Flows", AIAA–85–0247.

[18] Roe P.L., "Approximate Riemann Solvers, Parameter Vectors, and Difference Schemes", J. Comp. Phys., 43, 1981.

[19] Steger J.L., Warming R.F., "Flux Vector Splitting of Inviscid Gasdynamics with Application to Finite Difference Methods", J. Comp. Phys., 40, pp. 263–293, 1980.

[20] Van Leer B., "Flux-Vector Splitting for the Euler Equations", ICASE Report 82–30, September 1982.

[21] Glaister P., "An Approximate Linearized Riemann Solver for the Three Dimensional Euler Equations for Real Gases Using Operator Splitting", J. Comp. Phys., 77, 1990.

[22] Liou M.S., Van Leer B., Shuen J.S., "Splitting of Inviscid Fluxes for Real Gases", J. Comp. Phys., 87, 1990.

[23] Montagné J.L., Yee H.C., Klopfer G.H., Vinokur M., "Hypersonic Blunt Body Computations Including Real Gas Effects", NASA TM–10074, 1988.

[24] Yee H.C., "A Class of High-Resolution Explicit and Implicit Shock-Capturing Methods", NASA Technical Memorandum 101088, 1989.

[25] Grasso F., Bellucci V., "Numerical Solution of Viscous Hypersonic Flows in Chemical Non Equilibrium", 9th GAMM Conference on Numerical Methods in Fluid Mechanics, 25–27 September 1991, Lausanne, Switzerland.

Table 1 - Formation enthalpies, vibrational temperatures and molecular weights.			
Species	Δh_q^0 (J/Kg)	θ_q^v (K)	W_q
O	1.5425×10^7		15.9994
N	3.3614×10^7		14.0067
NO	2.9919×10^6	2739	30.0061
O_2	0	2273	31.9988
N_2	0	3393	28.0134

Table 2 - Degeneracies and electronic temperatures.[*]			
Species	i	$g_{q,i}$	$\theta_{q,i}^e$ (cm^{-1})
O	1	9	78
	2	5	15868
N	1	4	0
	2	10	19228
	3	6	28840
NO	1	4	0
	2	8	38807
O_2	1	3	0
	2	2	7918
	3	1	13195
	4	1	33057
	5	6	34690
	6	3	35398
	7	10	39279
N_2	1	1	0
	2	3	50204

[*]1 Kelvin = 1.439 cm^{-1}.

Species	$A_{q,1}$	$A_{q,2}$	$A_{q,3}$	$A_{q,4}$	$A_{q,5}$
O	.28236 E+1	-.89478 E-3	.83060 E-6	-.16837 E -9	-.73205 E-13
	.25421 E+1	-.27551 E-4	-.31028 E-8	.45511 E-11	-.43681 E-15
	.25460 E+1	-.59520 E-4	.27010 E-7	-.27980 E-11	.93800 E-16
	-.97871 E-2	.12450 E-2	-.16154 E-6	.80380 E-11	-.12624 E-15
	.16428 E+2	-.39313 E-2	.29840 E-6	-.81613 E-11	.75004 E-16
N	.25031 E+1	-.21800 E-4	.54205 E-7	-.56476 E-10	.20999 E-13
	.24820 E+1	.69258 E-4	-.63065 E-7	.18387 E-10	-.11747 E-14
	.27480 E+1	-.39090 E-3	.13380 E-6	-.11910 E-10	.33690 E-15
	-.12280 E+1	.19268 E-2	-.24370 E-6	.12193 E-10	-.19918 E-15
	.15520 E+2	-.38858 E-2	.32288 E-6	-.96053 E-11	.95472 E-16
NO	.35887 E+1	-.12479 E-2	.39786 E-5	-.28651 E -8	.63015 E-12
	.32047 E+1	.12705 E-2	-.46603 E-6	.75007 E-10	-.42314 E-14
	.38543 E+1	.23409 E-3	-.21354 E-7	.16689 E-11	-.49070 E-16
	.43309 E+1	-.58086 E-4	.28059 E-7	-.15694 E-11	.24104 E-16
	.23507 E+1	.58643 E-3	-.31316 E-7	.60495 E-12	-.40557 E-17
O_2	.36146 E+1	-.18598 E-2	.70814 E-5	-.68070 E -8	.21628 E-11
	.35949 E+1	.75213 E-3	-.18732 E-6	.27913 E-10	-.15774 E-14
	.38599 E+1	.32510 E-3	-.92131 E-8	-.78684 E-12	.29426 E-16
	.34867 E+1	.52384 E-3	-.39123 E-7	.10094 E-11	-.88718 E-17
	.39620 E+1	.39446 E-3	-.29506 E-7	.73975 E-12	-.64209 E-17
N_2	.36748 E+1	-.12081 E-2	.23240 E-5	-.63218 E -9	-.22577 E-12
	.32125 E+1	.10137 E-2	-.30467 E-6	.41091 E-10	-.20170 E-14
	.31811 E+1	.89745 E-3	-.20216 E-6	.18266 E-10	-.50334 E-15
	.96377 E+1	-.25728 E-2	.33020 E-6	-.14315 E-10	.20333 E-15
	-.51681 E+1	.23337 E-2	-.12953 E-6	.27872 E-11	-.21360 E-16

Table 3 - Constants for polynomial curve fits of thermodynamic properties.[*]

[*]Five values of the constants are given, corresponding to the following temperature ranges: 300 K – 1000 K; 1000 K – 6000 K; 6000 K – 15000 K; 15000 K – 25000 K; 25000 K – 30000 K

Table 4 - Collision integrals.[*]				
Species	$A^{1,1}$	$A^{2,2}$	$B^{1,1}$	$B^{2,2}$
O	-14.11	-14.71	-14.14	-14.79
	-14.76	-14.69	-14.86	-14.80
	-14.66	-14.59	-14.74	-14.66
	-14.69	-14.62	-14.76	-14.69
	-14.63	-14.55	-14.72	-14.64
N	-14.76	-14.69	-14.86	-14.80
	-14.08	-14.74	-14.11	-14.82
	-14.66	-14.67	-14.75	-14.66
	-14.66	-14.59	-14.74	-14.66
	-14.67	-14.59	-14.75	-14.66
NO	-14.66	-14.59	-14.74	-14.66
	-14.66	-14.67	-14.75	-14.66
	-14.58	-14.52	-14.64	-14.56
	-14.59	-14.52	-14.63	-14.56
	-14.57	-14.51	-14.64	-14.56
O_2	-14.69	-14.62	-14.76	-14.69
	-14.66	-14.59	-14.74	-14.66
	-14.59	-14.52	-14.63	-14.56
	-14.60	-14.54	-14.64	-14.57
	-14.58	-14.51	-14.63	-14.54
N_2	-14.63	-14.55	-14.72	-14.64
	-14.67	-14.59	-14.75	-14.66
	-14.57	-14.51	-14.64	-14.56
	-14.58	-14.51	-14.63	-14.54
	-14.56	-14.50	-14.65	-14.58

$$[*] A^{k,k} = \log_{10} \left(\pi \, \overline{\Omega}_{q,r}^{(k,k)} \right)_{2000}; \quad B^{k,k} = \log_{10} \left(\pi \, \overline{\Omega}_{q,r}^{(k,k)} \right)_{4000}$$

Table 5 - Coefficients for species viscosity.			
Species	A_{μ_q}	B_{μ_q}	C_{μ_q}
O	.0205	.4257	-11.5803
N	.012	.593	-12.3805
NO	.0452	-.0609	-9.4596
O_2	.0484	-.1455	-8.9231
N_2	.0203	.4329	-11.8153

Table 6 - Constants for Park's model.				
Reaction	$C_{f,r}$	$n_{f,r}$	$E_{f,r}/\kappa$	M
O_2+M_1	$2\ 10^{21}$	-1.5	59500	NO, O_2, N_2
O_2+M_2	10^{22}	-1.5	59500	O, N
N_2+M_1	$7\ 10^{21}$	-1.6	113200	
N_2+M_2	$3\ 10^{22}$	-1.6	113200	
$NO+M_1'$	$5\ 10^{15}$	0	75500	O_2, N_2
$NO+M_2'$	$1.1\ 10^{17}$	0	75500	O, N, NO
N_2+O	$6.4\ 10^{17}$	-1	38400	
$NO+O$	$8.4\ 10^{12}$	0	19400	

Table 7 - Constants defining the equilibrium constant.						
Reaction	B_1	B_2	B_3	B_4	B_5	M
O_2+M	2.855	0.988	-6.181	-0.023	-0.001	O, N, NO, O_2, N_2
N_2+M	1.858	-1.325	-9.856	-0.174	-0.008	
$NO+M$	0.792	-0.492	-6.761	-0.091	-0.004	
N_2+O	1.066	-0.833	-3.095	-0.084	0.004	
$NO+O$	-2.063	-1.480	-0.580	-0.114	0.005	

Reaction	$C_{f,r}$	$n_{f,r}$	$E_{f,r}/\kappa$	$C_{b,r}$	$n_{b,r}$	$E_{b,r}/\kappa$	M
O_2+M_1	$3.61\ 10^{18}$	-1	59400	$3.01\ 10^{15}$	-.5	0	O,N,NO,O_2,N_2
N_2+M_2	$1.92\ 10^{17}$	-.5	113100	$1.09\ 10^{16}$	-.5	0	O,NO,O_2,N_2
N_2+N	$4.15\ 10^{22}$	-1.5	113100	$2.32\ 10^{21}$	-1.5	0	
$NO+M_1$	$3.97\ 10^{20}$	-1.5	75600	$1.01\ 10^{20}$	-1.5	0	
$NO+O$	$3.18\ 10^{9}$	1.	19700	$9.63\ 10^{11}$	.5	3600	
N_2+O	$6.75\ 10^{13}$	0	37500	$1.5\ 10^{13}$	0	0	

Table 8 - Constants for Blottner's model.

10° Wedge		
altitude	(km)	61
u_∞	(m/s)	8100
T_∞	(K)	252.6
p_∞	(Pa)	20.35
ρ_∞	(kg/m^3)	$.281\ 10^{-3}$
T_w	(K)	1200
M_∞		25.4
Re/m		$.5\ 10^6$
Ref. length	(m)	1

Table 9 - Flow conditions

Test Case	Model			Wall Conditions
	TR	TC	CK	
1	2	2	3	non catalytic/fully catalytic
2	2	1	3	non catalytic/fully catalytic
3	2	2	2	non catalytic/fully catalytic
4	2	2	1	non catalytic/fully catalytic
5	1	1	3	non catalytic/fully catalytic

Table 10 - Description of test cases for 10° wedge simulation.

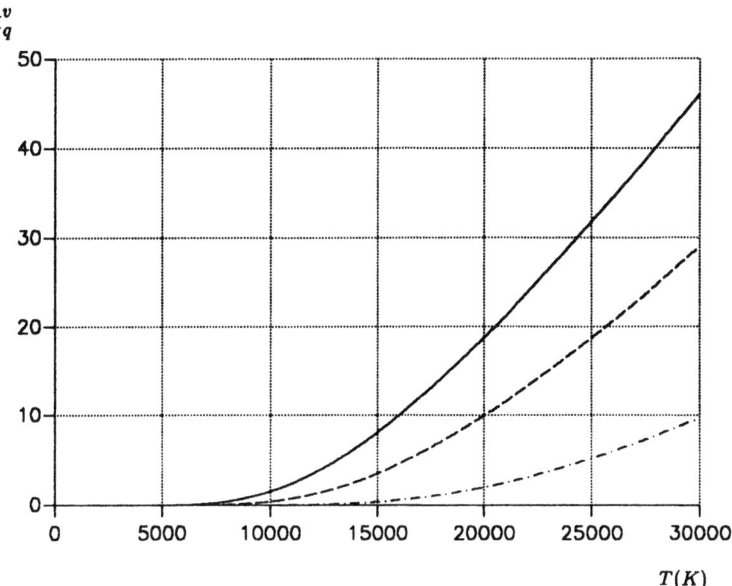

Fig. 1 – Percentage error (for vibrational energy estimation) vs. T:
——— O_2 ; – – – NO ; – · – N_2 .

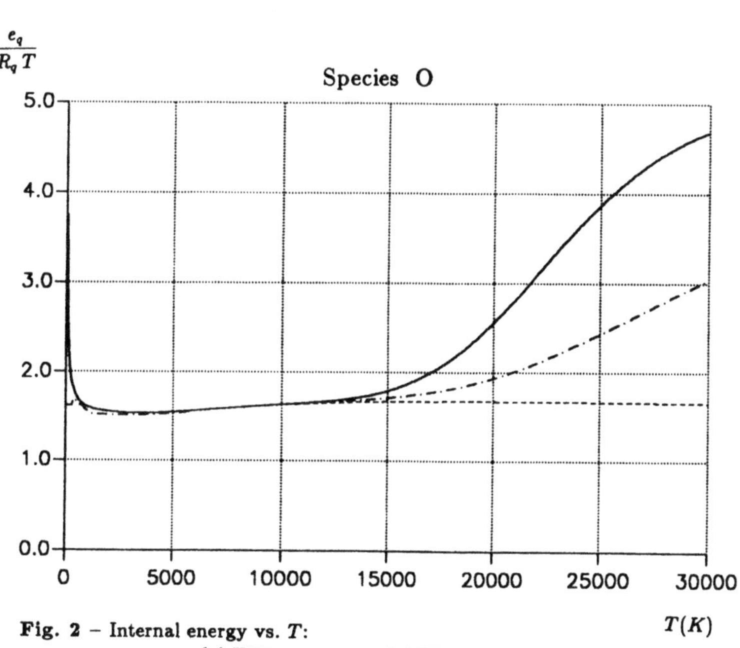

Fig. 2 – Internal energy vs. T:
– · – model TR2 ; ——— model TR1 with all electronic levels;
– – – model TR1.

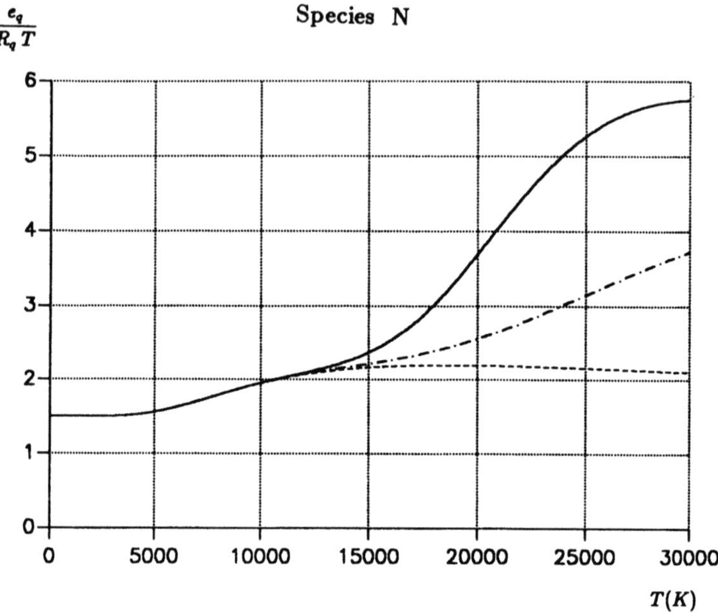

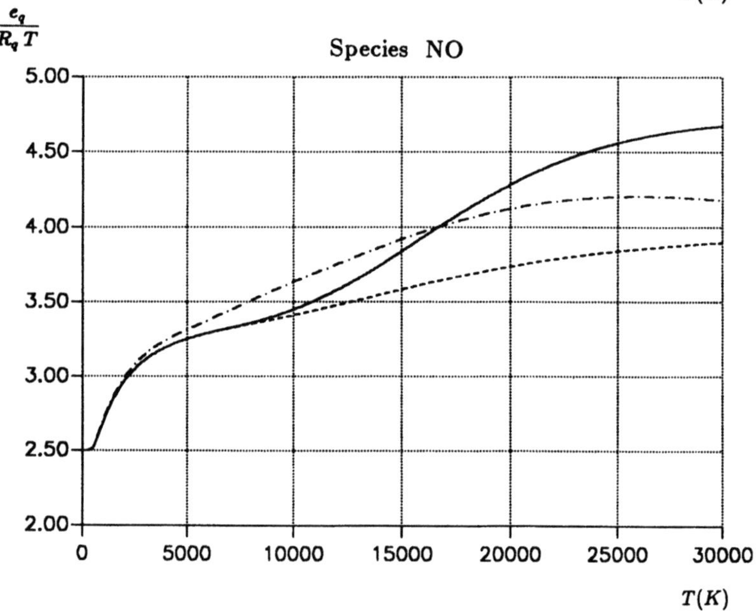

Fig. 2 – continued.

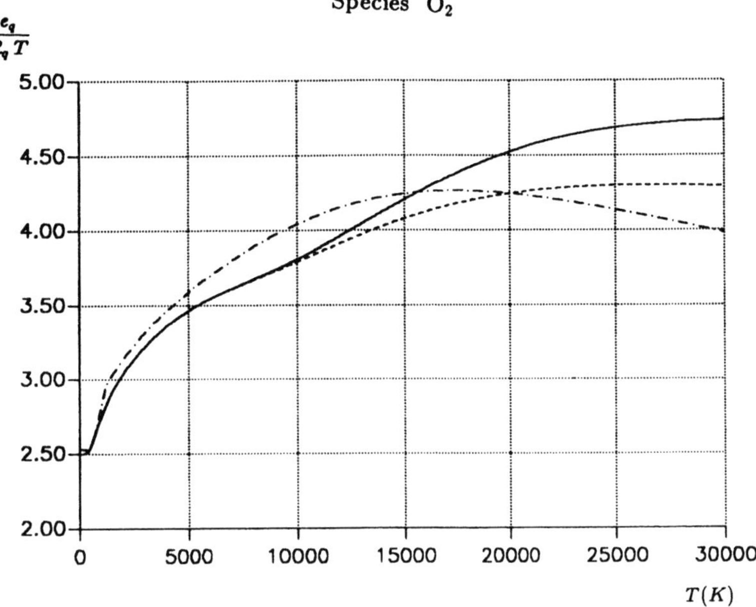

Species O_2

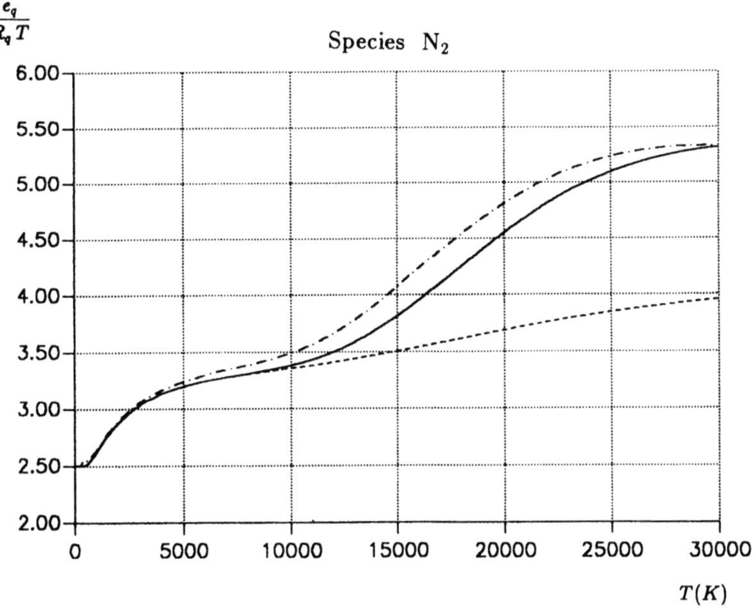

Species N_2

Fig. 2 – continued.

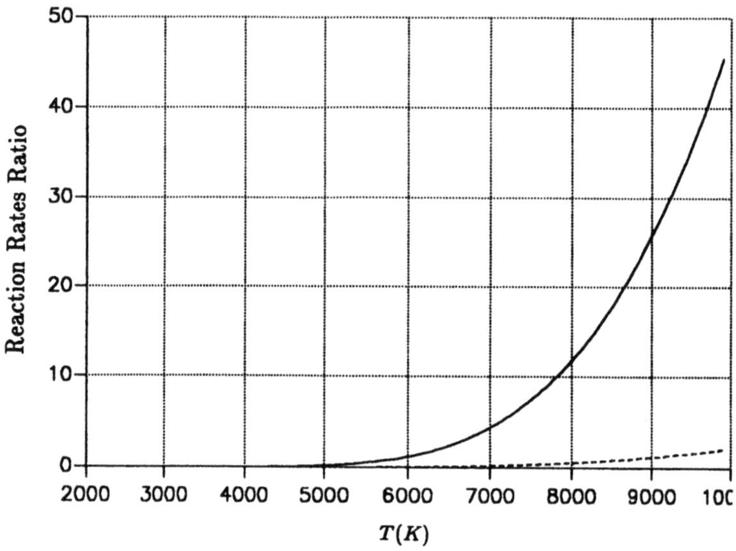

Fig. 3 – Ratio of exchange and thermal dissociation reaction rates of NO vs.
——— colliding species O, N, NO ; – – – colliding species O_2, N_2 .

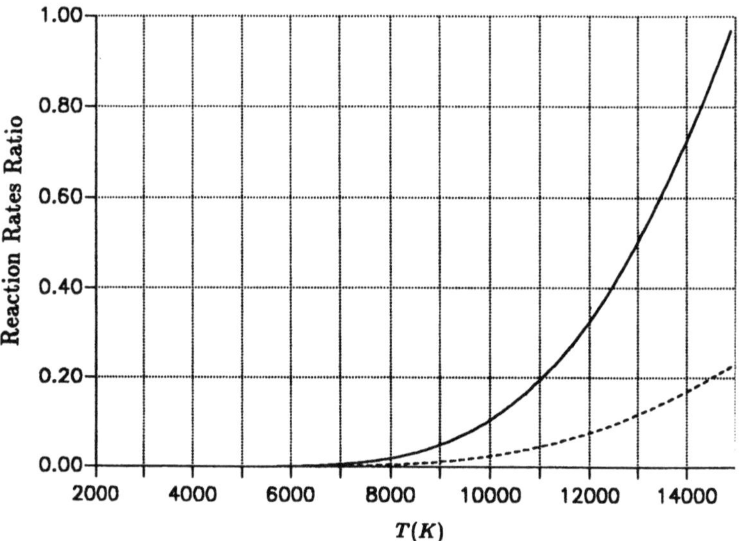

Fig. 4 – Ratio of exchange and thermal dissociation reaction rates of N_2 vs. T:
——— colliding species O, N ; – – – colliding species NO, O_2, N_2 .

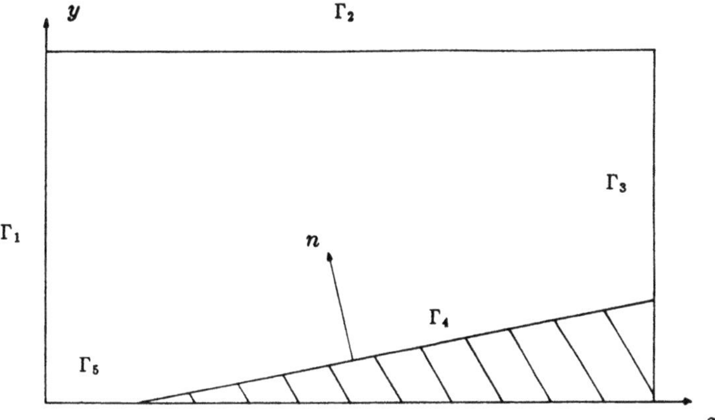

Fig. 5 – Geometry of wedge test case.

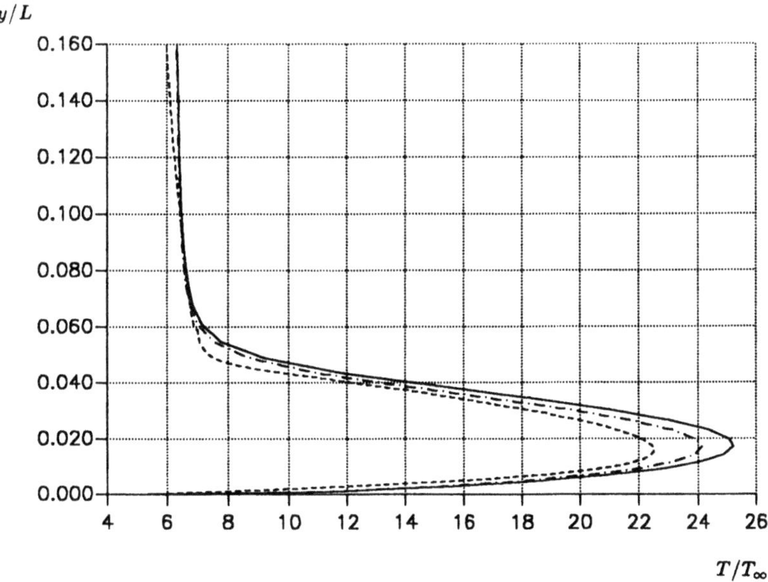

Fig. 6 – Influence of molecular transport with non catalytic wall:
——— Test case 1; $- \cdot -$ Test case 2; $- - -$ Ref.[5].

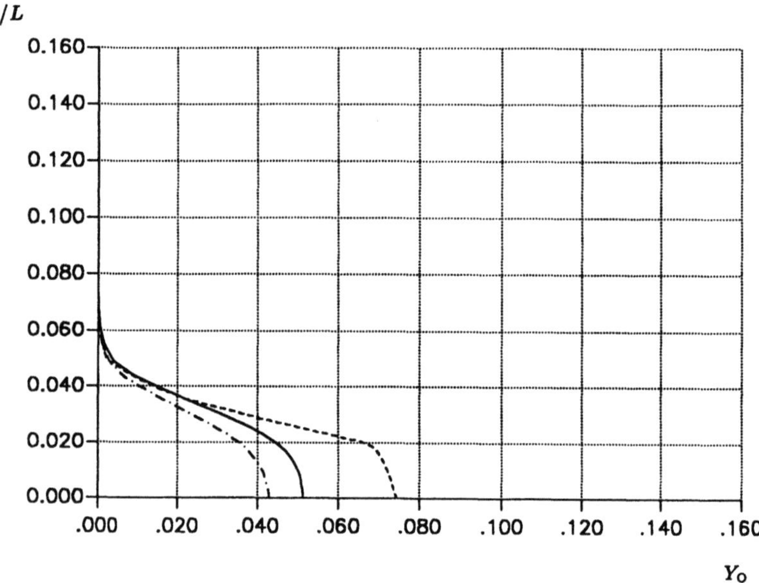

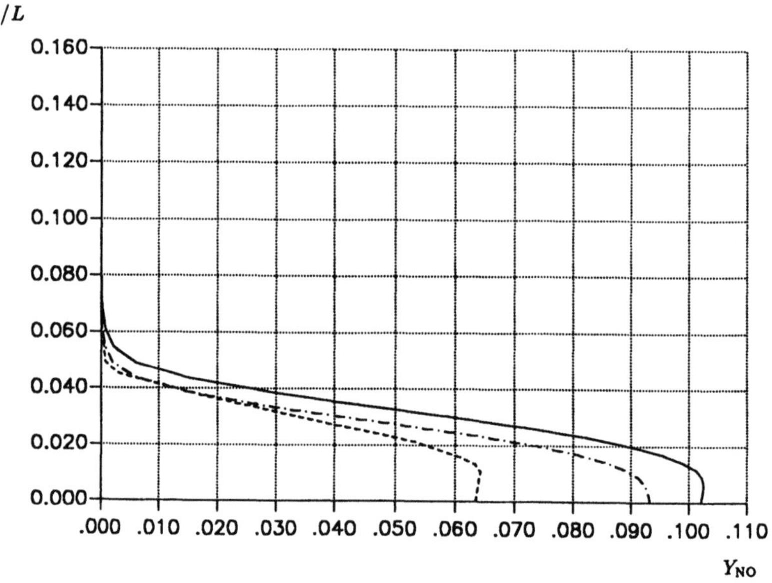

Fig. 6 – continued.

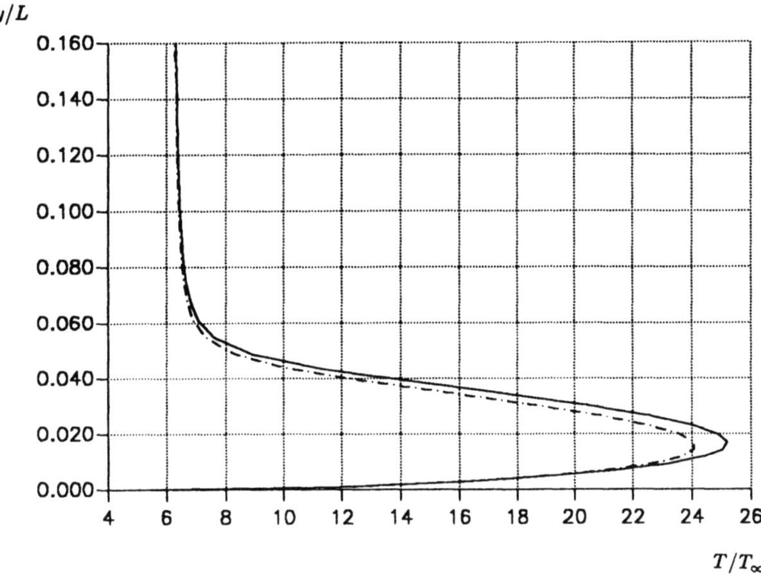

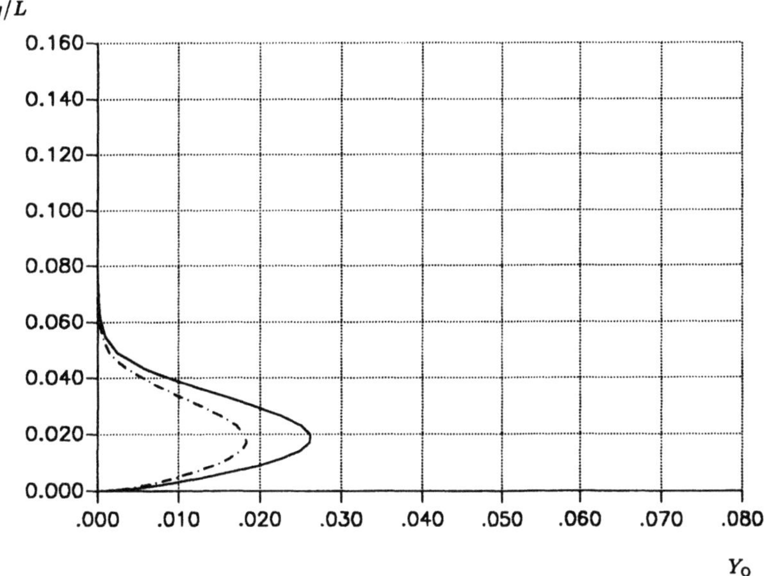

Fig. 7 – Influence of molecular transport with fully catalytic wall:
——— Test case 1; — · — Test case 2.

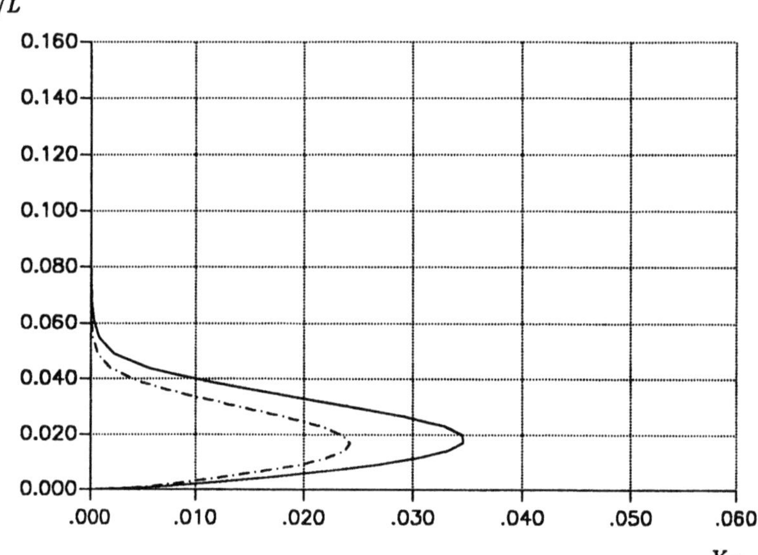

Fig. 7 – continued.

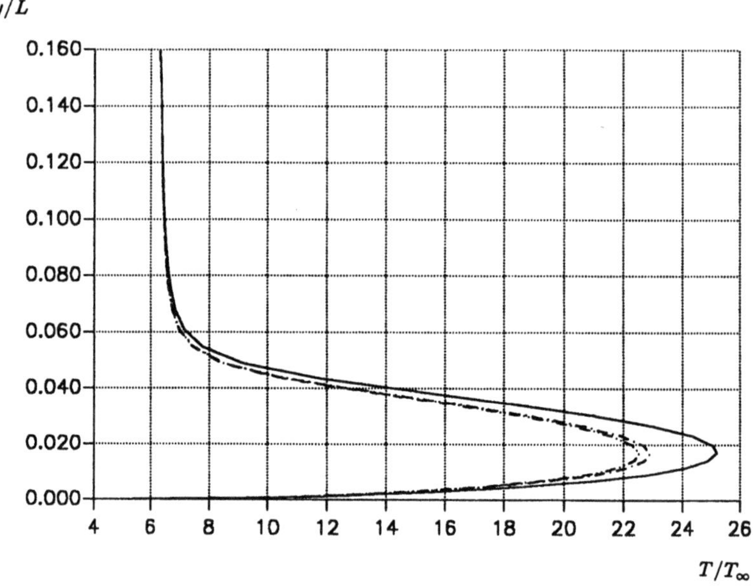

Fig. 8 – Influence of chemical kinetics mechanisms with non catalytic wall:
——— Test case 1;   — · — Test case 3;   — · · — Test case 4.

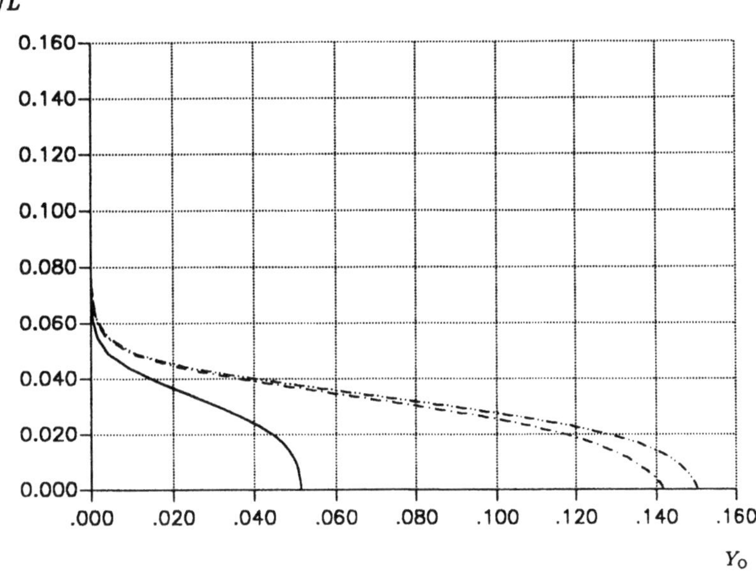

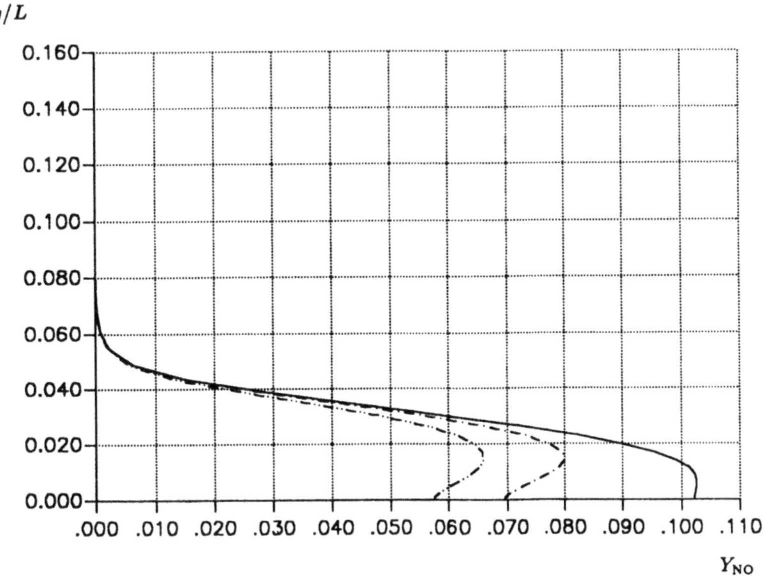

Fig. 8 – continued.

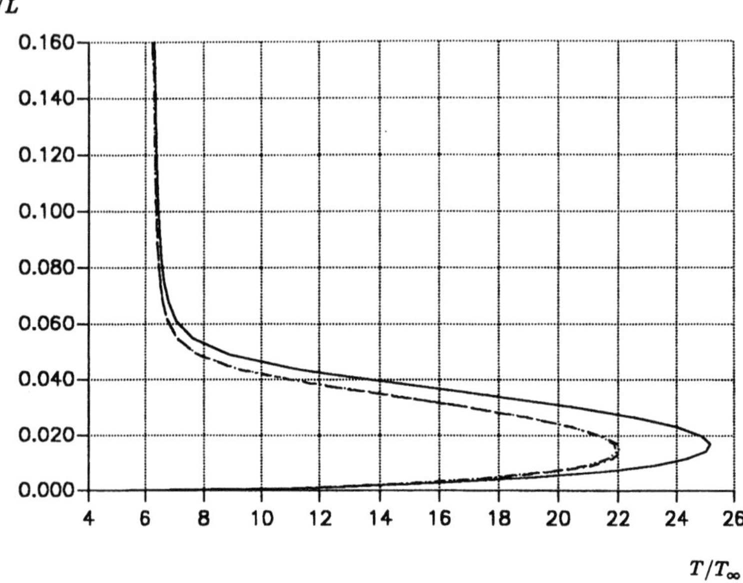

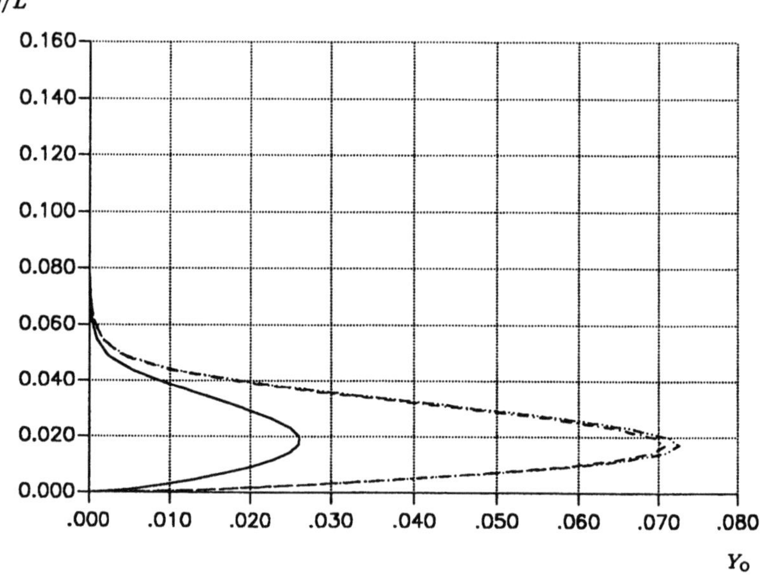

Fig. 9 – Influence of chemical kinetics mechanisms with fully catalytic wall: ——— Test case 1; — · — Test case 3; — · · — Test case 4.

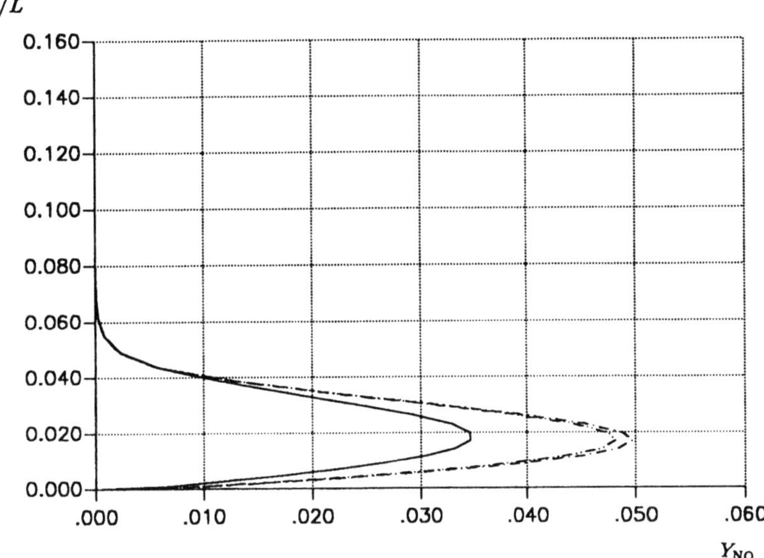

Fig. 9 – continued.

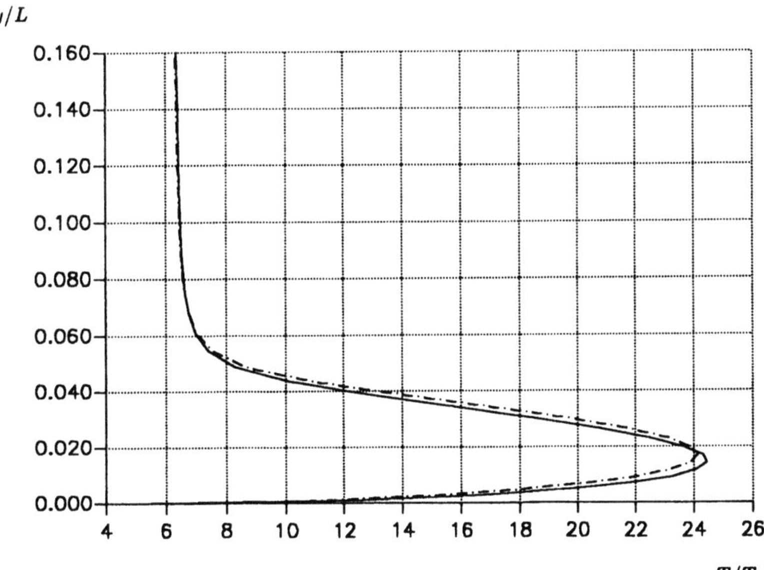

Fig. 10 – Influence of thermodynamic relations with non catalytic wall:
———— Test case 5; — · — Test case 2.

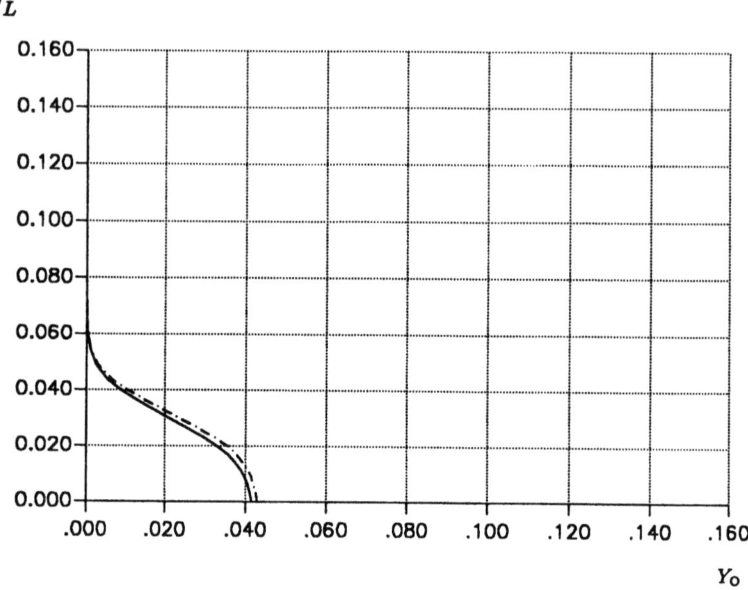

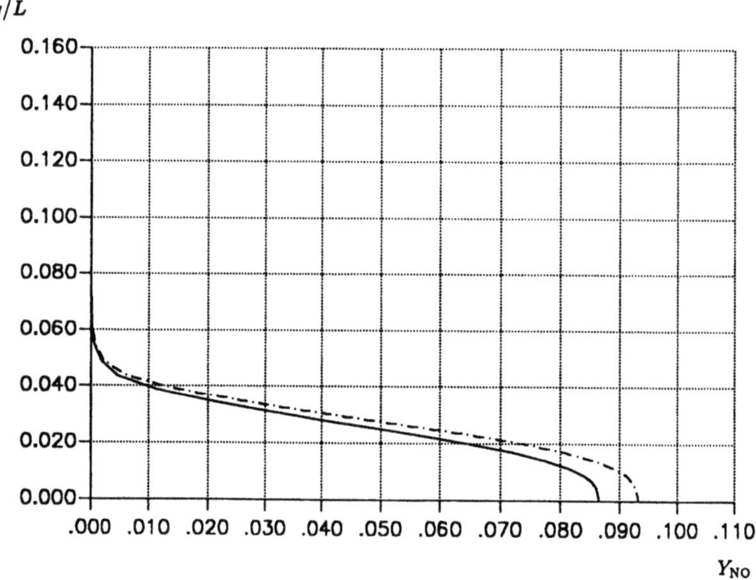

Fig. 10 – continued.

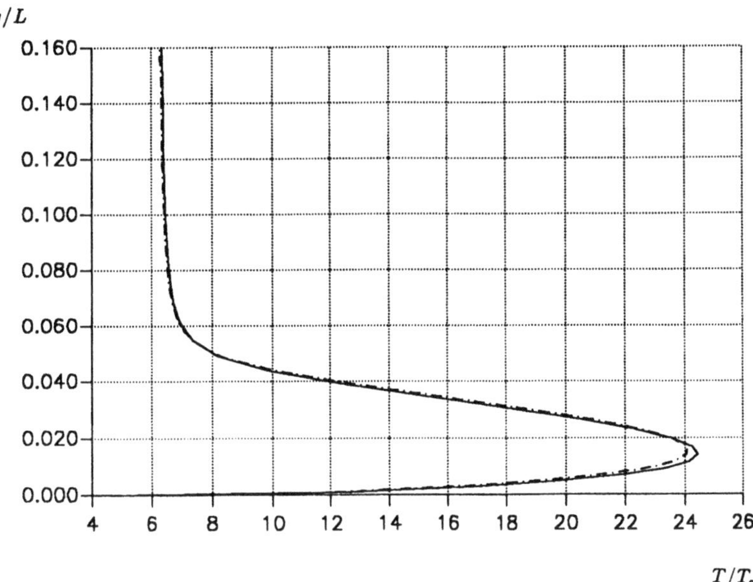

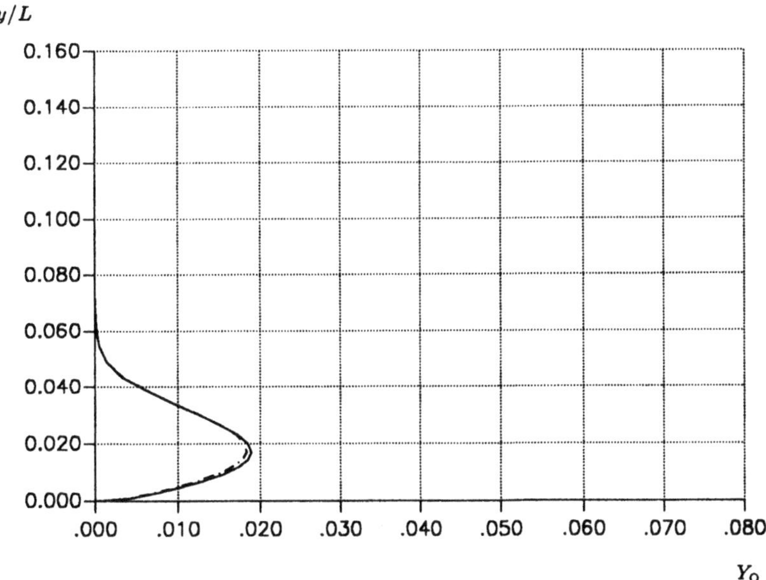

Fig. 11 − Influence of thermodynamic relations with fully catalytic wall:
———— Test case 5; — · — Test case 2.

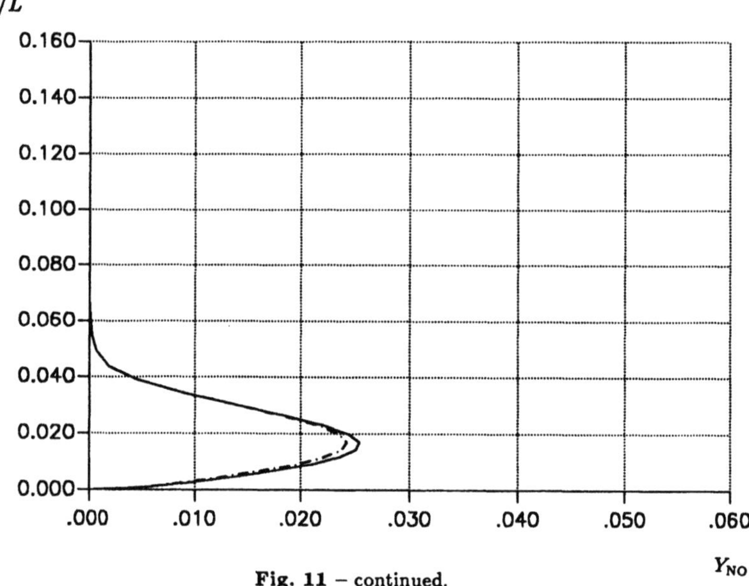

Fig. 11 – continued.

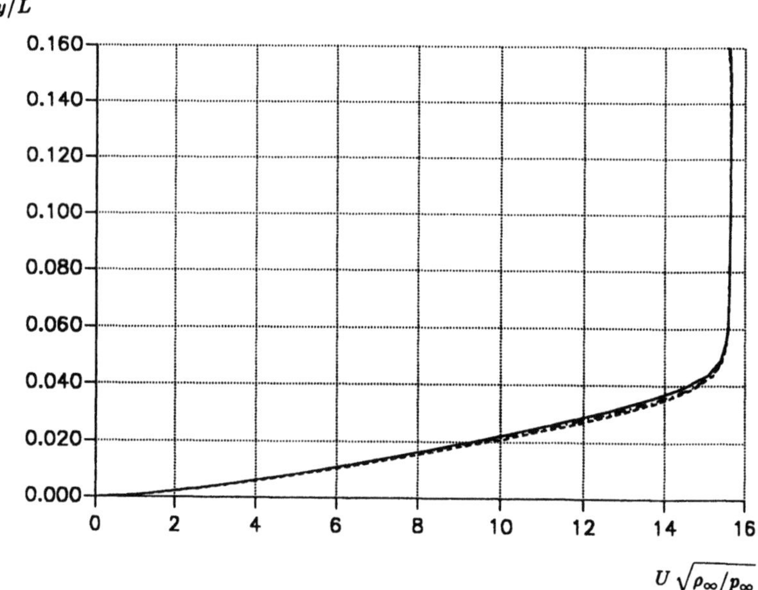

Fig. 12 – Influence of physical models on velocity field with non catalytic wall: —— Test case 1; — · — Test case 2; — — — Test case 4.

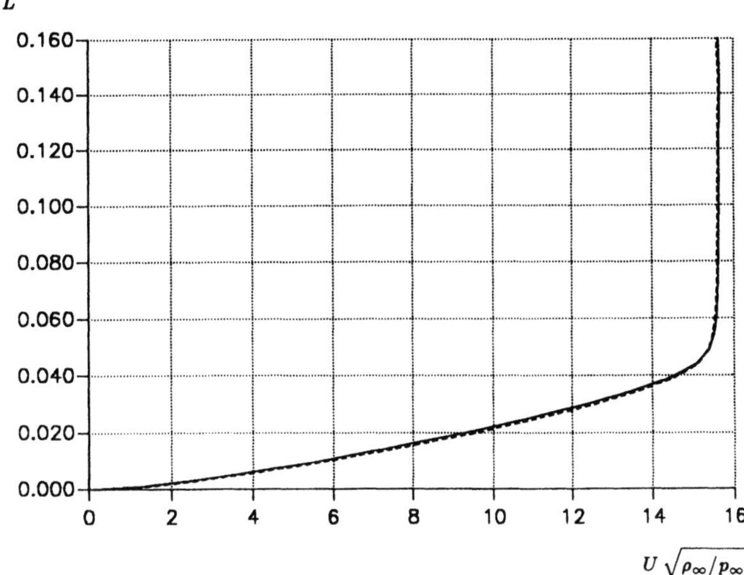

Fig. 13 – Influence of physical models on velocity field with fully catalytic wall: —— Test case 1 non catalytic; − − − Test case 1 fully catalytic.

WALL CATALYTIC RECOMBINATION AND BOUNDARY CONDITIONS IN NONEQUILIBRIUM HYPERSONIC FLOWS - WITH APPLICATIONS

Carl D. Scott
NASA Johnson Space Center
Houston, Texas 77058

ABSTRACT

The lecture discusses the meaning of catalysis and its relation to aerodynamic heating in nonequilibrium hypersonic flows. The species equations are described and boundary conditions for them are derived for a multicomponent gas and for a binary gas. Slip effects are included for application of continuum methods to low density flows. Measurement techniques for determining catalytic wall recombination rates are discussed. Among them are experiments carried out in arc jets as well as flow reactors. Diagnostic methods for determining the atom or molecule concentrations in the flow are included. Results are given for a number of materials of interest to the aerospace community, including glassy coatings such as the RCG coating of the Space Shuttle and for high temperature refractory metals such as coated niobium.

Methods of calculating the heat flux to space vehicles in nonequilibrium flows are described. These include two-layer methods (inviscid/boundary layer), viscous shock layer, parabolized Navier-Stokes, and Navier-Stokes calculations. These methods are applied to the Space Shuttle, the planned Aeroassist Flight Experiment, and a hypersonic slender vehicle such as a transatmospheric vehicle.

NOMENCLATURE

A,B...	chemical species
C_a	number of surface adsorption sites per unit area
C_i	mass fraction of species i
C_p	specific heat at constant pressure
D	dissociation energy
D	thermal desorption energy
D_{ij}	diffusion coefficient for multicomponent mixtures
D_{ij}	binary diffusion coefficient
E	activation energy per mole
f	stream function u/u_e
f	velocity distribution function
g	normalized enthalpy in boundary layer equations

$g12$	relative velocity in molecular collision
h	Planck constant
h	enthalpy
h_{Di}	energy per mass of dissociation
$h_{ic}, h_i{}^c$	enthalpy of formation of species i
H_e	total enthalpy at edge of boundary layer
j	index indicating axisymmetric or 2-dimensional
k	Boltzmann constant
k_w	catalytic recombination rate
L	length of Orbiter
Le	Lewis number
m	mass of particle
$\mathbf{M}_i$	net mass flux of ith species
$\mathbf{M}_i{}^{\downarrow}$	incident mass flux of ith species
M_i	normal mass flux of ith species
n	number density
P_{Ay}	normal stress tensor of species A
p	pressure
p	chemical reaction order
Pr	Prandtl number
q_c	chemical energy flux
R_N	nose radius
r	radial coordinate
Re	Reynolds number
Sc	Schmidt number
s	streamwise coordinate
T	temperature
t	time
u	velocity
u,v,w	velocity components
V_i	diffusion velocity of ith species
W	molecular weight
X	distance
y	coordinate normal to surface
z	normalized mass fraction C_a/C_{ae}

Greek symbols

β	chemical energy accommodation coefficient
ε	emittance
ε	rarefaction parameter $\sqrt{Re}$
η	normalized normal boundary layer coordinate
γ_i	catalytic recombination coefficient
κ	metric coefficient
λ	thermal conductivity

σ cross section
μ viscosity
ρ density
σ Stefan-Boltzmann constant
τ shear stress tensor
ω_i species production rate
ξ boundary layer steamwise coordinate

Subscripts

A,B,a,b   chemical species
a,m   atom, molecule
e, eq   equilibrium
e   edge of boundary layer
FC   fully catalytic
w   wall

Superscripts

s   edge of Knudsen layer
s   stagnation point

INTRODUCTION

Heating to hypersonic vehicles during flight at high altitudes is governed by the chemical or species boundary conditions as well as the state of the flow around the vehicle. At these low Reynolds number conditions the flow is in a state of chemical nonequilibrium; and these boundary conditions for the species equations and the energy equation account for chemical energy diffusion effects that are not included in equilibrium flow or in ideal gas flow. If the surface chemically reacts with the air the surface may be consumed or transformed such as by combustion or ablation. However, in this section we are not going to consider a surface that undergoes chemical changes, instead we will consider surfaces that are catalysts with respect to reactions of the ambient species in the nonequilibrium flow adjacent to the vehicle.

For the most part we will be considering flows that are in the continuum regime where the Navier-Stokes equations or some subset of them applies. However, to extend the applicability of the Navier-Stokes model to higher Knudsen numbers we will also consider slip flow model boundary conditions for reacting gases. On the other hand we will also consider the situation where the gas is not as rarified — the boundary layer regime.

Boundary conditions will be discussed for the simplified binary gas approximation and for the full multicomponent case for both the continuum regime and also in the lower density regime where slip effects may be present.

The objectives of these notes are to present the species equations and the boundary conditions associated with them. We will also consider the form of the energy equation boundary condition since that equation leads to our interest in this whole topic of catalysis and nonequilibrium flows — *aerodynamic heating*. It is hoped that these notes will serve to define catalysis and give the student and practitioner an understanding of the role of catalysis in aeroheating and physical intuition about it. Since catalytic recombination is a significant phenomenon that influences the heat flux to a vehicle in hypersonic flight, it will be important to know how to determine the catalytic reaction rates. Arc jet, flow reactor, and atomic beam experiments for determining catalytic reaction rates and some associated quenching rates will be discussed. Also, some theoretical developments that have been used to gain understanding of catalytic reactions important to aeroheating will be discussed briefly.

Of major relevance is the application of catalysis and nonequilibrium flow field solutions to the prediction of aeroheating to such vehicles as the Space Shuttle Orbiter, the aeroassisted orbital/space transfer vehicles (AOTV/ASTV), the Aeroassist Flight Experiment (AFE), and slender hypersonic vehicles such as aerospace planes (NASP).

At high temperatures the thermal and chemical characteristics of air in the shock layer of a hypersonic vehicle are altered in ways that depend on the atomic and molecular structure of the air constituents, e. g., nitrogen and oxygen atoms and molecules and ions. A similar statement can be said of the species in supersonic combustion in a scram jet, but combustion will not be considered here. The microscopic structure of these species affect the ways in which energy may be distributed and therefore affects the specific heat, chemical reaction rates, and transport properties. These properties, in turn, influence the character of shock waves and flow expansions — hence the pressure, temperature, and velocity distributions. The transport properties affect the boundary layer structure — hence the heat flux and shear stress; and the chemical composition can also affect the chemical energy that is diffused to the surface — hence the heat flux. For a review of the thermodynamics and transport properties associated with nonequilibrium hypersonic flows the reader should consult good texts on thermodynamics and transport properties. A summary of them are found in the notes for the First Joint Europe/US Short Course in Hypersonics.[1] Also found there is a discussion of gas phase chemical kinetics and the species equations and discussions of various flow regimes and the influence on the nature of the shock layer chemistry.

CATALYTIC ATOM RECOMBINATION

When atoms strike the surface of a material they may react to form molecules and in the process liberate some or all of their heat of dissociation. The process is referred to as *catalytic atom recombination.* In a nonequilibrium flow this reaction may also affect the heat transfer to the surface material.

The term *catalytic* is a chemical term associated with certain types of reactions. A mixture of reactants may be far from equilibrium, yet the reaction proceeds so slowly that, for all practical purposes, it does not proceed at all. However, if another species is added to the mixture which makes the reaction proceed at a much faster rate, but does not change chemically itself, then the added species is called a *catalyst.* Catalysts are common in the oil refining industry to enhance the speed of certain reactions for the production of gasoline and other petrochemical products. Every modern American car has a catalytic converter in its exhaust system to convert nitrogen oxides to simpler and less offensive products, e. g. N_2 and O_2. The catalyst in the converter is not consumed, although it may be *poisoned* or rendered ineffective by impurities such as lead.

In the following reaction the molecule denoted M is the catalyst because it remains the same molecule on both sides of the reaction equation.

$$2N + M \;\rightarrow\; N_2 + M \qquad\qquad (1\text{-}1)$$

Here the catalyst is a gas phase catalyst. However, in the case of the automobile catalytic converter, the catalyst is a surface species such as a metal compound.

The reaction equation would then read

$$2NO_2 + s \;\rightarrow\; N_2 + 2O_2 + s \qquad\qquad (1\text{-}2)$$

where s refers to a surface site.

Different materials act differently to enhance the rate of reactions. Therefore, one chooses the catalyst for the particular reaction of interest. Some catalysts simply serve as a third body to carry away excess energy as in reaction (1-1). Others serve as templates to hold molecules in certain positions long enough for there to be a rearrangement of the molecular structure, or for other molecules to find their way to the held molecule or atom. Some catalysts act by causing a rearrangement of the reactant molecules such that their potential energy surfaces are modified, particularly by lowering the potential energy barrier, thus changing the activation energy of the reaction. Whatever the mechanism, catalysts are very useful in the chemical industry.

When it comes to aeroheating during reentry or hypersonic flight, catalytic atom recombination is a *disadvantage*. When molecules are formed on the surface they may give up their latent heat of dissociation to the surface, increasing the heating of the skin. Additional heating is not what vehicle designers like. Therefore, spacecraft designers would like to choose a skin material that is a *poor* catalyst.

Although alluded to above, it should be emphasized that catalysis is a means of enhancing the rate of reactions or changing the equilibrium constant. However, if the gas in the boundary layer of the vehicle is in equilibrium then the rates of formation are already sufficiently fast that a surface catalyst will not have any effect on the formation of molecules. If the flow field around a hypersonic vehicle is in a state of equilibrium then as atoms approach the "cool" surface, the equilibrium will shift toward a undissociated state. In this process the gas is cooled by the wall and the gas phase atoms recombine and liberate their energy of dissociation to the gas in the boundary layer. This added heat tends to increase the heat flux to the surface via thermal conduction. We regard the equilibrium condition as resulting in the maximum heat flux q_{eq}. The equilibrium heat flux has been used as the reference condition in much of the state of the art. As a matter of terminology one also sees the term *fully catalytic*. This means that all atoms that strike the surface recombine on the surface, thus releasing all of their energy of dissociation to the surface. We shall see that the heat flux to a fully catalytic surface q_{FC} in nonequilibrium flow is approximately the same as if the flow were in chemical equilibrium. Therefore, we sometimes see both the terms *fully catalytic* as well as equilibrium wall heating used as a reference. Often the terms *fully catalytic wall* and *equilibrium wall* are used interchangeably. We must understand, however, that the two concepts are distinctly different.

MODELS OF CATALYTIC ATOM RECOMBINATION.

In the following we will discuss catalytic atom recombination from a phenomenological rather than a chemical mechanism approach. However, we will discuss the subject of catalysis mechanisms and the use theoretical of mechanistic and computational chemistry approaches to determine catalytic reaction rates in the section, Reaction Rate Theory for Catalytic Recombination. For now we will be content to deal with the subject in terms of reaction rates inferred from measurements and phenomenological models. We will first consider a simple model of gas/surface interactions then address additional features concerning energy accommodation.

Binary Interaction Model with Full Energy Accommodation. The simplest model of atom recombination on a surface is one in which we presume the atoms that diffuse to a surface either specularly reflect from the surface and neither gain nor lose energy at the surface - or, the atoms stick to the surface and are fully accommodated. Of those that stick some fraction may recombine with other atoms either on the surface or by direct collision with an impinging atom. See Fig.1. The molecules that are formed are presumed to lose all of their latent heat of dissociation to the surface. This model has been assumed in almost all of the calculations of aeroheating up to this time.

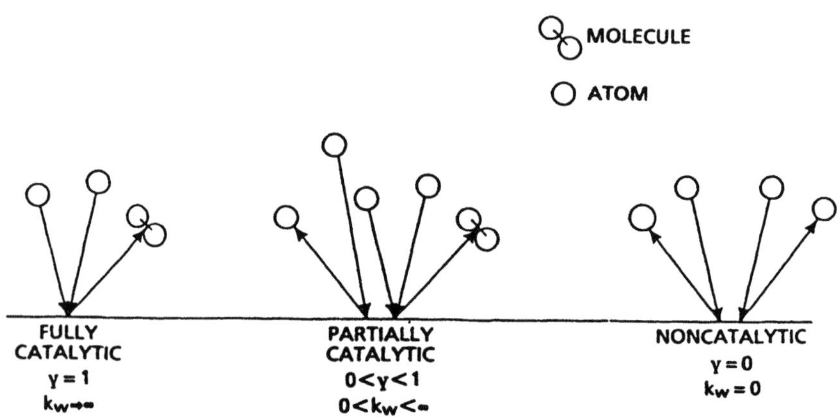

Figure 1.- Simple catalytic surface atom recombination model.

The fraction of incident atoms impinging on the surface that recombine is defined as γ;

$$\gamma_i \equiv \frac{|\mathbf{M}_i^{\uparrow}|}{|\mathbf{M}_i^{\downarrow}|} \qquad (1\text{-}3)$$

where $\mathbf{M}_i^{\downarrow}$ is the mass flux of the incident atoms "i" and $\mathbf{M}_i^{\uparrow}$ is the mass flux of recombining atoms or the net mass flux of atoms to the surface. This ratio γ is often referred to as the atom *recombination coefficient* or the atom *recombination probability*. This number depends on the particular atom and surface involved, and is a temperature dependent quantity. It is usually considered to be independent of the pressure or density because recombination on surfaces is usually a first order reaction.

Binary Interaction Model With Partial Energy Accommodation. A refinement to the previous model is made by considering the energy accommodation to be incomplete. Melin and Madix[2] and Halpern and Rosner[3] have measured recombination and the energy transfer independently and have found that the energy transferred to the surface less than total. That is, they found that even though a large fraction of the atoms recombine, the energy transferred per molecule to many materials is much less than the dissociation energy. This implies that the molecules formed on the surface must leave the surface in excited states or with a velocity much higher than would be expected based on the wall temperature. A fraction β called the chemical energy accommodation coefficient, has been defined in Ref. 2. Specific measurements of β for a number of metals is given in Ref. 3. This ratio of chemical energy transferred to the surface compared to the available energy from recombination is

$$\beta \equiv q_c/M_i h_{Di}$$

where q_c is the chemical energy flux, M_i is the net mass flux of atoms and h_{Di} is the energy of dissociation per unit mass. The disposition of dissociation energy may occur by molecules leaving the surface in excited states. They may quench in the boundary layer or by re-striking the wall and giving up that excitation energy. See Fig. 2. In the case of tile materials such as used on the Space Shuttle there is some evidence that glassy materials also have incomplete energy accommodation. The measurements of Breen, et al.[4] in which numbers of atoms recombining were measured and those of Scott[5], who determined the recombination coefficients from heat fluxes, are compared in Fig. 3. It can be seen that the recombination based on number of atoms is about a factor of 5 greater than that based on heat flux. This implies that β may be on the order of 0.2 for tile materials. One note of caution is that the coatings in the two cases was not exactly the same — one being Lockheed's LI-0042 and the other being reaction cured glass (RCG). A theory of the effects of incomplete energy accommodation and quenching on heat flux was developed by Rosner and Feng.[6]

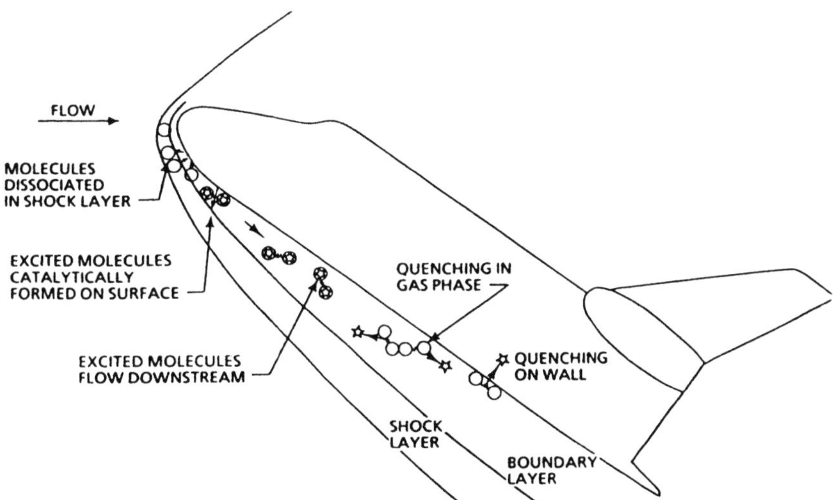

Figure 2.- Catalytic recombination model involving excited state molecule production and quenching.

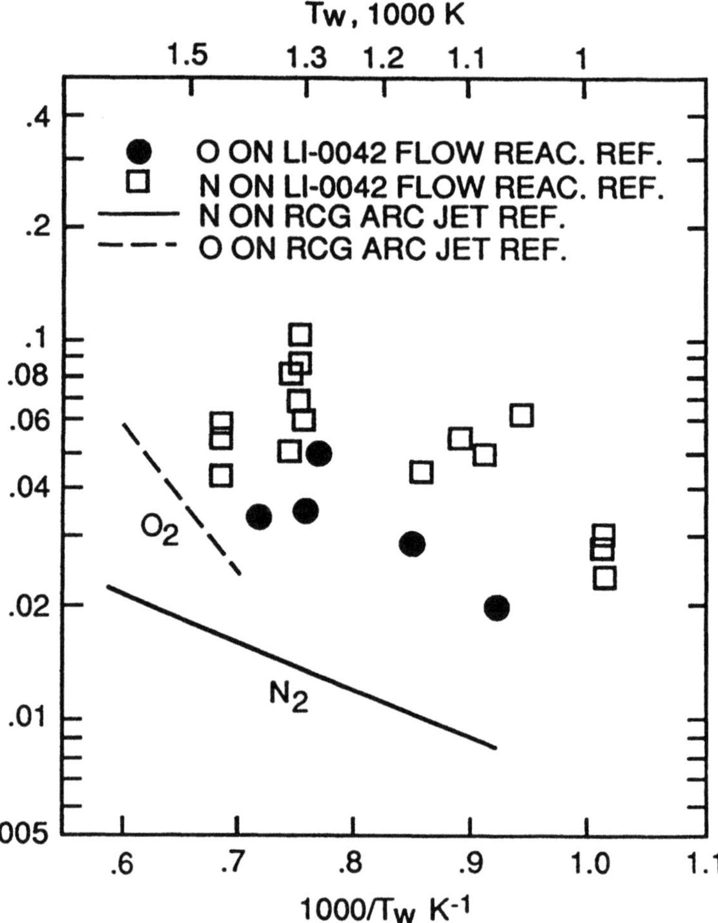

Figure 3.- Comparison of recombination coefficients inferred from atom density measurements and heat flux method.

However, since very little is known about β for real space vehicle surfaces, this refinement in the concept of catalytic heating has not been implemented in heat transfer calculations for vehicles. There is a small effect as seen in Fig. 4 which is a plot of heat flux ratio versus β. As β increases for a constant $\gamma = \gamma'\beta$ the heat flux ratio increase slowly. In this section γ' refers to recombination of numbers of atoms as opposed to energy transfer catalytic recombination coefficient γ. Except where otherwise noted, we will consider the effective energy transfer catalytic recombination coefficient $\gamma'\beta$ to be the same as γ. In other words, we will assume that the recombination coefficient determined from heat transfer measurements is based on the simple binary interaction model with full energy accommodation.

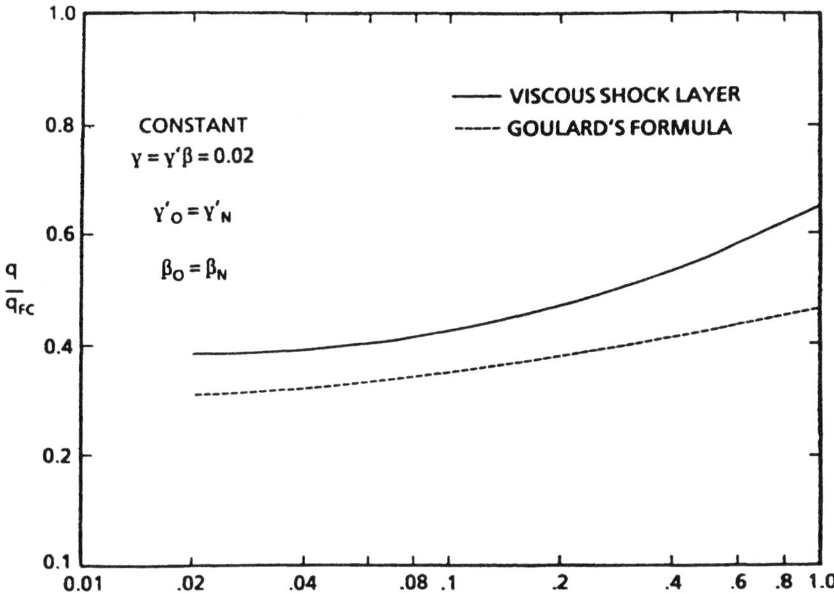

Figure 4.- Stagnation point heat flux variation with incomplete energy accommodation with constant $\gamma = \gamma'\beta$.

SURFACE RECOMBINATION COEFFICIENT AND REACTION RATE. In the literature concerning catalytic effects on heat transfer two terms are often used to express the rate at which atoms recombine on surfaces. We have already defined the recombination coefficient, but another quantity is in common usage, the catalytic recombination rate constant or speed k_w. If we equate the net mass flux of atoms to the surface with the density of atoms at the surface times a reaction rate constant, in analogy with a gas phase reaction rate, then the surface recombination rate k_w is the constant factor by which we multiply the gas phase density of atoms at the surface n_i to obtain the rate that atoms are converted to molecules on the surface per unit time and per unit area. By conservation of atoms the latter rate is equal to the net flux of atoms to the surface. In terms of the mass flux, k_w is defined by

$$M_i = (\rho C_i)^p \, k_{wi} \tag{1-4}$$

where M_i is the normal component of the net mass flux and p is the chemical order of the reaction (usually one.) Now k_w can be related to γ by equating expressions for the fluxes. By substituting (1-4) into (1-3) Scott[7] has shown that

$$k_{wi} = 2\gamma_i/(2-\gamma_i) \sqrt{kT/2\pi m_i} \tag{1-5}$$

for the no slip case, when the first order perturbation to the Maxwell-Boltzmann velocity distribution is assumed. The first order perturbation is consistent with the Navier-Stokes equations. If slip effects are considered then another more exact expression is obtained

$$k_{wi} = \gamma_i \sqrt{kT/2\pi m_i} \qquad (1\text{-}6)$$

We would get this same expression (the Hertz-Knudsen relation) for the no slip case if perturbation terms in the velocity distribution function are not included. Both expressions have been reported in the literature. When using expressions (1-5) and (1-6) in catalysis work one should use values of k_w in a manner that is consistent with the definition used when the original data was obtained.

ENERGY TRANSFER AND CATALYSIS. In a chemically reacting flow the heat transfer to a surface is composed of the usual conduction term plus the term resulting from diffusion of species to the surface. We can write this as

$$q = -\lambda \partial T/\partial y + \sum_{i=1}^{N_S} \beta_i h_{ci} M_i \qquad (1\text{-}7)$$

where q is the heat flux, λ is the thermal conductivity, T is the temperature, y is the coordinate normal to the surface, h_{ci} is the enthalpy of formation of the ith species, β_i is the chemical energy accommodation coefficient, and M_i is the normal mass flux of the ith species

$$M_i = \rho V_i = \rho \sum_{j \neq i}^{ns} D_{ij} \frac{m_i}{m} \left(\frac{\partial C_j}{\partial y} + C_j \frac{\partial \ln m}{\partial y} \right) \qquad (1\text{-}8)$$

where thermal diffusion has been neglected.

The problem for the aerothermodynamicist is to determine these quantities. Most authors have assumed that $\beta_i = 1$. In principle, finding the heat flux requires solving the flow equations with appropriate boundary equations for the particular geometry of interest. The heat flux q is then calculated from the solution using equation (1-7). The flow field equations are discussed in the section, Theoretical and Analytical Methods.

BOUNDARY CONDITIONS FOR THE SPECIES EQUATIONS. The general expression for the boundary equation for the atomic species equations is obtained by equating the mass flux relation (1-8) with the definition of the catalytic speed, equation (1-4)

$$[(\rho C_i)^p]_w \, k_{wi} = \left[\rho \frac{m_i}{m} \sum_{\substack{j=1 \\ j \neq i}}^{ns} D_{ij}\left(\frac{\partial C_j}{\partial y} + C_j\frac{\partial \ln m}{\partial y}\right)\!\frac{\partial C_j}{\partial y} + C_j\frac{\partial \ln m}{\partial y}\right)\right]_w \qquad (1\text{-}9)$$

Note that these mixed boundary equations are coupled in the sense that each C_{i_w} depends on the gradient of all the other mass fractions C_j. For first order reactions at the wall we have

$$C_{i_w} \, k_{wi} = \left[\frac{m_i}{m} \sum_{\substack{j=1 \\ j \neq i}}^{ns} D_{ij}\left(\frac{\partial C_j}{\partial y} + C_j\frac{\partial \ln m}{\partial y}\right)\right]_w \qquad (1\text{-}10)$$

or in terms of γ_i

$$C_{i_w} \, \gamma_i \sqrt{kT/2\pi m_i} = \left[\frac{m_i}{m} \sum_{\substack{j=1 \\ j \neq i}}^{ns} D_{ij}\left(\frac{\partial C_j}{\partial y} + C_j\frac{\partial \ln m}{\partial y}\right)\right]_w \qquad (1\text{-}10a)$$

Example for an Eleven Species Model . A complete set of species boundary conditions for an 11 species reacting model follow. The assumptions made are that the surface is fully catalytic for all ions and is noncatalytic for NO. The mass fluxes for each neutral species is given by

$$M_O = - C_{O_w} \, \gamma_O \sqrt{kT/2\pi m_O} - M_{O^+}$$

$$M_{O_2} = C_{O_w} \, \gamma_O \sqrt{kT/2\pi m_O} - M_{O_2^+}$$

$$M_N = - C_{N_w} \, \gamma_N \sqrt{kT/2\pi m_N} - M_{N^+}$$

$$M_{N_2} = C_{N_w} \, \gamma_N \sqrt{kT/2\pi m_N} - M_{N_2^+}$$

$$M_{NO} = - M_{NO^+}$$

where the ion mass fluxes at the wall are given by

$$M_{O^+} = \frac{\rho m_{O^+}}{m} \sum_{\substack{j \\ j \neq O^+}} C_j \, D_{O^+j}\left(\frac{\partial C_j}{\partial y} + C_j\frac{\partial \ln m}{\partial y}\right)$$

$$M_{O_2^+} = \frac{\rho m_{O_2^+}}{m} \sum_{j \neq O_2} C_j \, D_{O_2^+ j} \left(\frac{\partial C_j}{\partial y} + C_j \frac{\partial \ln m}{\partial y} \right)$$

$$M_{N^+} = \frac{\rho m_{N^+}}{m} \sum_{j \neq N^+} C_j \, D_{N^+ j} \left(\frac{\partial C_j}{\partial y} + C_j \frac{\partial \ln m}{\partial y} \right)$$

$$M_{N_2^+} = \frac{\rho m_{N_2^+}}{m} \sum_{j \neq N} C_j \, D_{N_2^+ j} \left(\frac{\partial C_j}{\partial y} + C_j \frac{\partial \ln m}{\partial y} \right)$$

$$M_{NO^+} = \frac{\rho m_{NO^+}}{m} \sum_{j \neq NO^+} C_j \, D_{NO^+ j} \left(\frac{\partial C_j}{\partial y} + C_j \frac{\partial \ln m}{\partial y} \right)$$

$$M_{e^-} = \frac{\rho m_{e^-}}{m} \sum_{j \neq e^-} C_j \, D_{e^- j} \left(\frac{\partial C_j}{\partial y} + C_j \frac{\partial \ln m}{\partial y} \right)$$

If the wall is fully catalytic then $\gamma_i = 1$ for atoms and the wall flux is limited by diffusion. However, in the literature one occasionally sees $k_w \to \infty$ as the limiting reaction rate for a fully catalytic wall. Then, the corresponding boundary condition for the atom becomes

$$C_{i_w} = 0 \qquad\qquad (1\text{-}11)$$

Although not strictly correct this simplifies the math and may be a reasonable approximation. With ions assumed to be fully catalytic one may approximate the boundary condition for them as

$$C_{N^+} = C_{N_2^+} = C_{O^+} = C_{O_2^+} = 0$$

The more exact boundary condition for fully catalytic species is equation (1-10a) with $\gamma = 1$.

Another limiting case is that of a noncatalytic wall, where none of the atoms striking the wall recombine, that is, $\gamma = k_w = 0$. In that case the net flux of atoms to the surface is zero and equation (1-9) for a multicomponent mixture becomes

$$\left[\rho\, \frac{m_i}{m} \sum_{j \neq i}^{ns} D_{ij}\left(\frac{\partial C_j}{\partial y} + C_j \frac{\partial \ln m}{\partial y}\right) \right]_w = 0 \qquad (1\text{-}12)$$

For a binary gas and a first order reaction (1-9) reduces to

$$(\rho_a)_w\, k_{wa} = -[\, \rho D_{am} \partial C_a/\partial y]_w \qquad (1\text{-}13)$$

for the atomic species. Since $C_m = 1 - C_a$ we do not need a corresponding equation for the molecule.

Frequently, a multicomponent mixture such as air is assumed to take on a binary gas character and equation (1-13) is used for the boundary conditions.[8,9,10,11] The binary diffusion coefficient D_{am} is chosen for the dominant pair of species. Little work has been done to assess the accuracy of this approximation.

In the case of a noncatalytic wall in a binary mixture the species concentration gradient is zero at the wall.

$$\partial C_a/\partial y = -\partial C_m/\partial y = 0 \qquad (1\text{-}14)$$

Contrast this with equation (1-12) for a multicomponent mixture.

SLIP BOUNDARY CONDITIONS. In rarefied flow such as for small vehicles or at high altitudes the density is so low that the Navier-Stokes (N-S) equations begin to break down in regions of the flow field where there is a steep gradient in flow properties such as in the boundary layer near a wall. The gas at the wall can no longer be described by a Maxwell-Boltzmann distribution function, nor is the Navier-Stokes approximation used in the Chapman-Enskog theory adequate. To extend the continuum theory to the rarified regime an approximation is made in which the flow at the wall is not assumed to be fully accommodated to the wall conditions as in the fully continuum theory. The slip model assumes that there is a discontinuity in velocity and termperature between the wall and the continuum zone. The distance of this *Knudsen layer* over which the continuum flow is not valid is of the order of a few mean free paths. Researchers[9] have shown that the slip model can improve the agreement of the Navier-Stokes method to higher altitudes where the Direct Simulation Monte Carlo (DSMC) method is appropriate. See Fig. 5 where the Stanton number versus Knudsen number is compared for three methods, one of which is N-S with slip. One can see that the agreement with the DSMC method is improved when wall slip is included.

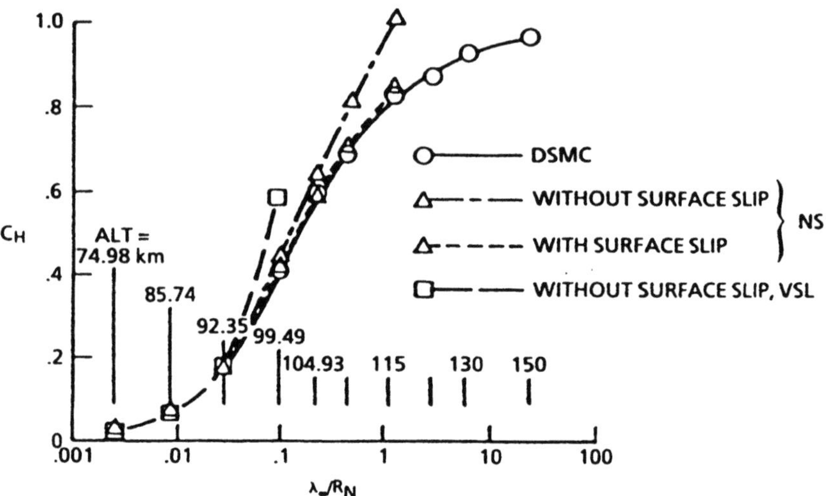

Figure 5.- Heat transfer coefficient versus Knudsen number predicted from different methods. From Ref. 9.

Species Boundary Condition With Slip. The slip boundary condition for recombining atoms is[10]

$$M_A^s = -\frac{\gamma_A}{(2 - \gamma_A)} \sqrt{\frac{kT_s}{2\pi m_A}} \left(\frac{P_{Ay}}{P_A^s} + 1\right)\rho C_i \qquad (1\text{-}15)$$

where, for a multicomponent mixture the net mass flux of species "i", neglecting thermal diffusion, is

$$M_i^s = \frac{\rho^s m_i}{m^s} \sum_{j \neq i}^{NS} D_{ij}^s \left(\frac{\partial C_j}{\partial y} + C_j \frac{\partial \ln m}{\partial y}\right)_s \qquad (1\text{-}16)$$

The superscript s refers to the outer edge of the Knudsen layer, or the inner boundary of the continuum gas near the wall. For zero net mass flux at the wall, the mass flux of atoms must balance the mass flux of molecules, thus

$$M_A = -M_M \qquad (1\text{-}17)$$

where $A = O$ for $M = O_2$ and $A = N$ for $M = N_2$. For all other species M_i is usually assumed to be zero since surface reactions and their rates are not known. P_{Ay} is the (partial) normal stress tensor for atom A. Contrast expression (1-15) with the non-slip case (1-10a).

190

Temperature Jump.

$$\frac{T_s}{T_w} = \left[-\frac{\sqrt{\pi}}{n_s} \sum_{i=1}^{ns} \frac{M_{iy}}{m_i} - \frac{\sqrt{\pi}}{n_s} \sum_{diatomics} \frac{M_{iy}}{m_i} + \frac{1}{2}\left(\frac{P_s}{p^s} + 1\right) \sum_{diatomics} \frac{m^s}{m_i} C_i^s \sqrt{\frac{kT_s}{2\pi m_i}} \right.$$

$$\left. + \frac{1}{2}\left(\frac{P_s}{p^s} + 1\right) \sum_{i=1}^{ns} \frac{m^s}{m_i} C_i^s \sqrt{\frac{kT_s}{2\pi m_i}} \right] \frac{1}{A} \qquad (1\text{-}18)$$

where

$$A = \left[-\sqrt{\pi}\frac{2-\theta}{\theta}\left(\frac{1}{2}\frac{\lambda}{\rho}\frac{\partial T}{\partial y} - \frac{5}{4}\sum_{i=1}^{ns}\frac{M_{iy}}{nm_i}\frac{1}{n}\sum_{diatomics}\frac{M_{iy}}{m_i}\right) \right.$$

$$+ \frac{1}{4}\left(3\frac{P_y}{\rho^s} + 1\right)\sum_{i=1}^{ns}\frac{m^s}{m_i}C_i^s\sqrt{\frac{kT_s}{2\pi m_i}}$$

$$\left. + \frac{1}{2}\left(\frac{P_s}{\rho^s} + 1\right)\sum_{diatomics}\frac{m^s}{m_i}C_i^s\sqrt{\frac{kT_s}{2\pi m_i}} \right] \qquad (1\text{-}19)$$

Here it is assumed that the internal energy modes are fully accommodated with the wall and that the vibrational and rotational modes are fully excited. The sticking coefficient θ is the fraction of the incident molecules that stick or diffusely reflect.

Velocity Slip. The tangential slip velocity for a multicomponent species is given by the relation[10]

$$v_{0x}^s = \left\{ \sqrt{\frac{\pi}{2}}\frac{2-\theta}{\theta}\left[\frac{\mu}{\sqrt{kT}}\left(\frac{\partial v_{0x}}{\partial y} + \frac{\partial v_{0y}}{\partial x}\right)\right] + \frac{1}{5}\left(\frac{1}{kT}\frac{\partial T}{\partial x}\sum_{i=1}^{ns}\frac{n_i\lambda_i}{n}\sqrt{m_i}\right) \right.$$

$$\left. - n_s\sum_{i=1}^{ns}\sqrt{m_i}\sum_{j\neq i}^{ns}D_{ij}^s\left(\frac{\partial C_j}{\partial x} + C_j\frac{\partial \ln m}{\partial x}\right)_s \right\} \cdot \left\{\sum_{i=1}^{ns}n_i^s\sqrt{m_i}\right\}^{-1} \qquad (1\text{-}20)$$

The velocity component in the z-direction is similar.

The general slip theory as well as slip boundary conditions for simple gases and for other coordinate systems can be found in Reference 10. The effects of slip flow and the improvement in agreement with a more exact analysis for low density flows (DSMC) is seen in Fig. 5 (taken from Ref. 9). There we see that as the density decreases or the Knudsen number and altitude increase the non-slip heating deviates more and more from the DSMC method results.

REACTION RATE THEORY FOR CATALYTIC RECOMBINATION. Most of the reaction rates for catalytic atom recombination are obtained from measurements in either arc jets or flow reactor experiments. Our ability to predict the reaction rates from chemical kinetic theory computational chemistry, or other reaction rate theory is very limited. However, some researchers have taken up the task and have achieved some very limited results. It is expected that as the field of computational chemistry continues to develop and with improvements in super computers we will see more analytical predictions of reaction rates. At present the theory of surface catalytic reaction rates depends heavily on empirical data for inputs. With the theories and continued accumulation of empirical information we will begin to understand the mechanisms involved and may finally be able to calculate the reaction rates from first principles.

Chemical Kinetics Model of Gas-Surface Interaction. Jumper, et al[12,13] have worked out an analytical model for fluorine atom recombination on nickel and applied it to a laser nozzle wall. Seward[14] applied their theoretical model to oxygen atom recombination on silica. In an appendix of that same dissertation Seward also presents hydrogen atom recombination coefficients on SiO_2. The model is semi-empirical and describes the physical and chemical microscopic details that may be occurring on the surface when atoms recombine and desorb as molecules. Both first order and second order reactions are accounted for. The model accounts for adsorption of atoms, migration on the surface, recombination, and finally desorption of the molecules. The final form of the recombination coefficient expression is

$$Y = \frac{2\, S_{0_O} N_O \delta_{O_2}\, P \exp\left(-\dfrac{E}{kT}\right)}{U}$$

where

$$U = S_{0_O} N_{O_2} \delta_{O_2} + \delta_O S_{0_{O_2}} N_{O_2} + \delta_O \delta_{O_2}$$

$$+ \left(N_O S_{0_{O_2}} N_{O_2} + N_O \delta_{O_2}\right) P \exp\left(-\frac{E}{kT}\right) \qquad (2\text{-}1)$$

where E is the activation energy of the recombination reaction, S_{0_O} and $S_{0_{O_2}}$ are the initial or clean surface sticking coefficients for O and O_2 respectively, N_O and N_{O_2} are the incident number fluxes of O and O_2 respectively, P is the steric factor for recombination, and δ_O and δ_{O_2} are the thermal desorption rate for O and O_2 respectively. These latter quantities can be written as

$$\delta = C_a \frac{kT}{h} \exp(-\frac{D}{kT}) \tag{2-2}$$

where C_a is the number of adsorption surface sites per unit area, h is Planck's constant, and D is the thermal desorption energy.

Seward used available data for some of the parameters involved in the expression (2-1). He then calculated the recombination coefficient for oxygen recombination on silica. His results, calculated with parameters given in Table 1, are shown in Fig. 6. These results have a behavior similar to the measurements of Kolodziej and Stewart[15] and Scott[5], however, the peak recombination rate predicted by Seward for SiO_2 is at much too low a temperature compared with the arc jet measurements for HRSI. This indicates that the molecular oxygen bond energy is higher for HRSI. Certain other parameters also may not be appropriately modeled, particularly the number of active sites on the surface. Further work is needed to satisfactorily use Seward's reaction kinetics model to predict atom recombination on the RCG coating of the Shuttle tile material. Seward and Jumper[16] have postulated that the recombination sites on silica are silicon atoms, which they believe improves the model by increasing the number of active sites. Validation of the model awaits detailed measurements of the surface physical parameters and recombination probability over a wide range of temperature.

Table 1 Computational Parameters for Oxygen Atom Recombination on Silicon Dioxide Assuming Atomic Oxygen Chemisorption Only

<u>PARAMETER</u>	<u>VALUE</u>
P, Steric factor	$2.24 \times 10^{-5} \exp(0.00908\ T)$ (maximum value = 0.1)
E, Activation energy	1 kcal/mole = 6.949×10^{-21} J/atom
S_0, Initial Sticking Coefficient	$0.05 \exp(-0.002\ T)$
C_a, Surface Sites	5×10^{14} sites/cm^2
D, Thermal Desorption Energy	81 kcal/mole = 5.629×10^{-19} J/atom

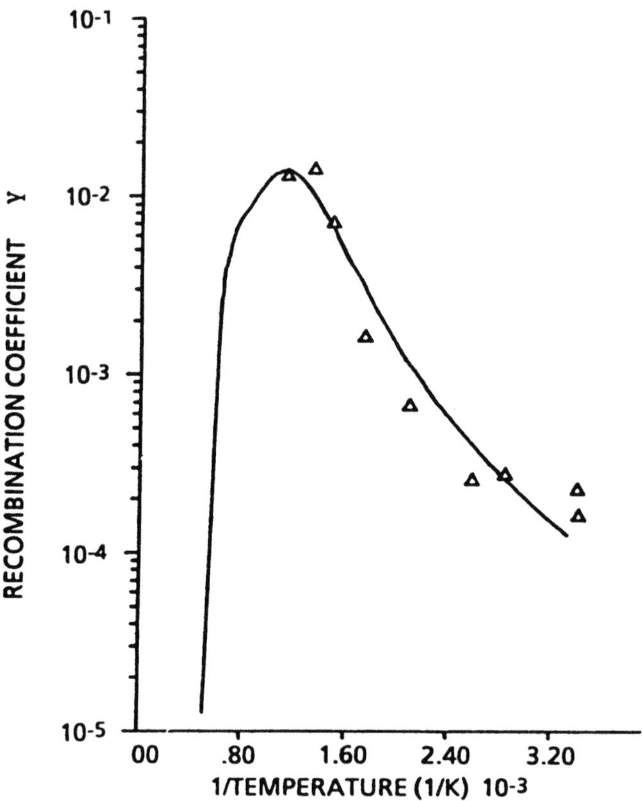

Figure 6.- Recombination coefficient for oxygen on silicon dioxide.
From Ref. 14.

Work by R. J. Willey,[17] using Seward's theoretical model as applied to HRSI, is compared with Kolodziej and Stewart's[15] measured recombination coefficients which are shown in Fig. 7. The trend with pressure tends not to agree, implying that the measurements are inaccurate or the theory needs refining.

Computational Chemistry with Interaction Potential and Chemical Dynamics Model. Theoretical dynamical methods to study the collision processes that may participate in the dynamics of atom recombination and quenching on thermal protection materials are being developed. An initial study was begun by Swaminathan, et al.[18] Electronic energy transfer between gas phase oxygen atoms and a surface consisting of SiO_2 was investigated using semiclassical collision theory. The energy transfer was described within the Self-Consistent Eikonal Approximation method (SCEM). Participation of the vibrational modes of the surface was included via the Generalized Langevin Equation (GLE) method. These investigators were able to model the interaction process in a

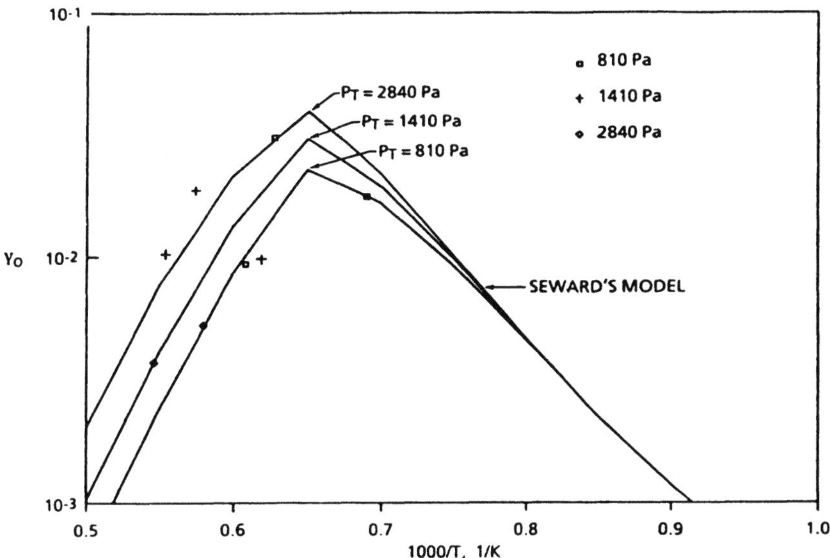

Figure 7.- Comparison of measured recombination coefficients of oxygen on S_iO_2 with Seward's model. (Courtesy of R. J. Willey).

somewhat statistical way, the surface acting as a heat bath for the atoms and absorbing and giving up energy to the electronic state of the atom. Swaminathan, et al. applied this technique to collisions of O with SiO_2 where $O(^1D)$ and $O(^3P)$ species can participate as two electronic states leading to quenching and relaxation phenomena. The dynamics of the collision is given in Fig. 8 where shown are the total energy of gas atom E_{gas}, the probability for quenching the the electronically excited state of the atom P_{12}, the surface atom position r_0, and the atom-surface separation distance R. Note that energy is given up upon collision, but most of it is recovered when the atom leaves the surface. This is a consequence of the relatively constant and finite value of the transition probability when the atom is very near the surface.

For catalytic recombination a similar calculation is needed to model the recombination process and the accommodation of chemical bond energy by the surface. The transfer of energy to internal modes of the formed molecule also must be modeled. It is anticipated that such research will be forthcoming in the next two or three years.

CATALYTIC RECOMBINATION MEASUREMENTS. Several techniques have been used to measure the catalytic recombination coefficient or recombination rates on a variety of materials. There is a large body of literature on techniques near room temperature for many combinations of gas species and

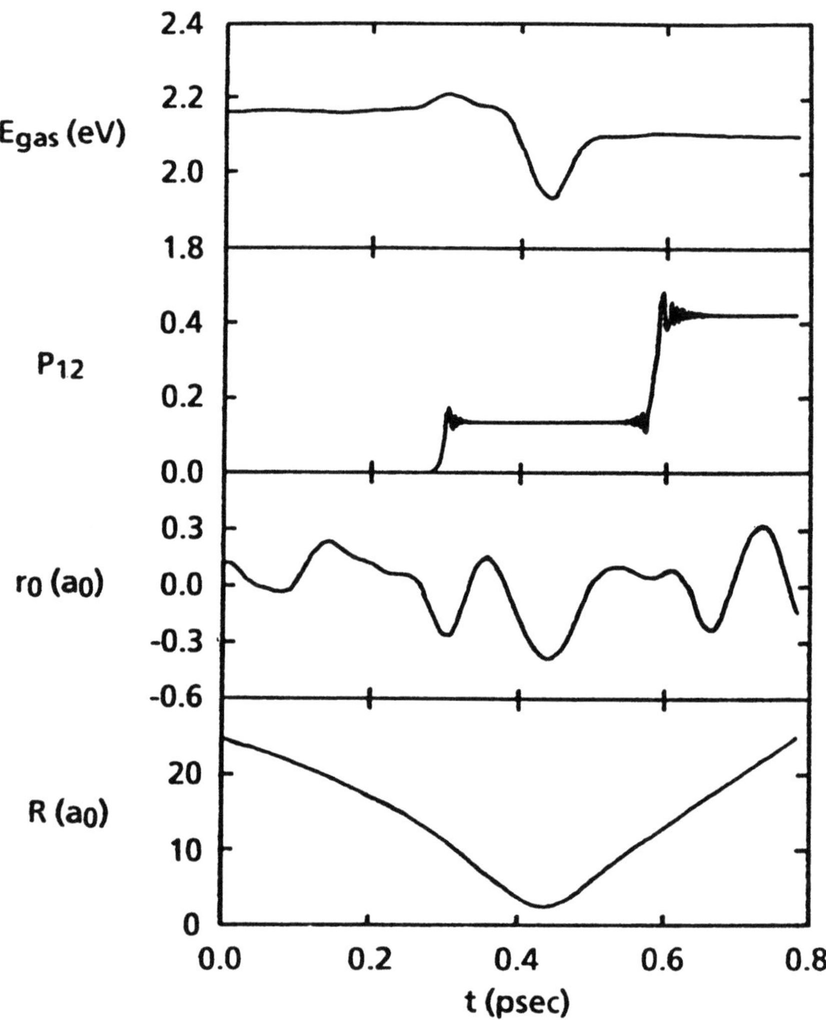

Figure 8.- Solution to interaction equations for oxygen on silica. From Ref. 18.

surface materials. In this section we will confine ourselves to oxygen and nitrogen atom recombination.

Techniques differ in both the method used and in the basic information determined. As mentioned earlier, there are recombination coefficients based on direct atom loss measurements; and there are measurements based on heat transfer measurements. All of the atom loss measurements are done in flow tube reactor experiments. Both the method of detection of atoms and the range of temperature differ in the experiments. Some methods of detecting and measuring species are: chemiluminescent titrations, side-arm oxygen atom detection with Wrede-Hartek or other atom sensitive devices, laser induced fluorescence of molecules, laser multiphoton ionization of molecules, and spontaneous emission. Usually, atoms such as nitrogen and oxygen are generated in a microwave discharge device. However, they may be generated in a high power radio-frequency discharge. The atoms are allowed to flow over a specimen made of the material of interest; and the change in atom concentration with distance is a measure of the rate of atom loss on the surface due to recombination.

Some references of atom recombination from heat flux measurements have been done in flow reactor experiments,[19,20] while others have been done in arc jets.[21,22,5,23,24] An example for the Shuttle tile material, RCG, is shown in Fig. 9. It was usually assumed that the energy transfer due to recombination involved all the dissociation energy. That is, for each recombination event on the surface of the material, one unit of dissociation energy was absorbed by the surface. This assumption ignored the possibility of less than complete energy accommodation, i. e., $\beta < 1$. Both atom loss and heat flux measurements have been done independently in flow reactors to determine both the recombination coefficient γ and the chemical energy accommodation coefficient β. Melin and Madix[2] obtained data for oxygen on metals, whereas, Halpern and Rosner[3] did the same for nitrogen recombination on metals over a wide range of temperature.

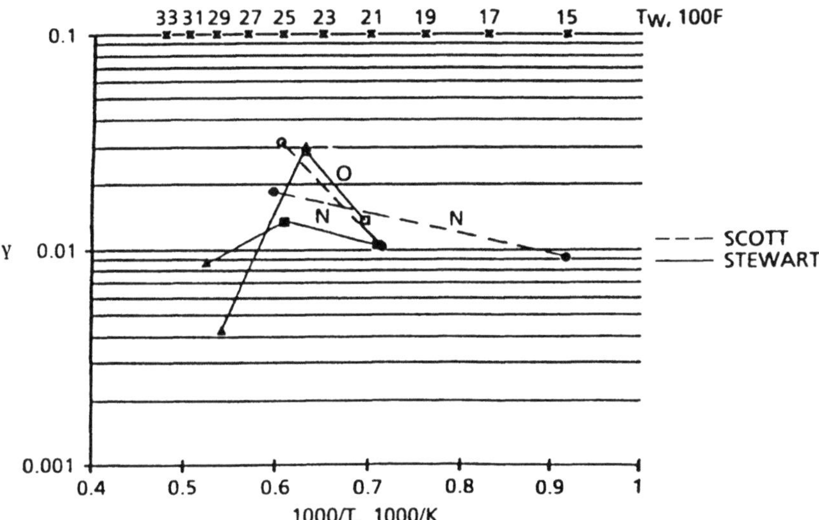

Figure 9.- Measured energy flux recombination coefficients.

Attempts have been made to determine the oxygen recombination coefficient for the Space Shuttle tile material from flight heat flux measurements.[25] Such inferences require accurate heating and flow field predictions as well as accurate measurements. Whereas the measurements are believed to be reliable, the heat flux prediction techniques are not sufficiently accurate to obtain accurate recombination coefficients. Moreover, the unjustified assumption in Ref. 23 that the ratio $k_{wN}/k_{wO} = 0.3$ renders those inferred recombination rates for oxygen unreliable. However, the more recent measurements of Kolodziej and Stewart[15] should be more reliable.

Evaluation of the large amount of data[4,20,21,22,24,26,27,28,29,30,34,36] reveals that recombination coefficients are difficult to measure accurately, since surface condition and possible contaminants significantly affect the measurements. Measurements by different investigators are given in Table 2 for nitrogen and oxygen recombination on room-temperature nickel. It can be seen that there is a large uncertainty in the data. Most likely, the scatter is due to surface preparation and partly due to measurement technique.

Table 2 Reported Recombination Coefficients

Gas/surfaces species	γ,γ' (a)	k_w cm/s (a)	$\gamma'\beta$	Reference temperature, K	Technique	Reference
N/Ni	0.000065	(1.12)		300	Side arm	Rahman and Linnett[28]
N/Ni	0.022	(400)		350		Rojinsky as reported by Goulard[36]
N/Ni	0.1	(1900-2700)		400-800	Atom balance flow tube	Wood and Wise[29]
N/NiO	0.014	(258)		358	Side arm	Dickens and Sutcliff[30]
N/Ni(NiO)	(0.1) (0.4)	2200 1200		500 100	Arc jet	Anderson[21]
N/Ni	(0.29)	520		350	Arc jet	Pope[22]

Table 2 Concluded

Gas/surfaces species	γ,γ' (a)	k_w cm/s (a)	$\gamma'\beta$	Reference temperature, K	Technique	Reference
N/NiO	0.0013	(22)		300	Flow Reactor	Marinelli[20]
O/Ni	0.0085	(150)			Flow reactor heat flux	Myerson[34]
O/Ni	0.028	(500)		298	Side arm	Greaves and Linnett[26]
O/NiO	0.0077	(160)			Side arm	Greaves and Linnett[26]
O/NiO	0.0089	(180)			Side arm	Greaves and Linnett[26]
O/NiO	0.01	(180)		300	Flow reactor	Breen,et al.[4]
O/Ni	0.017	(50)	0.0027		Flow reactor heat flux	Melin and Madix[2]
Air/Ni	(0.09)	1700	400		Arc jet heat flux	Anderson[21]
Air/Ni	(0.017)	300	350		Arc jet heat flux	Pope[22]
O/NiO	0.00096	(17)		300	Flow reactor	Marinelli[20]

[a]Parentheses indicate values calculated from reported values.

Flow Reactors. Techniques for measuring catalytic recombination coefficients fall into two basic categories: atom loss experiments and heat flux experiments. The atom loss experiments fall into two basic types: side arm flow reactor experiments and duct reactor experiments. In each the loss of atoms is measured and related by simple theory to the recombination coefficient.

Side arm reactors. One of the earliest techniques was a side arm technique developed by Smith[31] and adapted by Greaves and Linnett[26]. The technique makes use of a side arm duct coated with the material of whose recombination coefficient is to be determined. A moveable probe which is sensitive to the atoms of interest is inserted into the side arm. The probe response as function of distance will yield the recombination coefficient via a solution of the diffusion equation for a catalytic wall tube having zero convection. The conditions must be such that gas phase recombination is negligible. The solution to this equation is given in Ref. 26. An example of the physical arrangement for a side arm reactor is shown in Fig. 10a. Similar techniques have been used by other researchers who used various detector techniques, such as, Wrede-Hartek gages, afterglow emission measurements and titrations, electron magnetic resonance spectroscopy, etc. Many detection techniques may be used, depending on the species of interest.

Straight flow reactors. Straight flow reactors have been used in which the wall of the tube is lined with the material of interest and similar detection techniques have been used as above. Another application of flow reactors is one in which the sample of interest is inserted into the flow. Halpern and Rosner[3] used a fast flow reactor to measure both γ and β for nitrogen recombination on metal wires. Their apparatus is shown in Fig. 10b. Recently, Marinelli, et al.[32] used laser induced fluorescence to measure the atom and molecular nitrogen and oxygen production or loss rates on the Shuttle tile RCG coating at near room temperature. Instead of the wall being coated with the RCG, a coated rod was inserted into the flow tube as shown in Fig. 11. Some techniques rely on the measurement of the heat flux to determine the recombination coefficient. The results of this technique may result in data that is more applicable to aerothermodynamics applied to the reentry spacecraft because the question of incomplete energy accommodation is bypassed somewhat. At least, it is an effect that already may be accounted for approximately in the measurements. Flow reactor experiments that have provided nitrogen recombination on platinum and copper were done by Prok[33]. Myerson[34] measured the oxygen recombination coefficient on noble metals and titanium dioxide; whereas, Hartunian, et al.[35] measured oxygen recombination on silver. Most of these measurements were made for surfaces at room temperature. For temperatures of hypersonic vehicle interest heating devices need to be used in conjunction with the atom loss or heat flux measurements. In some cases the heating is provided externally, such as with a heating coil or oven or even radiatively heated with a high power cw laser. Another technique to be discussed in the next section is the arc jet, which is particularly appropriate for thermal protection materials.

<u>*Arc Jets.*</u> In arc jet flows the samples for which γ is to be determined are subjected to heating in a dissociated flow environment. The energy flux catalytic atom recombination coefficient is derived from measurements of the heat flux via

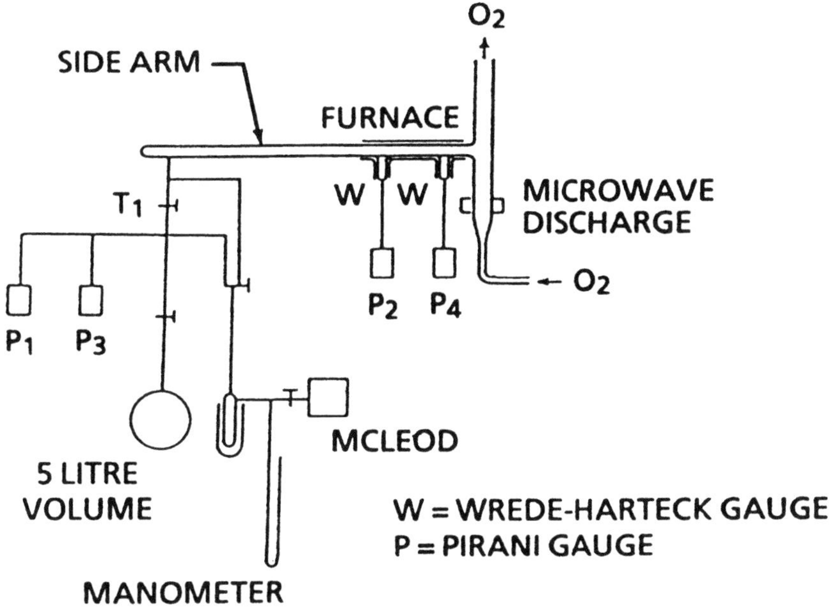

Figure 10a.- Diagram showing experimental lay-out when using microwave-maintained discharge of side arm reactor. From Ref 26.

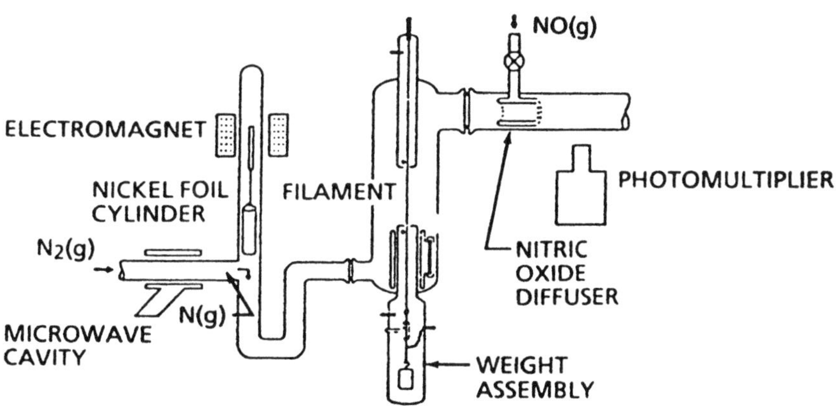

Figure 10b.- Schematic diagram of coaxial filament flow reactor (CFFR) and flow system. From Ref. 3.

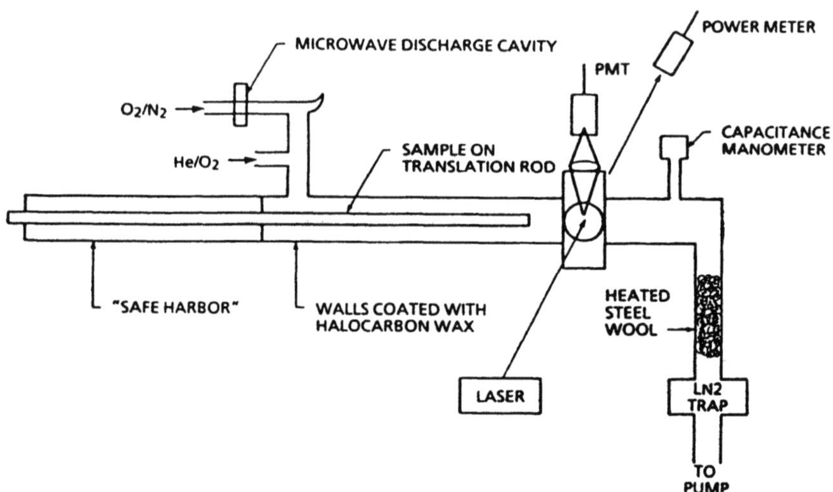

Figure 11.- Discharge flow reactor with sample on internal rod. From Ref. 20.

a theory such as that of Goulard[36] or some other heating prediction such as viscous shock layer or boundary layer equation solutions. One difficulty of using arc jets is that the free stream flow has not been well characterized. Approximations have to be made to estimate the atom fractions, since the kinetics of the flow and the initial conditions upsteam of the nozzle throat are not well known.

Determinations of energy flux recombination coefficients have been made by Pope[22], who measured nitrogen and air recombination on copper, nickel silver, and their oxides, gold platinum, tungsten, silicon monoxide, and teflon at discrete temperatures. Anderson[21] studied nitrogen recombination on nickel, platinum, copper, gold, and silica. His measurements were made at wall temperatures of 400, 500, and 1000 K. Scott[37] measured room temperature energy transfer recombination rates on chrome oxide, siliconized carbon/carbon, hafnium/tantalum carbide on carbon/carbon, and niobium silicide. Scott[5,23] also measured temperature dependent energy transfer recombination coefficients on the Space Shuttle tile RCG coating and an overcoating used on the Orbiter Experiment (OEX) wall catalysis experiment of Stewart et al.[38] Subsequently, Kolodziej and Stewart[15] have extended these measurements to higher temperatures. These results are compared in Fig. 9.

Zwan, et al.[39] measured the recombination speed of "air" atoms on silicon carbide and nickel at high temperatures in an arc jet facility. These materials were applied to a graphite cylinder/wedge test article by chemical vapor deposition. They used a viscous shock layer code to predict the heat flux to a cylinder whose axis was normal to the flow axis. Their results for SiC were reasonably consistent with the heating measured on the Space Shuttle Orbiter. Their approach has the deficiency that the results assumed that the oxygen is fully dissociated and no account was taken for different recombination coefficients for oxygen and nitrogen separately. In spite of this deficiency, their approach illustrates a fairly careful experimental approach using arc jet facilities with then current diagnostic capabilities to determine the flow properties. More modern techniques are being developed using spectroscopy and lasers at the NASA Johnson Space Center[40] which can improve the knowledge of the flow and lead to more accurate heating predictions, and hence, better accuracy of the catalysis measurements.

Recently Cunnington, et al.[41] measured the catalytic recombination coefficent for several coatings applicable for hypersonic vehicles in an arc jet facility. Since most structural materials may degrade in oxidizing environments, they must be coated to prevent oxidation. These coatings should have high emittance and low catalycity to maintain low temperatures. Several candidate coatings for protecting titanium aluminide alloys were tested in the NASA Langley Research Center arc jet to determine their catalytic characteristics at temperatures from 300 and 1255 K. Chemical vapor deposited boron-oxygen-silicon and aluminum-boron-oxygen-silicon compositions were applied to Alpha 2 and Gamma titanium-aluminide alloys. The emittance, catalytic recombination coefficients, and weight change (to determine oxidation resistance) as a function of arc jet exposure time (at 1255 K) were determined from the tests. The recombination efficiencies they determined are given in Table 3.

Table 3 Recombination Efficiencies of Coated and Uncoated
Titanium-aluminides at 1255 K

	Recombination Efficiency after Exposure		
	0.5 hr	1.0 hr	5.0 hr
Uncoated Ti$_3$Al	0.073	0.071	-
Uncoated TiAl	0.038	0.050	-
SiB$_x$, B-O-Si	0.002	0.002	0.005
Al-B-O-Si, B-O-Si	0.005	0.013	0.015
Sermalloy W	0.012	0.025	-

Shock Tubes. Dissociated flow may be obtained in other high temperature facilities such as shock tubes and shock tunnels. These facilities frequently produce flow that is near chemical equilibrium. However, if care is taken to operate at low densities and pressure the flow in the shock layer of a blunt body immersed in the flow may be out of equilibrium and therefore be capable of use in determining the atom recombination probablity, much as in the case of arc jet flow. In fact this has been done in the case of oxygen recombination on nickel and copper by McCaffrey and East.[42] Their method of determining the recombination coefficient is to compare the heating on these metals with that on silica using a differential thin film heat flux gage coated with the various materials. Their analysis used the Goulard[36] stagnation point heat flux relation

203

and the Fay and Riddell[43] heating analysis to be discussed later. These shock tube measurements yielded recombination coefficients of 0.13 for copper and $0.018 < \gamma < 0.022$ for nickel.

Atomic Beams. None of the techniques already discussed have addressed the issue of incomplete energy accommodation and the production and quenching of excited species on thermal protection materials. One such technique being developed by W. J. Marinelli at Physical Sciences, Inc. is an atomic beam technique in which a beam of atoms is generated in an RF discharge then impinges on a sample in an ultrahigh vacuum chamber. The atomic beam apparatus of Marinelli, et al. is shown in Fig. 12. The incident flux of atoms and the resultant recombined molecules are measured using mass spectrometry and laser multiphoton ionization. Detection of the ions is done by a channel plate electron multiplier having a large number of detection elements. This technique should be able to determine the energy state of the molecules produced using a retarding potential in front of the plate. The experiment will also be used to determine the probability of formation of the excited molecules. The samples may be "prepared" in the ultrahigh vacuum chamber by flowing suitable gasses over the surface of the sample prior to the measurement. The sample also may be rotated and translated to position it for various measurements.

CLASSIFICATION OF MATERIALS. Materials of interest for the spacecraft thermal protection designer are ones that have low recombination efficiencies. These generally fall into the category of glasses, silica, and perhaps silicon carbide. Highly catalytic materials are usually found among the metals and metal oxides. In between fall some salts and metal carbides. One should consult the references cited above for specific values for the catalytic rates. To this author's knowledge fused silica has the lowest values of γ_N and γ_O at room temperature, while silver has the highest values.

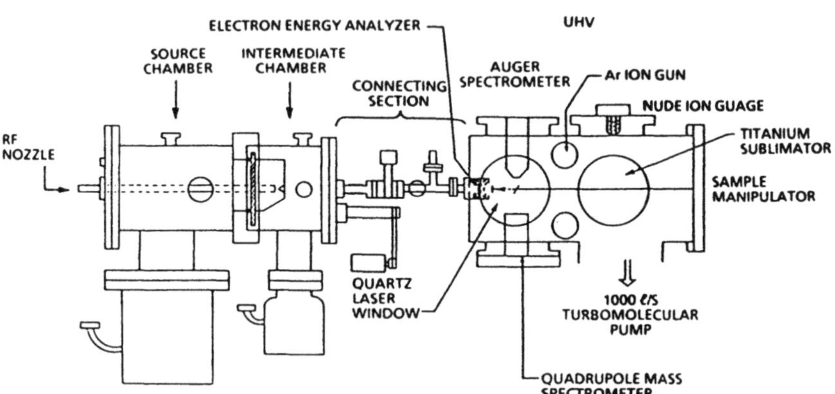

Figure 12.- Overall view of atomic beam apparatus.

To obtain the effect of finite rate or noncatalytic atom recombination on surfaces for applications of aerospace interest the flow field equations are solved with the appropriate boundary conditions discussed in the sections, BOUNDARY CONDITIONS FOR THE SPECIES EQUATIONS and <u>Example for an Eleven Species Model</u>. The flow field equations are usually the Navier-Stokes equations or some subset of them. Solutions techniques for these equations are discussed by others in this short course and, therefore, will not be discussed in detail here. However, for completeness the equations will be described briefly; and references to solution methods and examples will be given. The equations will be discussed in the order of generality from the most general equations, the Boltzmann equations, to the more specific Euler, viscous shock layer, and boundary layer equations.

THE FLOW EQUATIONS. From a microscopic viewpoint the basic starting point of flow equations is the Boltzmann equation which describes the evolution of states in phase space, i. e., ensembles of molecules having a given set of velocities, locations, and internal energy states. The Boltzmann equation for a single species can be expressed as

$$\partial f(\mathbf{x},\mathbf{v})/\partial t + \mathbf{v} \cdot \nabla_{\mathbf{X}} \, f(\mathbf{x},\mathbf{v}) + \mathbf{F} \cdot \nabla_{\mathbf{V}} \, f(\mathbf{x},\mathbf{v})$$

$$= \iiint [f'_1 f'_2 - f_1 f_2] g_{12} d\sigma dv_z$$

where $f(\mathbf{x},\mathbf{v})$ is the velocity distribution function, $f_{1,2}$ and $f'_{1,2}$ are the distribution functions of molecules 1 and 2 before and after collision respectively, g_{12} is the relative velocity and $d\sigma$ is the differential collision cross section. The solution to this equation is not simple nor does it yield the flow quantities such as the velocity, density, temperature, etc. directly. One must evaluate the velocity moments of the distribution function to obtain these latter quantities. The usual flow equations, the *Navier-Stokes equations* can be derived from the Boltzmann equation by calculating moments of the Boltzmann equation. Moments of the right hand side then become the terms containing the transport of mass, momentum and energy, e. g. diffusion, stress, and heat conduction.

<u>*The Navier-Stokes Equations*</u>. The Navier-Stokes equations expressed in conservation form can be written as follows

Continuity

$$\partial\rho/\partial t + \nabla\cdot\rho\mathbf{u} = 0 \qquad\qquad (3\text{-}1)$$

Momentum

$$\partial\rho\mathbf{u}/\partial t + \nabla\cdot(\rho\mathbf{u}\mathbf{u} + \Pi) = 0 \tag{3-2}$$

Energy

$$\partial\rho(e + \rho\,\mathbf{u}\mathbf{u}) + \nabla\cdot[\rho\mathbf{u}(e + \rho\,\mathbf{u}\mathbf{u}) + \Pi\cdot\mathbf{u} + \mathbf{q}] = 0 \tag{3-3}$$

Species Diffusion

$$\partial\rho C_1 + \nabla\cdot(\rho\mathbf{u}C_1 + \mathbf{M}_1) = \rho\omega_1 \tag{3-4}$$

where ρ is the density, $\mathbf{u}$ is the velocity vector, Π is the stress tensor, e is the internal energy, $\mathbf{q}$ is the heat flux vector, $\mathbf{M}_1$ is the species diffusion vector, C_1 is the species mass fraction, ω_1 is the species production rate. The flux quantities can be written as follows:

Stress Tensor

$$\Pi = p\delta_{ij} + \mu\,(\partial u_i/x_j + \partial u_j/x_i) - \frac{2}{3}\mu\,\nabla\cdot\mathbf{u}\,\delta_{ij} \tag{3-5}$$

Heat Flux Vector

$$\mathbf{q} = -\lambda\nabla T + \sum_1 \mathbf{M}_1 h^c_1 \tag{3-6}$$

Mass Diffusion Vector

$$\mathbf{M}_1 = \rho W_1/W \sum_m D_{1m}[\nabla C_m + C_m\nabla\ln W] \tag{3-7}$$

where p is the static pressure, μ is the viscosity, δ_{ij} is the Kroneker delta, λ is the thermal conductivity, h^c_1 is the heat of formation of species 1, W_1 and W are the molecular weights of species 1 and the mixture, respectively, and D_{1m} is the diffusion coefficient for the 1,m pair of species.

The Euler equations, the viscous shock layer equations, and the boundary layer equations are subsets of the Navier-Stokes equations in which various approximations are made to simplify the equations for an appropriate application.

The Navier-Stokes (N-S) equations are valid over a wide range of flows of hypersonic interest. However, the N-S approximation does tend to break down at very high altitudes where the density is very low. In such cases it is necessary to either solve the Boltzmann equation, which is an exceedingly formidable task, or use the direct simulation Monte Carlo (DSMC) method.[44] The DSMC method is outside the scope of these notes. For discussions of the DSMC method one

should see, for example, works by G. A. Bird, J. N. Moss or H. A. Hassan and
their coworkers. At intermediate altitudes or densities it is sometimes possible
to apply slip boundary conditions to extend the range of validity of the N-S equations. Slip boundary conditions for the N-S equations may be found in Ref. 10.

Euler Equations. The Euler equations are a subset of the Navier-Stokes
equations in which all the viscous terms have been dropped. These equations
apply to the inviscid layer around bodies in hypersonic flight and to the inviscid
core of internal flows. Solving the Euler equations is much simpler than the
full N-S equations. Reductions in computational time and increased accuracy
where viscous effects are negligible may be obtained. To determine the heat flux
and effects of finite rate catalytic recombination one must couple the solution of
the Euler equations with a viscous calculation — the boundary layer equations.
The combined method is the *two layer method.* One uses the solution of the
Euler equations for outer boundary conditions for the boundary layer equations.
Usually one can solve the three dimensional Euler equations for complex bodies
and apply the boundary conditions to the axisymmetric boundary layer equations
and thereby save much computational effort over solutions of the full Navier-
Stokes equations.

Boundary Layer Equations. The boundary layer equations are a subset of the
Navier-Stokes equations in which the normal pressure gradient terms have been
neglected and in which a similarity transformation may be performed to the
partial differential equations and transform them to ordinary differential equations.
For chemically reacting flows the equations cannot truly be decoupled except in
an approximate sense where iterative solutions are performed. The equations are
written in body oriented coordinate system as follows:

Continuity

$$\frac{\partial \rho u r^k}{\partial s} + \frac{\partial \rho v r^k}{\partial y} = 0 \tag{3-8}$$

Streamwise Momentum

$$\rho u \frac{\partial u}{\partial s} + \rho v \frac{\partial u}{\partial y} = \frac{1}{r^k} \frac{\partial}{\partial y}[\rho r^k (\nu + \varepsilon_M)\frac{\partial u}{\partial y}] - \frac{\partial P}{\partial s} \tag{3-9}$$

Energy

$$\rho u \frac{\partial h}{\partial s} + \rho v \frac{\partial h}{\partial y} = u \frac{\partial P}{\partial \xi} + \frac{1}{r^k} \frac{\partial}{\partial y}[r^k (-q)] + m(\frac{\partial u}{\partial y})^2 \tag{3-10}$$

Species

$$\rho u \frac{\partial C_i}{\partial s} + \rho v \frac{\partial C_i}{\partial y} = \frac{1}{r^k} \frac{\partial}{\partial y}[r^k (\rho \varepsilon_D \frac{\partial C_i}{\partial y} - M_{yi})] + w_i \tag{3-11}$$

When transformed using the transformation

$$\xi = \int_0^s \rho_e u_e \mu_e r^{2k_0} ds \qquad (3\text{-}12a)$$

$$\eta = \frac{r_0^k u_e}{\sqrt{2\xi}} \int_0^y \rho \, dy \qquad (3\text{-}12b)$$

and writing the dependent variable for the stream function $f' = \partial f / \partial \eta = u/u_e$ the boundary layer equations take the form

Streamwise Momentum Equation

$$ff'' + (Cf'')' = \frac{2\xi}{u_e} \frac{du_e}{d\xi}[(f')^2 - \frac{\rho_e}{\rho}] \qquad (3\text{-}13)$$

Energy Equation

$$(Cg'/Pr)' + fg' = \frac{2\xi f'g}{H_e} \frac{dH_e}{d\xi} + [\frac{u_e^2}{H_e}(1 - \frac{1}{Pr})Cff'']'$$

$$+ [\frac{C}{Sc}(\frac{1}{Le} - 1) \frac{(h_A - h_M)\alpha_e}{H_e} z_A']' \qquad (3\text{-}14)$$

Species Equation

$$(\frac{C}{Sc} z_A')' + f z_A' = \frac{2\xi f z_A}{\alpha_e} \frac{d\alpha_e}{d\xi} - \frac{2\xi w_A}{\rho \rho_e u_e^2 \mu_e r_0^{2k}\alpha_e} \qquad (3\text{-}15)$$

where $z_A(\eta) = \alpha/\alpha_e$, $g(\eta) = H/H_e$, and $C = \rho\mu/\rho_e\mu_e$. These equations have been written for a binary gas. For multicomponent gases there are additional species equations for as many species as are added. Likewise, the last term in the energy equation becomes a sum over all species less one.

These equations must be solved numerically for the general problem. Fay and Riddell[43] solved these equations for the stagnation point of a sphere and correlated the results for a fully catalytic surface by the formula

$$-q = 0.763 \, Pr^{-0.6} \sqrt{\rho_e\mu_e \frac{du_e}{dx}} \left(\frac{\rho_w\mu_w}{\rho_e\mu_e}\right)^{0.1} (H_e - h_w)[1 + (Le^{0.63} - 1)\frac{h_c}{H_e}] \qquad (3\text{-}16)$$

Goulard[36] solved in closed form the chemically frozen boundary layer equations for the stagnation point of the sphere in a binary gas flow. His result is the ratio of the finite catalytic to fully catalytic heating. The form of the equations that he solved are as follows:

Streamwise momentum

$$ff'' + (Cf'')' = \frac{2\xi}{u_e} \frac{du_e}{d\xi} [(f')^2 - \frac{\rho_e}{\rho}]$$ (3-17)

Energy Equation

$$(Cg'/Pr)' + fg' = [\frac{u_e^2}{H_e}(1 - \frac{1}{Pr})Cff'']'$$
$$+ [\frac{C}{Sc}(\frac{1}{Le} - 1)\frac{(h_A - h_M)\alpha_e}{H_e} z_A']'$$ (3-18)

Species Equation

$$(\frac{C}{Sc} z_A')' + f z_A' = 2\xi f \frac{dz}{d\xi}$$ (3-19)

Goulard applied the boundary conditions $f(0) = f'(0) = 0$, $f'(\infty) = 1$, $g(0) << 0$, $g(\infty) = 1$, and

$$\left(\frac{\partial z}{\partial \eta}\right)_w = \frac{\sqrt{2\xi}}{r_0 u_e} \frac{k_w}{\rho_w^{2-m} D_w} C_e^{m-1} z_w^m$$

 Goulard's result is

$$\frac{q}{q_{FC}} = \frac{1 + (Le^{2/3} \phi - 1) \frac{h_c C_e}{h_{se}}}{1 + (Le^{2/3} - 1) \frac{h_c C_e}{h_{se}}}$$ (3-20)

where Le is the Lewis number, the subscript e refers to boundary layer edge, and s refers to the stagnation point. The parameter ϕ is given by

$$\phi = \frac{k_w}{k_w + S}$$ (3-21)

and

$$S = \frac{0.47 \, Sc^{-2/3}}{\rho_w} \left\{ \frac{\mu_{es}\rho_{es} \, 2u_\infty}{R_N} \left[\frac{\rho_\infty}{\rho_{es}} \left(2 - \frac{\rho_\infty}{\rho_{es}} \right) \right]^{1/2} \right\}^{1/2}$$ (3-22)

where Sc is the Schmidt number, μ_{es} is the viscosity, u_∞ and ρ_∞ are the free stream velocity and density respectively, and R_N is the nose radius.

The analytical techniques such as that of Goulard have been extended by reducing the number of assumptions and also extending the solutions to beyond the stagnation point by quadrature integrations of the boundary layer equations, especially the species equation. Nonequilibrium and arbitrary catalytic boundary conditions have been accounted for in the recent work of Inger.[45] That work accounts for viscous dissipation heating of the boundary layer such as might exist on slender vehicles and provides a means for calculating the catalytic heating effects from the stagnation point to the aft of the vehicle.

Viscous Shock Layer Equations. In flow regimes in which there is significant interaction between the inviscid outer flow and the viscous flow it is convenient to approximate the Navier-Stokes equations with the _viscous shock layer_ equations. These are similar to the boundary layer equations, but no transformation to similarity variables is made. The equations are usually normalized with respect to the conditions immediately behind the shock wave. These conditions behind the shock are found using modified Rankine-Hugoniot relations. The viscous shock layer equations for the stagnation point were first solved by Blottner[46], whereas, Davis[47] subsequently formulated the VSL equations for axisymmetric and two-dimensional bodies. The method was extended to multicomponent mixtures by Miner and Lewis[48] and by Moss[49]. The two-dimensional governing equations can be written in the following form:

Continuity

$$\frac{\partial}{\partial s}[(r + y\cos\phi)^j \rho u] + \frac{\partial}{\partial y}[(1 + \kappa y)(r + y\cos\phi)^j \rho v] = 0 \qquad (3\text{-}23)$$

s-Momentum

$$\frac{1}{1+\kappa y}\rho u \frac{\partial u}{\partial s} + \rho v \frac{\partial u}{\partial y} + \rho u v \frac{\kappa}{1+\kappa y} + \frac{1}{1+\kappa y}\frac{\partial P}{\partial s} = \varepsilon^2 \frac{\partial}{\partial y}\left[\mu\left(\frac{\partial u}{\partial y} - \frac{\kappa u}{1+\kappa y}\right)\right]$$

$$+ \varepsilon^2 \mu \left(\frac{2\kappa}{1+\kappa y} + \frac{j\cos\phi}{r + y\cos\phi}\right)\left(\frac{\partial u}{\partial y} - \frac{\kappa u}{1+\kappa y}\right) \qquad (3\text{-}24)$$

y-Momentum

$$\frac{\partial P}{\partial y} = \frac{\kappa u}{1+\kappa y}\rho u^2 - \frac{1}{1+\kappa y}\rho u \frac{\partial v}{\partial s} - \rho v \frac{\partial v}{\partial y} \qquad (3\text{-}25)$$

Energy

$$\frac{1}{1+\kappa y}\rho u\, C_p\frac{\partial T}{\partial s} + \rho v C_p\frac{\partial T}{\partial y} - \frac{1}{1+\kappa y}u\frac{\partial P}{\partial s} - v\frac{\partial P}{\partial y} =$$

$$\varepsilon^2\frac{\partial}{\partial y}\left(\kappa\frac{\partial T}{\partial y}\right) + \varepsilon^2\left(\frac{\kappa}{1+\kappa y} + \frac{j\cos\phi}{r+y\cos\phi}\right)\lambda\frac{\partial T}{\partial y} - \varepsilon^2\sum_{i=1}^{ns}J_i C_{pi}\frac{\partial T}{\partial y}$$

$$+\,\varepsilon^2\mu\left(\frac{\partial u}{\partial y} - \frac{\kappa u}{1+\kappa y}\right) - \sum_{i=1}^{ns}h_i\,w_i \tag{3-26}$$

Species

$$\frac{1}{1+\kappa y}\rho u\frac{\partial C_i}{\partial s} + \rho v\frac{\partial C_i}{\partial y} = \varepsilon^2\frac{\partial}{\partial y}(J_i) - \varepsilon^2\left(\frac{\kappa}{1+\kappa y} + \frac{j\cos\phi}{r+y\cos\phi}\right)J_i \tag{3-27}$$

where J_i is the diffusion mass flux of species i. Note that the VSL equations are uniformly accurate to order of ε^2 throughout the flow field — unlike the boundary layer equations. These equations are solved iteratively in the coordinate normal to the body and then marched downstream. Since the dependent variables are normalized with respect to the shock values it is necessary to know the shock shape or shock angle at each point downstream of the stagnation point. This is not known *a priori* and a first guess is required or some other knowledge of the shock shape. After solving the equations along the body it is usually necessary to iterate globally to account for the updated shock shape obtained from the previous iteration.

A three-dimensional nonequilibrium viscous shock layer code was developed by Kim, Swaminathan, and Lewis.[50,51] They applied the method to the space shuttle orbiter geometry for a noncatalytic wall. Subsequently, Thompson[52] made corrections to that code and showed comparisons with Shuttle measurements. Some of his results will be shown in the section on applications. The VSL equations cannot handle cross flow separation because of the solution method and the degree of approximation used. A higher order approximation that can handle cross flow separation (but not streamwise separation) is the Parabolized Navier-Stokes approximation which will be discussed next.

Parabolized Navier-Stokes (PNS). The Parabolized Navier-Stokes equations are a subset of the Navier-Stokes in which certain terms associated with normal gradients are neglected and other approximations are made which render the system of equations parabolic in the streamwise variable and elliptic in the other variables. The solution method is essentially a marching technique in the streamwise dimension as is the VSL method. However, the PNS method can handle cross flow separation, although not separation in the marching direction. Although the parabolized Navier-Stokes method was first developed many years ago, its application to problems of interest to catalytic recombination has lagged. Part of the reason is that the code is particularly applicable to conditions where nonequilibrium effects are not very important; and even in those cases where nonequilibrium is significant, catalytic effects on heating may not be very important. Current PNS codes that have been developed for nonequilibrium flow are those of Bhutta, Lewis, and Kautz,[53] Prabhu, Tannehill, and Marvin,[54,55] and Tannehill, et al.[56] The published applications of these codes do not display the effects of finite rate catalysis, therefore, these techniques will not be discussed further.

APPLICATIONS: EFFECTS OF CATALYSIS ON HYPERSONIC VEHICLES

In this section we will discuss the ideas developed in the previous sections concerning nonequilibrium flow and catalysis, but will relate these concepts to their influence, along with the effects of finite rate and noncatalytic atom recombination, on the heat flux to hypersonic vehicles. The examples to be discussed are the Space Shuttle Orbiter, Aeroassisted Orbital Transfer Vehicles, the Aeroassist Flight Experiment, and slender hypersonic transatmospheric vehicles (e. g. NASP). However, before discussing particular examples we should turn our attention to some general trends associated with nonequilibrium and catalytic phenomena in hypersonic flight.

TRENDS IN HYPERSONIC NONEQUILIBRIUM AND CATALYSIS.

Effects of Altitude and Recombination Rate. To get an appreciation of the influence of the surface finite rate catalytic recombination in atmospheric flight Inger[57] has calculated the stagnation point q/q_{FC} for a one-foot sphere at a velocity of 26 000 ft/s for various recombination rates. His results are shown in Figs. 13 and 14. It can be seen that at very high and very low altitudes the heat flux is almost independent of the recombination rate. At altitudes in between these extremes the value of q/q_{FC} decreases appreciably. At the high altitudes not much dissociation has occurred in the shock layer because the shock layer is tenuous and the collision frequency is low. Therefore, not much energy is absorbed by dissociation. Most of the energy is in translational modes. At the very low altitudes, the shock layer is so dense that the boundary layer is near equilibrium and the atoms recombine in the boundary layer before they have a chance to strike the wall. Hence, the boundary layer is heated and there is more potential for heat transfer by conduction. In Fig. 14 the q/q_{FC} is plotted against the Reynolds number. Again we see the influence of the density on the relative heating.

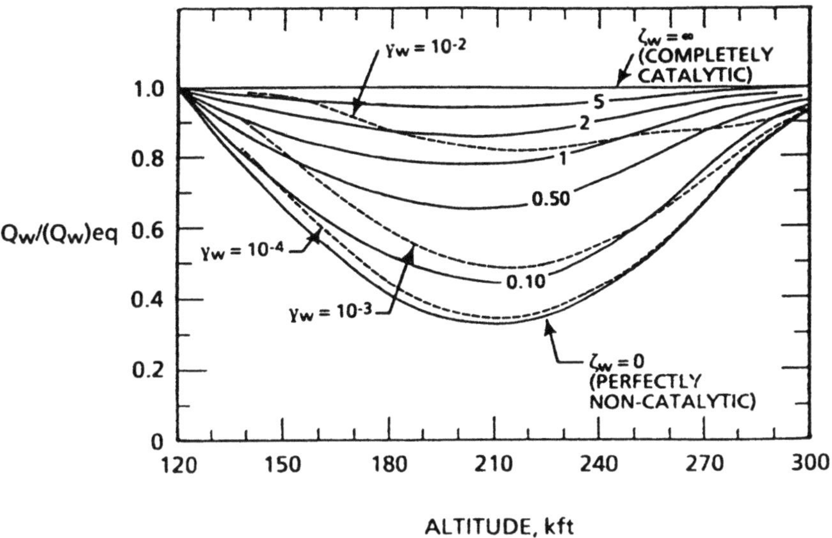

Figure 13.- Effect of surface catalycity on nonequilibrium heat transfer in nonequilibrium flowfield for a 1-ft radius sphere at a velocity of 26,000 ft/s. From Ref. 57.

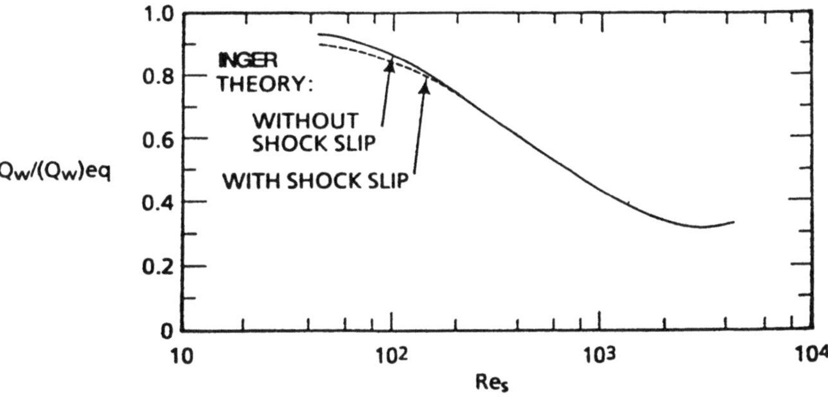

Figure 14.- Nonequilibrium heat transfer dependence on Reynolds number at high altitudes. From Ref. 57.

 The relative heating q/q_{FC} depends on the vehicle size, since the size affects the time the air requires to flow around the vehicle. The longer the flow time the closer the flow can come to dissociation equilibrium. This effect is seen in Fig. 15 where Scott, et, al.[58] calculations of the relative heat flux is plotted against the body radius for various altitudes and velocities. For almost all conditions the larger the body the lower the relative heating. This demonstrates the influence of the dissociation nonequilibrium since more flow time allows dissociation to go toward completion. At the lowest altitude the solution tends to not converge. As the flow approaches equilibrium the BLIMPK code has difficulty handling a reacting solution near equilibrium, presumably because the species production terms are differences between large terms and the relative accuracy decreases.

SPACE SHUTTLE ORBITER. This section relies heavily on a review article[11] and other work in a number of references to be cited.

The Shuttle Orbiter is a hypersonic glide re-entry vehicle that spends much of its entry time at relatively tenuous altitudes in which chemical nonequilibrium predominates in the shock layer. Calculations have shown that both dissociation nonequilibrium[59,60] and recombination nonequilibrium[11] exist. On the windward side, the dissociation nonequilibrium exists in the inviscid layer as was seen in Fig. 16 which compared the results of the reacting inviscid flow field calculation and an equilibrium calculation. Recombination nonequilibrium

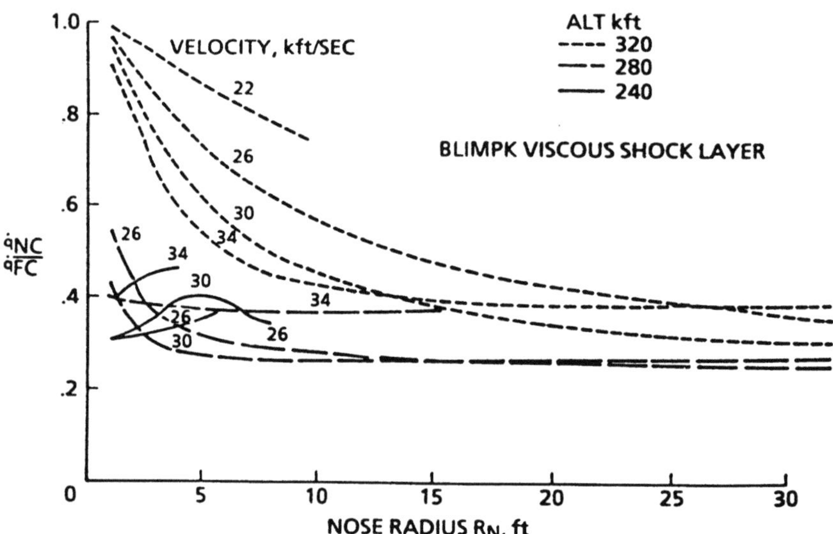

Figure 15.- Noncatalytic to fully catalytic heat flux ratio at stagnation point of sphere.

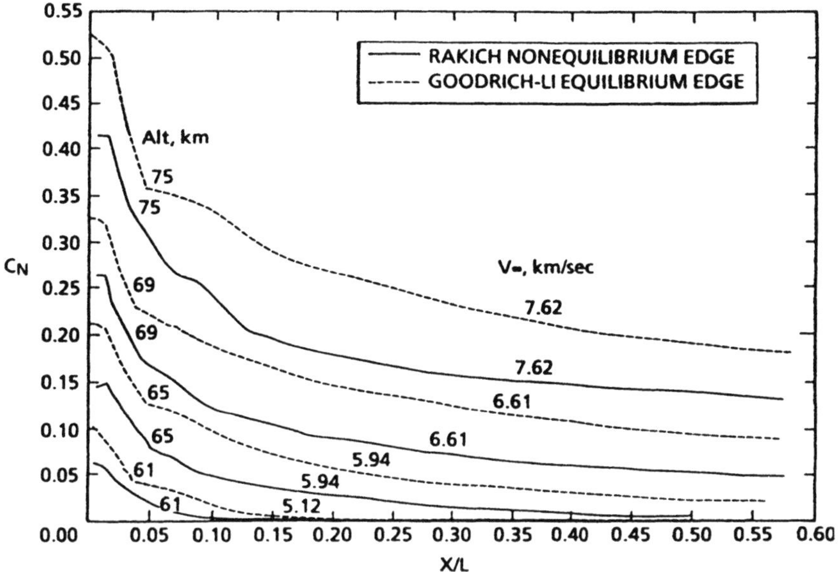

Figure 16.- Nitrogen atom mass fraction on centerline of Orbiter at boundary layer edge.

exists in the boundary layer as seen in Fig. 17 where at an altitude of 65 km the composition is almost frozen. A noncatalytic wall is assumed in the calculation so as not to mask equilibrium composition profile if it existed. It is also expected that nonequilibrium exists in the lee side flow, as is shown by Li[61] in another lecture in this short course. Li showed that significant nitrogen dissociation persists in the flow field far down stream of the nose above the canopy of the Orbiter. If the flow were in equilibrium there, the temperature would be low enough to quench almost all the nitrogen atoms. Some experimental evidence that supports this nonequilibrium will be discussed later. Verification of nonequilibrium phenomena has not been obtained directly from species concentration measurements. Instead it is inferred by comparing heat transfer measurements with the reacting flow field results.

Determining the presence of chemical nonequilibrium was made easier because the high-temperature reusable surface insulation (HRSI) tile coating, reaction cured glass (RCG), is relatively noncatalytic. Had the tile coating been highly catalytic, the heat flux to the tiles would have been near the equilibrium value and it would be difficult to discriminate between flow field and surface recombination effects. Also of great importance in demonstrating the nonequilibrium flow behavior is the catalytic surface effects experiment performed on several shuttle flights by Stewart, Rakich, and Lanfranco.[36] In their experiment, selected tiles were coated with a highly catalytic material, iron-cobalt-chromia

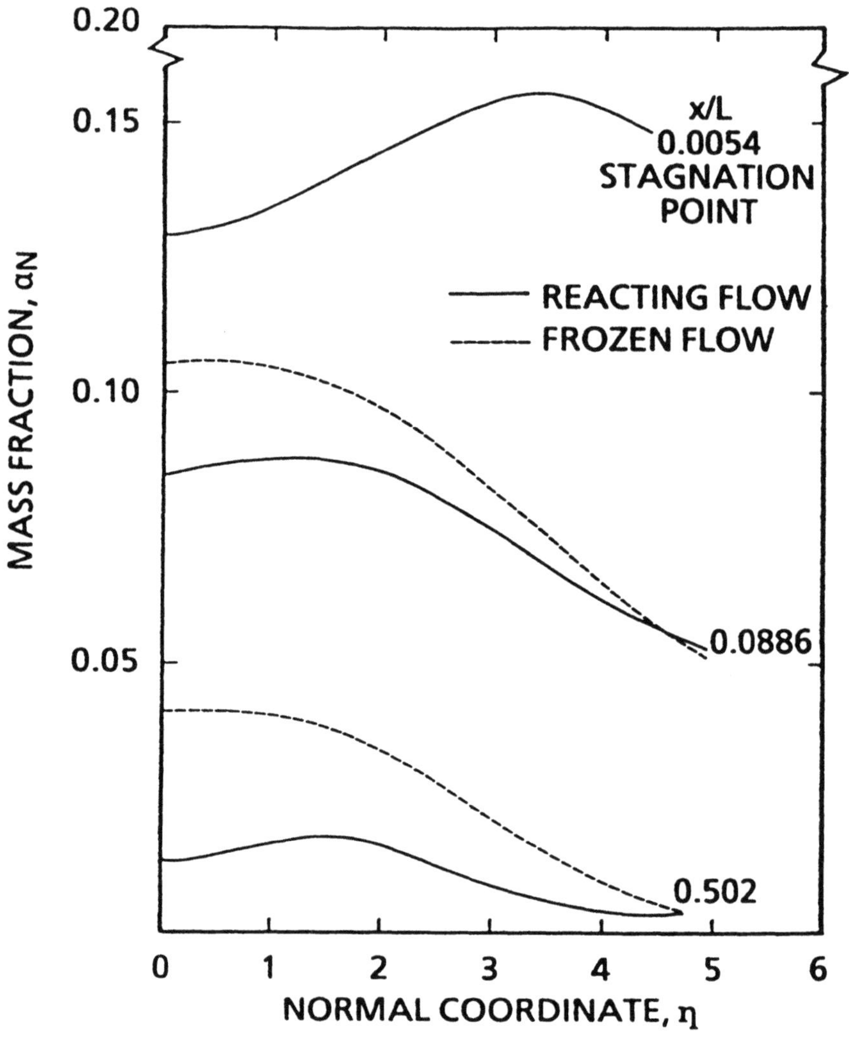

Figure 17.- Nitrogen atom mass fraction profiles normal to surface at various stations on the Orbiter centerline at an altitude of 65 km.

spinel. By a comparison of the heat flux to the over coated tiles and to the standard tiles one could demonstrate the relative catalycity of the two materials. The initial results of the experiment were reported by Rakich, et al.[62] Prior to the flight experiments, predictions of the noncatalytic nature were reported in Refs. 38, 59, and 60 based on arc jet measurements of the (energy transfer) catalytic recombination coefficients and flow field and boundary layer computations.

Measurements of Shuttle Heat Flux. Surface temperature measurements of several instrumented HRSI tiles, distributed along the lower surface of the Orbiter are considered here. Some of the locations are indicated in Fig. 18. The flights considered are STS-2, STS-3, and STS-5. Heat fluxes were inferred from measured temperature by a modified radiation equilibrium heat flux relation

$$q = 1.06 \, \varepsilon \sigma T_w^4 \qquad (4\text{-}1)$$

where ε is the emittance and σ is the Stefan-Boltzmann constant. The factor 1.06 accounts for conduction effects and was based on a thermal math model analysis of the tile with time dependent heating.[63] Measured flight heat fluxes along the windward centerline of the Orbiter are given in Fig. 19. It is apparent that there is some variation from flight to flight even though the fluxes are normalized to account for minor differences in the trajectories of the three flights. See Ref. 11 for details. It has been speculated that the shift toward higher heat fluxes is due to changes in the surface characteristics, especially the catalytic recombination coefficient, possibly due to a build-up of contaminants on the surface.

Calculations of the Shuttle Heat Flux. Until Li's[61] recent Navier-Stokes solutions nonequilibrium calculations of the flow field and heat fluxes on the windward centerline of the Orbiter during re-entry fell into three categories. These are the two-layer method, the axisymmetric viscous shock layer method, and a three-dimensional viscous shock layer method.

The earliest calculations[64,65] used the two-layer approach of Rakich and Lanfranco.[66] By this technique nonequilibrium solutions of the Euler equations were used as boundary conditions for a nonequilibrium nonsimilar boundary layer code,[67] which calculated the boundary layer profile and heat fluxes. Finite rate catalytic reaction rates were used in the wall boundary conditions.

Two sets of axisymmetric viscous shock layer results are presented that are based respectively on the reacting code of Moss[49] and the reacting code of Miner and Lewis,[48] both are modified to account for finite rate catalytic recombination coefficients. Shinn, Moss, and Simmonds[68] applied the Moss code[49] and Scott[10] applied the Miner and Lewis[48] code.

The three-dimensional viscous shock layer method was developed by Kim, Swaminathan, and Lewis,[50,48] who applied the method to the shuttle geometry and assumed a noncatalytic wall. Thompson[52] extended and corrected some errors in this code which will be discussed later.

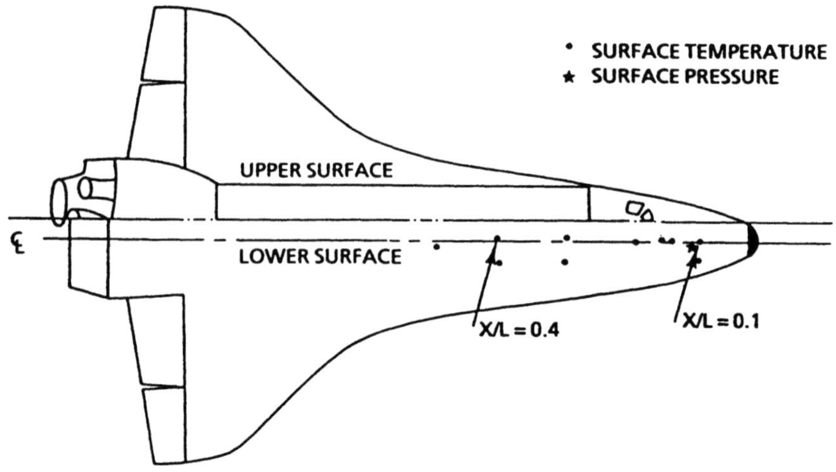

Figure 18.- Some windward centerline instrumented tile locations on the Space Shuttle Orbiter.

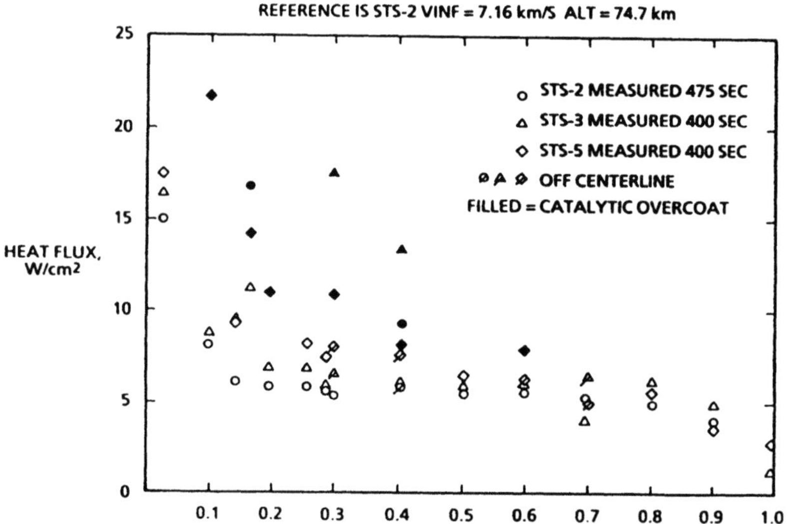

Figure 19.- Comparison of measured radiation equilibrium heat fluxes near windward centerline of Orbiter. $V_\infty = 7.16$ km/s and altitude of 74.7 km STS-2, STS-3, and STS-5.

<u>*Comparison of Methods.*</u> In the following, the three sets of methods are compared and evaluated.

In a previous chapter it was noted that it is often assumed that the heat flux predicted by equilibrium calculations and by nonequilibrium calculations with a fully catalytic wall are approximately equal. To verify this for the methods considered here, comparisons were made between equilibrium two-layer results of Goodrich, et al.[69] and the nonequilibrium two-layer method of Rakich and Lanfranco.[60] Also compared are the equilibrium and nonequilibrium viscous shock layer (VSL) results of Shinn, et al.[68] These results are seen in Fig. 20.

Shinn's equilibrium results are about 10 percent higher than her nonequilibrium fully catalytic results, whereas, Goodrich's equilibrium results are about 20 percent higher than the nonequilibrium fully catalytic (FC) wall results (BLIMPK) for the forward part of the vehicle. The agreement is somewhat better farther aft. One concludes from this that the methods overall agree with one another within about 20 percent. The differences in subtleties of the methods and the assumed geometries may account for the discrepancies.   Also shown in Fig. 20 is a comparison between the noncatalytic (VSL) results of Shinn, the two-layer (BLIMPK) results and the three-dimensional VSL results of Kim, et al.[50] We

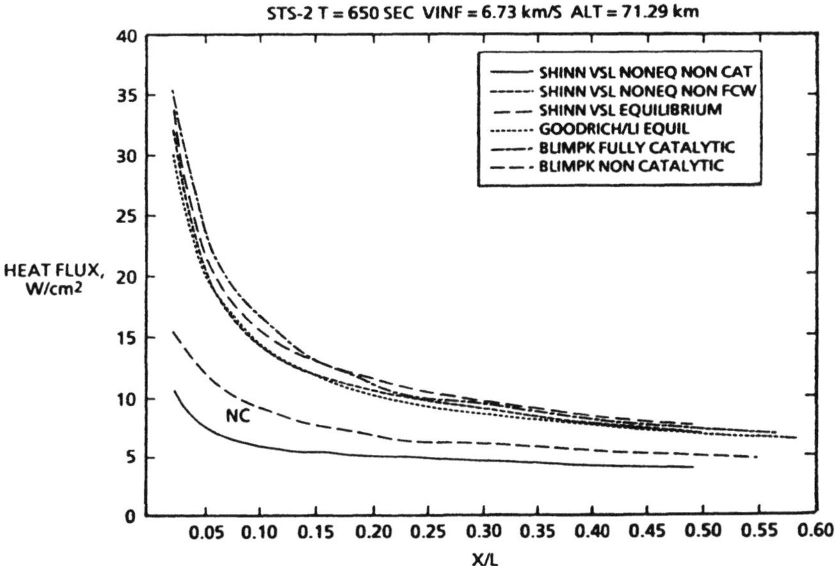

Figure 20.- Comparison of various calculation methods for nonequilibrium fully catalytic and noncatalytic boundary and for equilibrium flows for the Shuttle Orbiter.

see a large discrepancy in the results which is due in part to the fact that the VSL method resulted in a larger amount of dissociation than Rakich's inviscid method. The discrepancy could also be due to the differences in the geometry and the method of obtaining edge conditions. The good agreement between the axisymmetric VSL results and the three-dimensional results in the forward part of the vehicle may indicate that the geometries do not differ very much until farther downstream, where the vehicle deviates considerably from an axisymmetric shape. The bottom of the vehicle is almost flat aft of x/L = 0.1. As will be seen in the next section improvements in the three-dimensional VSL code may explain this discrepancy.

Comparison of Measured and Calculated Heat Flux. Windward side of orbiter. Attention is now turned to comparing the calculated and measured heat fluxes along the lower surface of the Orbiter. The calculations were made with two sets of catalytic recombination coefficient relations, the temperature dependent ones of Ref. 5 measured in an arc-jet and the constant ones of Rakich, et al.[62] A comparison is made in Figs. 21 and 22 between the measurements and several calculations for STS-2 at two times in the trajectories. The measurements are near the centerline of the vehicle except for a few points that are about 1.3 m off the centerline. In the higher altitude cases (Fig. 21), the viscous shock layer methods with the recombination coefficients of Scott[5] yield better agreement for X/L < 0.3. The two-layer method with $k_{w_O} = 80$ cm/s also agrees with the measurements at X/L > 0.5. At the lower altitude (Fig. 22), the two-layer methods yield better agreement for X/L > 0.2.

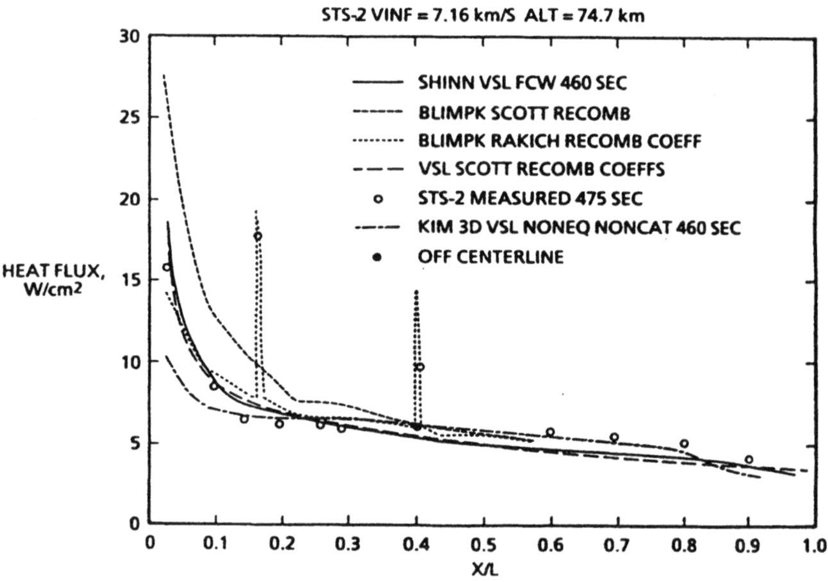

Figure 21.- Comparison of calculated and measured Shuttle heat fluxes for STS-2, t = 460 sec, V_∞ = 7.16 km/sec, and altitude of 74.7 km.

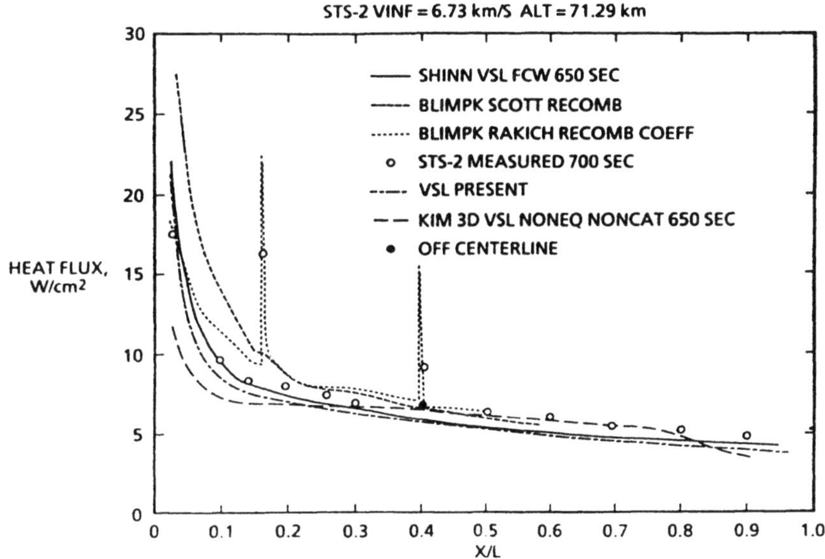

Figure 22.- Comparison of calculated and measured Shuttle heat fluxes for STS-2, t = 650 sec, V_∞ = 6.73 km/sec, and altitude of 71.29 km.

In the higher altitude cases, the nonequilibrium axisymmetric VSL methods with temperature-dependent γ_O and γ_N yield good agreement at X/L < 0.3 and the nonequilibrium two-layer with k_{w_O} = 80 cm/s yields good agreement for X/L < 0.5. For the lower altitude case, this two-layer approach yields better agreement for X/L > 0.2 than the axisymmetric VSL methods. The nonequilibrium two-layer method with temperature-dependent γ_O and γ_N results are about 30 percent higher than the measurements on the nose are for all cases presented here, but the agreement improves toward the middle of the vehicle and at lower altitude. It is apparent that the two-layer approach predicts higher heat fluxes for given wall boundary conditions than do the VSL approaches. This difference may be due in part to the VSL having a slightly higher level of dissociation as well as to differences in the flow field modeling.

In comparing the three-dimensional nonequilibrium calculations of Kim, et al.[51] with other noncatalytic predictions, one sees that the heat flux does not decrease as rapidly down the vehicle as do the axisymmetric VSL and two-layer calculations. This variation indicates a possible influence of geometry and cross flow that is more adequately accounted for in three-dimensional models. In Fig. 21, the 3-D VSL results tend toward better agreement with the measurements than the other calculations aft of X/L = 0.6. This three-dimensional approach was further investigated with appropriate finite rate recombination coefficients by Thompson[52]. He used the three-dimensional code which he adapted and corrected

221

from Kim, et al.[51] Thompson's calculations were for a simulated Orbiter geometry, which matched the bottom and extreme forward part of the upper surface. His computational geometry is shown in Fig. 23. He applied a seven species nonequilibrium gas model and simple binary approximation in the species equations boundary condition. The catalytic recombination rates he employed were from Scott's[5] measurements and oxygen rates from Zoby, Gupta, and Simmonds.[70] His results indicate improved agreement with flight measurements over the calculations of Kim, et al.[51] which Thompson attributes to corrections in the code and the inclusion of certain viscous terms not found in Kim's version. He also concluded that there is not much difference in the wind-ward centerline heat flux distribution over an axisymmetric method such as Shinn, Moss, and Simmonds.[68] See Fig. 24. Thompson[52] compared his three-dimensional method with some centerline STS-2 flight measurements in Fig. 25. The comparison is not uniformly good, but compares very well with most of the measurements. The advantage of the three-dimensional over two-dimensional method is the ability to obtain transverse heating distributions as seen in Fig. 26.

Catalytic surface effects experiment. As mentioned above several tiles of the Orbiter were coated with a highly catalytic coating of iron-cobalt-chromia spinel to compare with the normal base line RCG-coated tiles. The temperature on these over coated tiles exceeded that on the normal tiles by several hundred degrees. This indicated that the base line tiles have a low recombination coefficient. Since these over coated tiles were located as far aft as X/L = 0.6 along

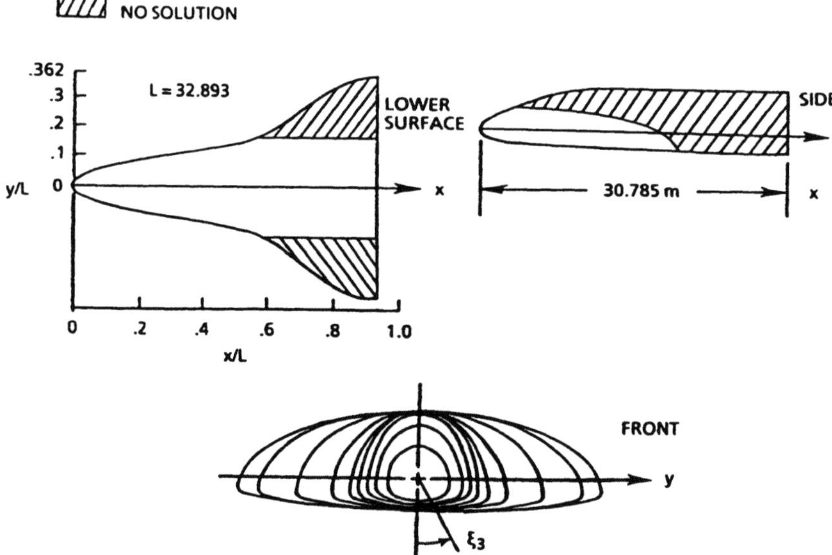

Figure 23.- Computational geometry approximating lower surface of orbiter. From Ref. 52.

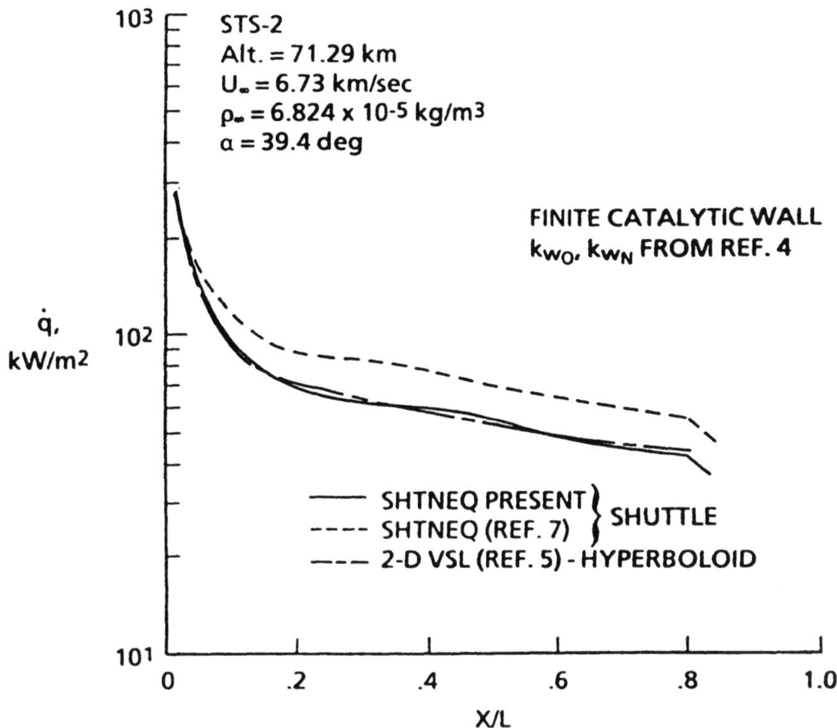

Figure 24.- Comparison of 2-D and 3-D viscous shock layer methods. From Ref. 52.

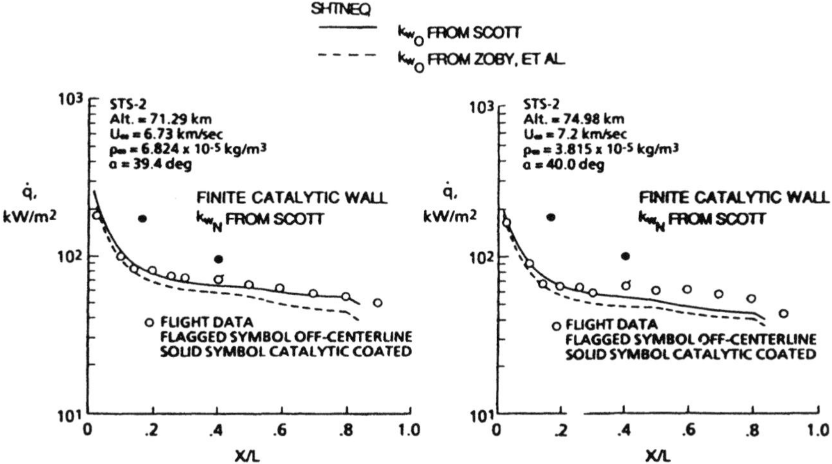

Figure 25.- Comparison of heat flux measurements with 3-D VSL calculations. From Ref. 52.

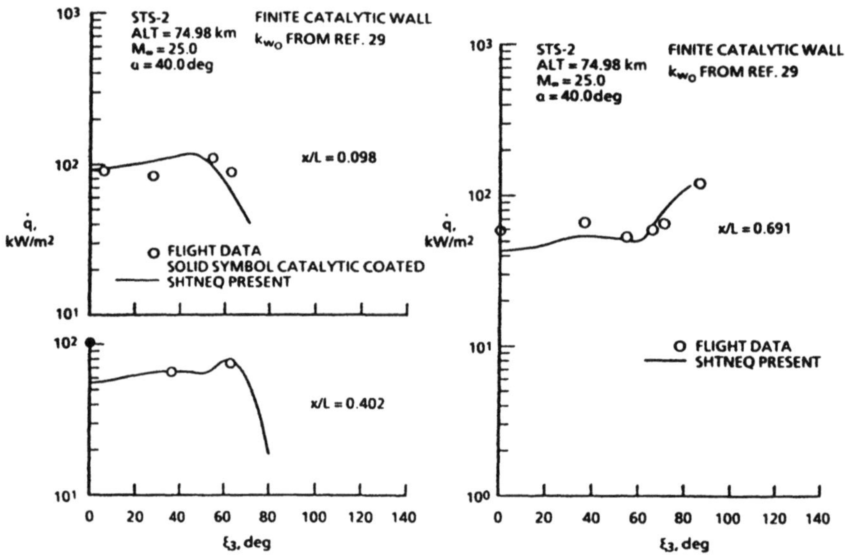

Figure 26.- Nonequilibrium heat-transfer predictions in circumferential direction. From Ref. 52.

the windward centerline of the Orbiter and showed the same high heating overshoot, as seen in Fig. 19, one realizes that the boundary layer flow is far from equilibrium, even that far aft on the vehicle. Similar measurements were made on the lower surface of the wings of the Orbiter, which showed similar results.[11] One might expect similar behavior on a more slender vehicle at low angle of attack. However, it would probably be less pronounced aft since one would expect less of the boundary layer to be dissociated. That is, effects of the normal shock in which the flow is greatly dissociated would not extend as far aft.

Calculations show that the overshoot in heat flux relaxes quickly after a discontinuity in surface recombination rate. However, the distance required to reach the heating value that would exist had the entire surface been highly catalytic is very long. So long that it is impractical to say that it is ever reached. Shuttle experience indicates that the relaxation zone is longer than 1.5 meters. However, the very high part of overshoot does not persist very far, only a few centimeters from the leading edge of the overcoat. This phenomenon can be seen clearly in Fig. 27. The implies that if one has a discontinuity in materials such as a ceramic-to-metal joint, then one might experience temperatures at the leading edge of the metal to overheat it if the overshoot is not properly accounted for in the design. The Orbiter nose cap is such a situation and will be discussed next.

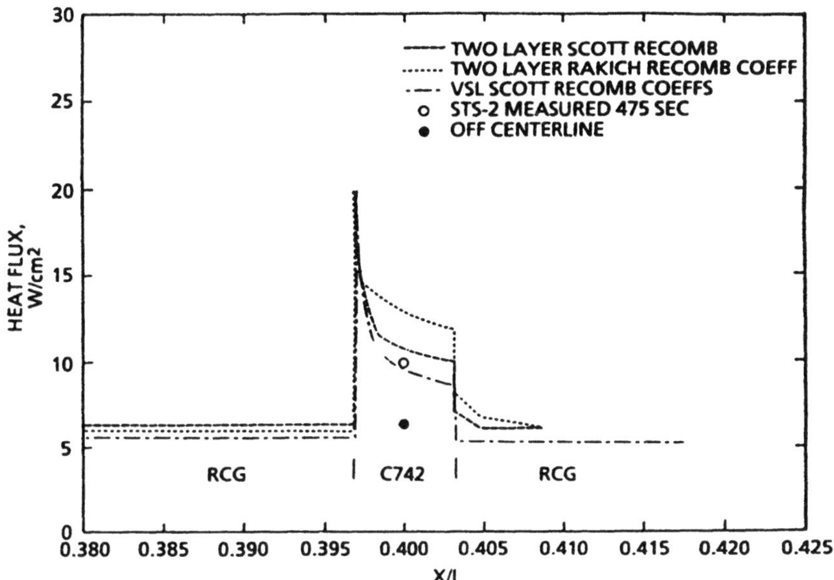

Figure 27.- Heat fluxes showing overshoot on C742-coated tile at X/L = 0.40, STS-2, t = 475 sec. VINF = 7.16 km/sec: ALT = 74.7 km.

Orbiter nose cap. The nose cap and wing leading edges of the Orbiter are made of a reinforced carbon/carbon (RCC) which has been coated with an oxidation inhibiting material. This coating is mostly silicon carbide, but also includes some alumina and silica. Arc jet experiments[38] as well as flight experience have shown that this coating has a low catalycity at the temperatures experienced by the Orbiter during reentry (about 1600 to 1750 K). Stewart, Rakich and Lanfranco[38] estimated the catalytic speed k_W of RCC to be about 110 cm/s at a temperature of 1600 K. A higher value of k_W = 700 cm/s was estimated from comparisons of calculations with flight measurements by Curry, et al.[71] and Ting, et al.[72]

The temperature of the RCC nose cap on the orbiter is measured with radiometers attached to the inner structure of the nose, behind the nose cap. From thermal analysis and a calibrated thermal math model the incident heating to the nose cap of the orbiter is inferred. In this work Ting, et al.[72] compared the viscous shock layer method of Miner and Lewis[48] with the BLIMPK boundary layer method and BLIMPK viscous shock layer option. As would be expected on the basis of previous work by Cheng[73] and others the boundary layer predictions were lower than the viscous shock layer calculations in the low Reynolds regime. See Fig. 28. At still lower $Re_C = \rho_s u_\infty R_N / \mu_s$ the VSL method dropped significantly below the boundary layer value. However, at these very low Reynolds numbers the flow is approaching free molecular and neither

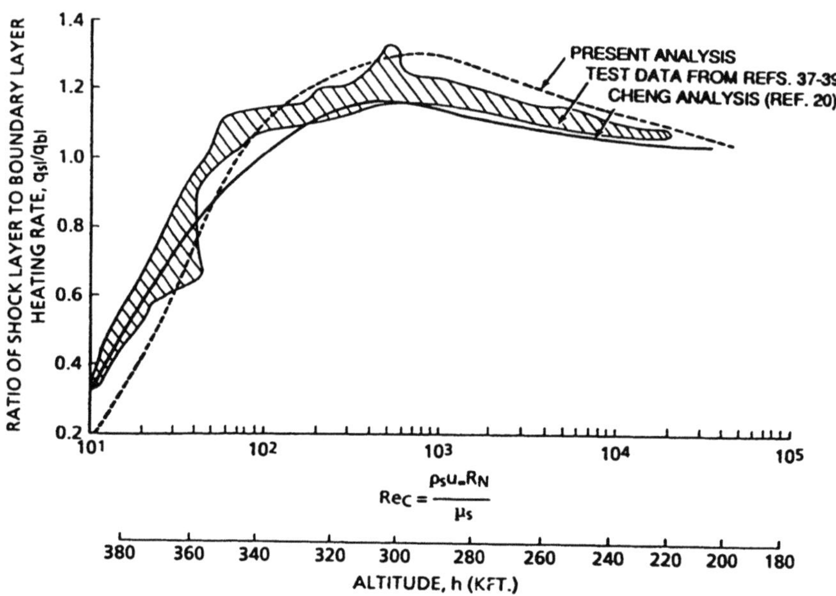

Figure 28.- Variation of shock layer to boundary layer heating rate as a function of Reynolds number. From Ref. 72.

method is applicable. Likewise, due to the different amount of dissociation and flux of chemical energy the two different methods agreed with the flight measurements only if the wall catalytic recombination coefficients differed. The best agreement with measurements for the boundary layer calculations was obtained with k_W = 700 cm/s, whereas, the best overall agreement with the viscous shock layer methods obtained when $k_W \approx$ 400 cm/s. See Fig. 29 where the Stanton number is plotted against $\sqrt{Re_S}$ and where $Re_S = \rho_\infty u_\infty R_N/\mu_S$. There it can be seen that the VSL methods compare very well with the flight stagnation point measurements. All reports of values of k_W for RCC quote an effective value which does not discriminate between nitrogen and oxygen recombination. Measurements of k_{w_N} and k_{w_O}, independently, and as functions of temperature for RCC have not yet been made.

Orbiter Nose Cap with Pressure Ports - In recent flights of the Orbiter, Columbia, the nose cap is fitted with pressure ports that provide pressure measurements during reentry. This system is the Shuttle Entry Air Data System (SEADS)[74]. The orifices of the pressure ports are made of a coated columbium (niobium) and have a fairly high catalytic recombination coefficient. Since the RCC around the ports have a fairly low recombination coefficient there is the possibility of a higher heat flux on the ports than on the surrounding RCC. In

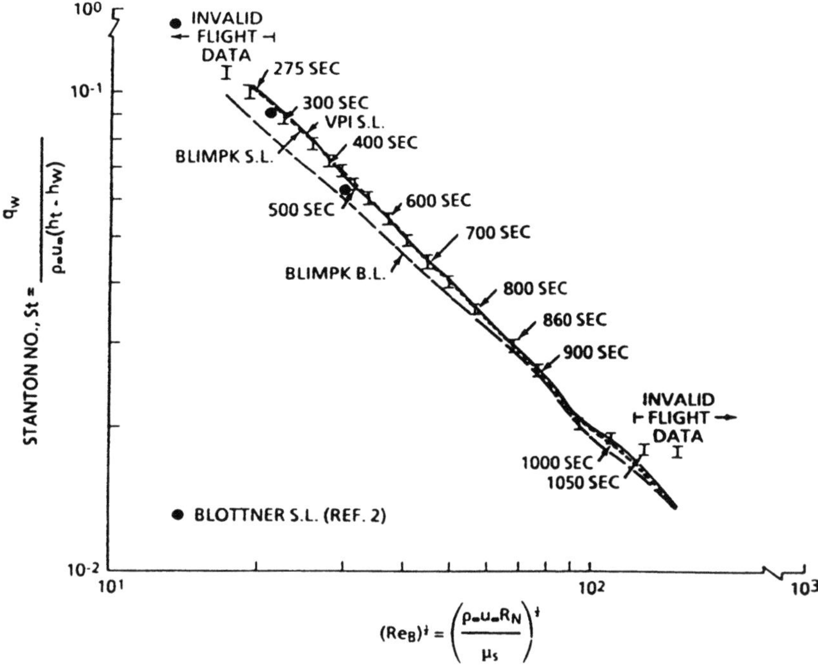

Figure 29.- Stanton number as a function of $(Re_B)^{1/2}$ in STS-5 BET for variable k_w. From Ref. 72.

fact, the heat flux may be higher than if the surface were fully catalytic everywhere as was the case in the previous section for the Catalytic Surface Effects Experiment. This effect was investigated by Ting, Rochelle, and Curry[75] who calculated the heat flux to the nose cap using a finite rate chemistry boundary layer code (BLIMPK). The layout of the orifices on the nose cap of the orbiter is shown in Fig. 30. They used catalytic rates $k_w = 700$ cm/s for the RCC and $k_w = 2500$ cm/s for the columbium. Their results are shown in Fig. 31 where we can see that the heat flux to the ports is significantly higher than to the RCC that surrounds the port. However, Ting, et al.[75] concluded from thermal analysis that the surface temperature of the ports and their surrounding RCC reached about the same temperature because of conduction and radiation heat transfer between the ports, the RCC nose cap, and the underlying structure. The nose cap is hollow, allowing internal radiative heat transfer.

An interesting result of their calculations is the distribution of the conductive heat flux

$$q_{cond} = -\lambda \frac{dT}{dy}$$

and the distribution of diffusive (chemical) heat flux

$$q_{chem} = \sum_{j=1}^{ns} h_i^{c} M_j$$

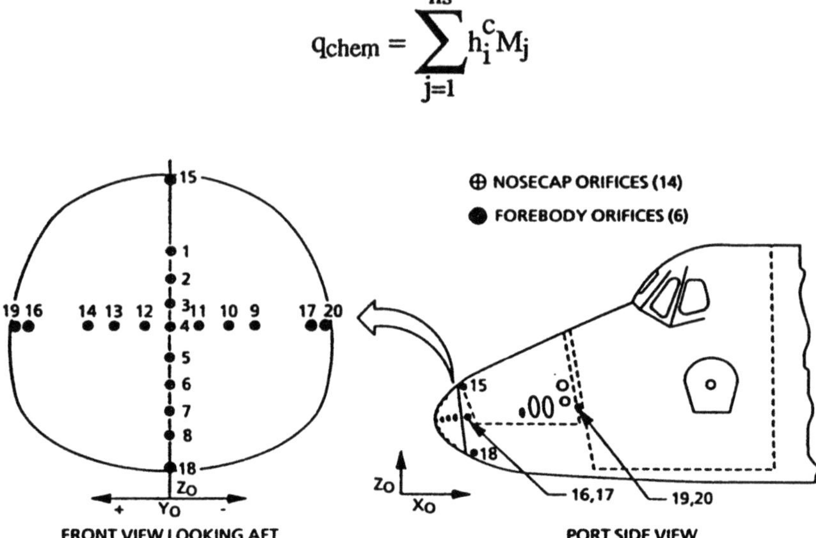

Figure 30.- SEADS flush orifice configuration. From Ref. 75.

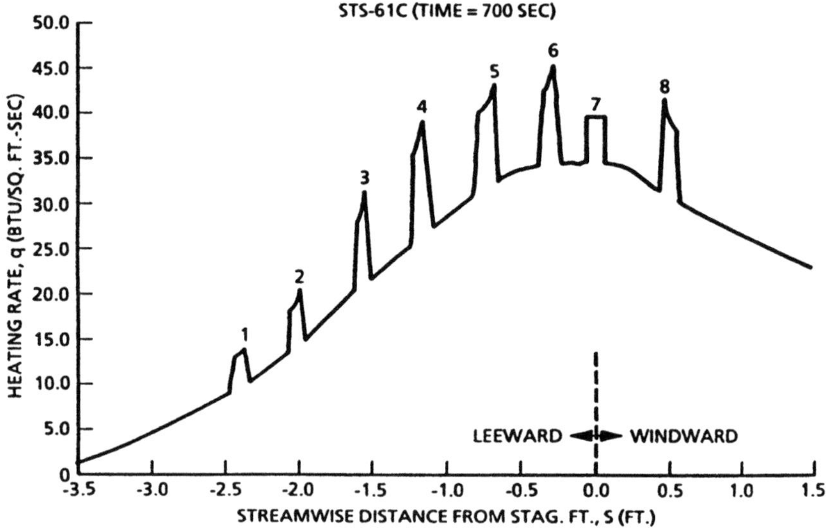

Figure 31.- Heating rate distribution around SEADS nose cap.
STS-61C (time = 700 sec). From Ref. 75.

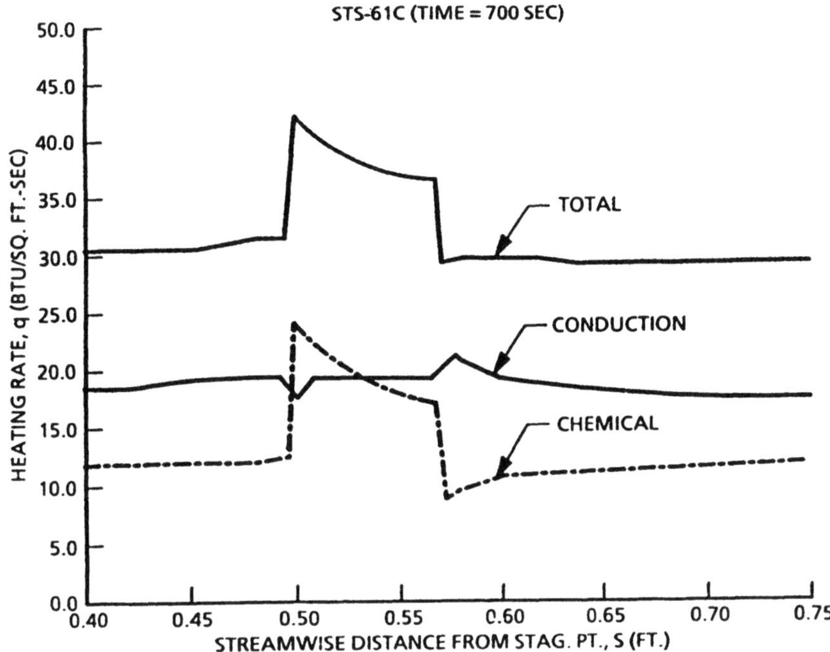

Figure 32.- Total, conduction and chemical heating rate at port 8.
STS-61C (time = 700 sec). From Ref. 75.

in the vicinity of the ports. It can be seen in Fig. 32 that the conductive heat flux is almost constant, whereas, the chemical heat flux increases sharply at the port. There is a slight perturbation in the conductive heat flux at the upstream and downstream edges of the port due mainly due to the discontinuity in temperature at the interface.

Leeward Side Nonequilibrium. This section has been primarily concerned with the nonequilibrium flow on the windward side of the Orbiter because it is for that region that almost all the prediction techniques have been developed to date. As three-dimensional computation capability advances, the leeward side and wake regions during reentry will be of increased interest. As an adjunct to the NASA-Ames catalytic surface effects experiments, one tile on the vertical stabilizer was coated with the highly catalytic spinel coating for flight STS-5. The tile was instrumented with a surface thermocouple, and, likewise, an uncoated tile on the opposite side of the stabilizer was instrumented with a surface thermocouple. The temperature history during the reentry is presented in Fig. 33. It is seen that the temperature of the spinel-coated tile was as much as 120 K hotter than the basic RCG-coated tile. This difference in temperature implies about a factor of

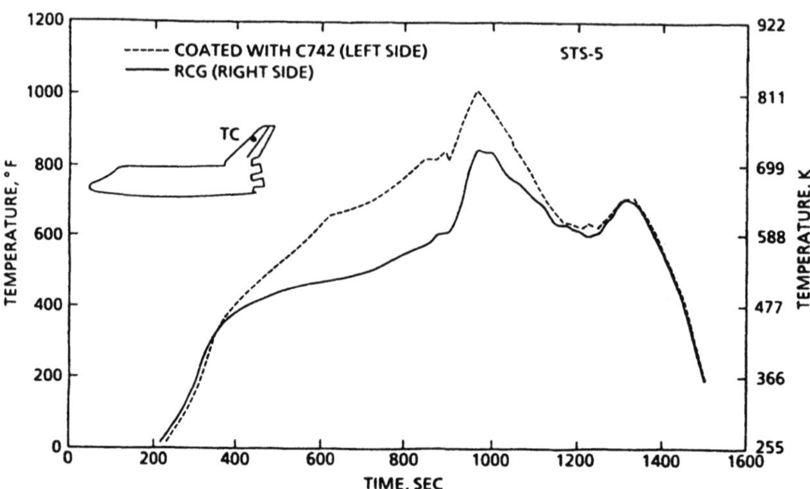

Figure 33.- Comparison of temperatures on vertical stabilizer on symmetrically located tiles. One tile is overcoated with highly catalytic iron-cobalt-chromia spinel (C742).

two difference in heat flux which results from a dissociated nonequilibrium leeward side flow field in the vicinity of the vertical stabilizer. Existence of nonequilibrium is not surprising, since the recombination reaction rate in the gas phase should be very small due to the low-pressure and low-temperature expanded flow around the leeward side of the vehicle. It is presumed that the nonequilibrium dissociation persists far into the wake of the vehicle. This theory is evidenced by the afterglow trail of a re-entry body such as a meteor and also observed during Orbiter re-entry.[76]

AEROASSISTED VEHICLES. The effects of nonequilibrium and related heating to surfaces having finite rate catalycity are particularly important for aerobraking or aeroassisted vehicles such as aeroassisted orbit transfer vehicles (AOTV) and the Aeroassist Flight Experiment (AFE). These vehicles use the atmosphere to reduce the energy of the orbit upon return from geosynchronous Earth orbit (GEO) or from the Moon or Mars. Similar vehicle concepts are visualized for aerocapture at Mars and for entry into the Mars atmosphere during descent to the surface. It is desired that the heating and the acceleration loads during these maneuvers be as benign as possible; therefore, the trajectories are designed such that the vehicles fly at as high an altitude as possible and still maintain sufficient aerodynamic control to allow for uncertainties in the aerodynamics characteristics, guidance, navigation, control, and the atmospheric density. The resultant high altitude flight results in high energy, low density flow around the vehicle; and the shock layer near the surface is significantly out of equilibrium. This nonequilibrium composition near the wall allows a noncatalytic surface to reject the energy associated with dissociation of the molecules, hence, the heating will be much less than to an entry vehicle that deeply penetrates the atmosphere, such as the Apollo command module or ballistic missiles. Of course all reentry vehicles spend some time at very low density conditions; but these deeply penetrating vehicles spend so little time there that their design is not affected by the high altitude part of their trajectories.

Aeroassisted Orbital Transfer Vehicles. One of the first analyses of the effects of nonequilibrium and catalysis on generic AOTV's was a study by Shinn and Jones[77] who calculated the heating to the hyperboloids using a viscous shock layer code adapted from Moss.[49] Their calculations were for several points in possible AOTV trajectories and they parametrically varied the recombination co-efficients, body size and streamline distance along the surface of the hyperboloid. Their results indicated that a significant reduction in heating could be realized for AOTV's having low catalycity surfaces. In Fig. 34 Shinn and Jones show the dependence of q/q_{FC} on altitude, steamwise distance, and γ for a 2.5-m nose

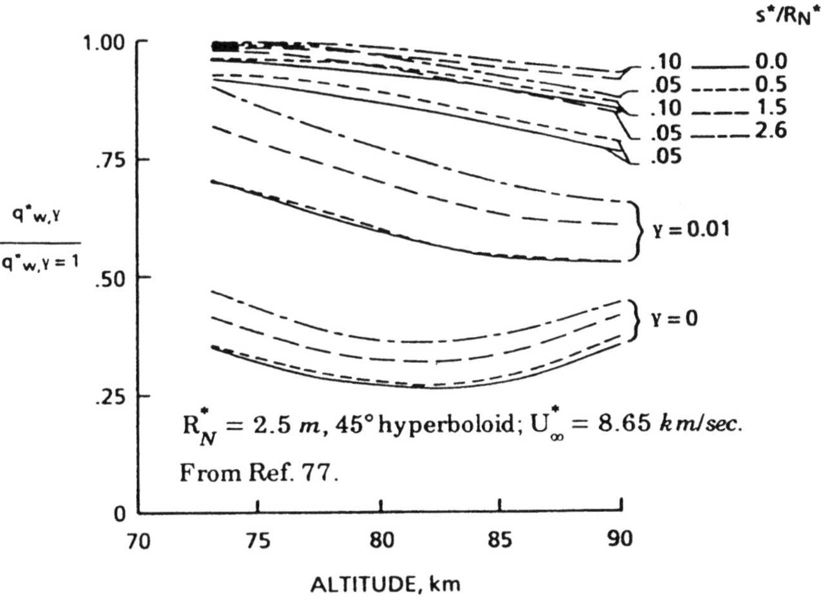

Figure 34.- Ratio of heating to the fully catalytic value as a function of altitude for several γ values and body locations.

radius, 45-degree hyperboloid body at a velocity of 8.65 km/s. Scott, et al.[78] presented a paper about a year later in which a blunted raked off cones were analyzed and they showed the dependence of heating on nose radius, altitude and velocity as was seen in Fig. 15. This calculation of q_{NC}/q_{FC} for the stagnation point of a sphere was obtained using the viscous shock layer option of the BLIMPK code.

Aeroassist Flight Experiment. Following the early feasibility studies of the AOTV's it was apparent that there were a number of technology issues that needed to be resolved before having sufficient confidence to go ahead with the development of AOTV's. In particular the gas chemistry and nonequilibrium effects on the radiation heating from the shock layer and the catalytic effects were very uncertain and important. Also, the ability to guide the vehicle through a varying uncertain atmosphere was questioned. Therefore, NASA recommended a flight experiment to help answer and resolve these and other issues. The flight experiment called the Aeroassist Flight Experiment (AFE) was approved for flight; and the current launch date is set for 1996. Currently there are 12 science experiments planned to fly on the AFE. One of the experiments is to measure the heat flux distribution over the vehicle. Another is similar to the Shuttle Orbiter experiment,[38] Catalytic Surface Effects experiment. The experiment on the AFE is called the Wall Catalysis Experiment.[79] The objective of the experiment is to assess experimentally the effects of nonequilibrium and catalysis on vehicles in AOTV-like trajectories. This will be done by comparing the heating to the standard thermal protection tile coating, reaction cured glass (RCG), with the heating to some reference material, such as a highly catalytic material as was used on the Orbiter experiment. Figure 35 taken from Ref. 79 shows the layout of the AFE aerobrake with locations indicated for the specially coated tiles. Preliminary calculations of the heat flux distribution in the vicinity of the coated tiles is given in Fig. 36, also taken from Ref. 79. The typical overshoot in heating on tiles coated with highly catalytic material compared with the low catalycity RCG can be seen. The method used here by Stewart and Kolodziej is a boundary layer calculation with the pressure distribution obtained using an inviscid method of integral relations.

AFE aeroheating environment methodology. The current methodology[80] for determining the aeroheating environment for the aerobrake of the Aeroassist Flight Experiment relies on boundary layer techniques. Although the AFE is not an axisymmetric shape (see Fig. 37) axisymmetric analog boundary layer techniques are employed due to the great number of calculations to be done for many points in the trajectory and for many points over the aerobrake surface. Higher fidelity techniques such as three-dimensional viscous shock layer or three-dimensional Navier-Stokes methods are not practical at this time for use as design tools. The VSL methods rely on shock shape data that requires either nonequilibrium Euler solutions or experimentally determined shock shapes. Neither are practical at this time. Likewise, the N-S techniques are too

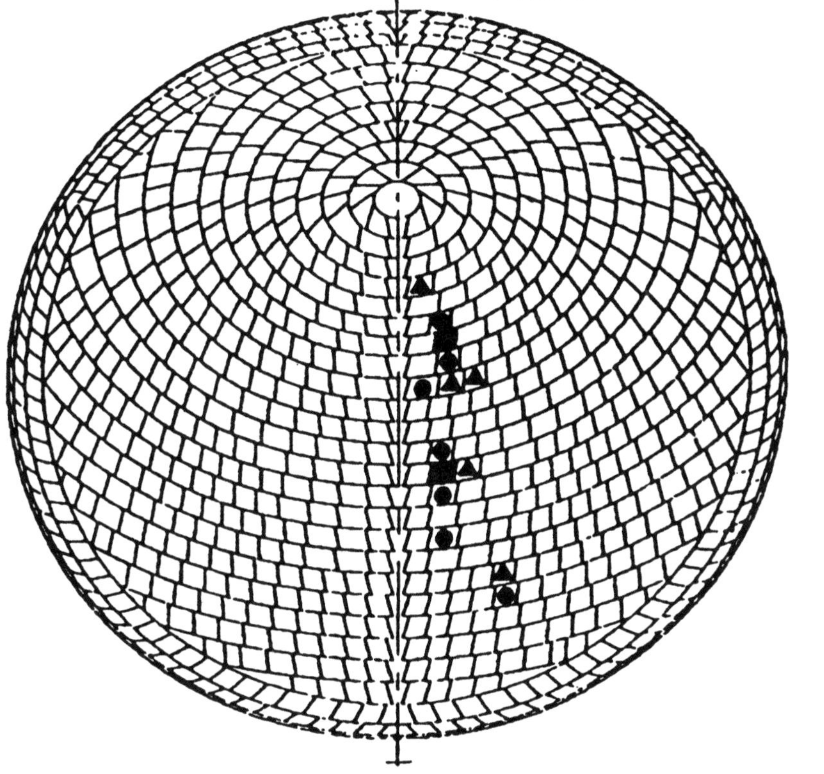

Figure 35.- Temperature profile across AFE wall catalysis experiment tile. From Ref. 79.

computationally intensive to be practical at this time. Therefore, we are content to design the AFE with the approximate two-layer approach for axisymmetric analog pressure distributions over the AFE windward surface. It was found that differences in the pressure distribution over the vehicle as a function of trajectory condition did not significantly affect the heating distribution over a range of altitudes from 85 km down to perigee (ca. 75 km) and back up to 85 km. The most significant heating occurs within this range. Moreover, the peak heating which occurs near perigee is the principle TPS design factor.

The heating for the AFE is calculated using pressure distributions determined from solutions to the three-dimensional Euler equations. These pressures are then input to a boundary layer code as boundary conditions along with

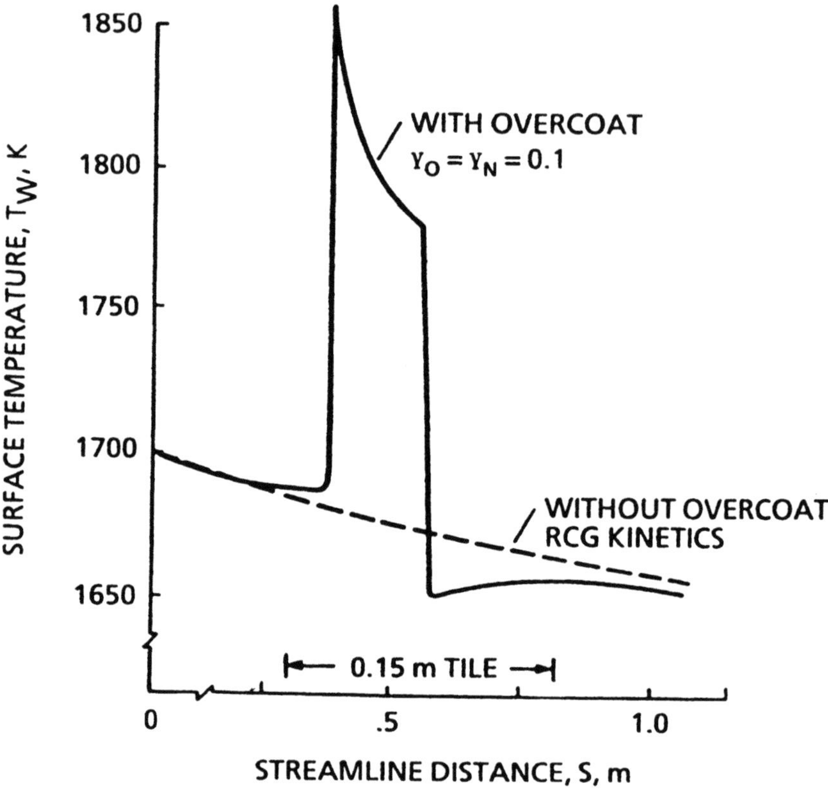

Figure 36.- Typical surface temperature profile across wall catalysis experiment tile. From Ref. 79.

geometrically determined metric coefficients. These metric coefficients are equal to the distance from the axis of the AFE to the body surface at each streamwise station. A 5-species nonequilibrium gas model is used in the calculations with two catalytic surface reactions, nitrogen and oxygen recombination. The interaction of NO with the surface is assumed to be noncatalytic.

The geometry of the AFE is shown in Fig. 37. The windward side of the aerobrake consists of an ellipsoid, an elliptic cone, and a skirt having a circular cross section in planes passing through the cone axis. The base of the cone where the skirt is attached is circular. The AFE trims such that the geometric stagnation point is at the center of the ellipsoid surface, and is also on the cone axis.

The boundary layer code is first run in its equilibrium chemistry mode to obtain the species concentrations required for input for the reacting mode. The inviscid

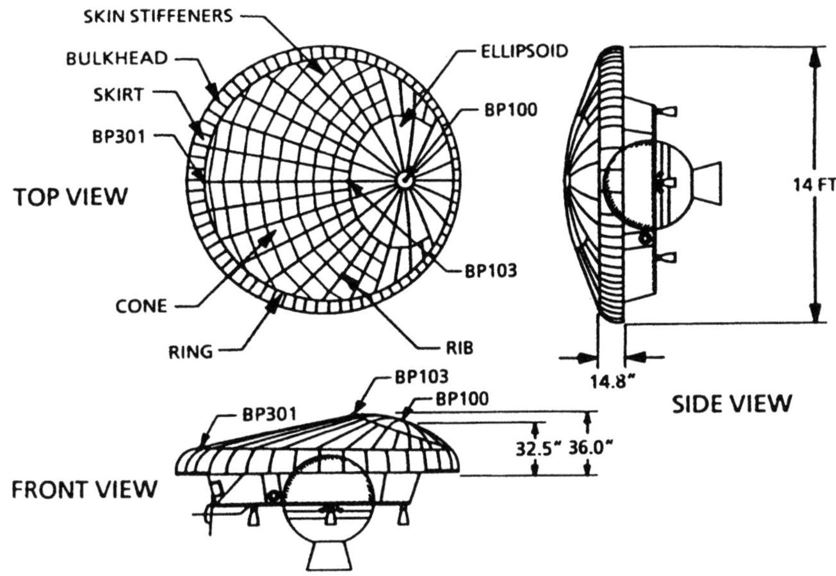

Figure 37.- Aeroassist flight experiment vehicle configuration.

flow around the very blunt configuration is assumed to be isentropic with normal shock entropy, therefore, entropy is not a required input. A perhaps more exact technique using the code is the viscous shock layer method, which requires not only the pressure distribution to be input, but also the shock shape. The VSL option is feasible for the stagnation point, but has not been implemented for the rest of the aerobrake. The stagnation point heating for the two methods differ by about 10 percent at peak heating as is seen in Fig. 38 for a fully catalytic surface. The percent difference is greater at higher altitudes.

The effects of different sets of recombination data is seen in Fig. 39 where the stagnation point heat flux for a fully catalytic surface is compared with values from Kolodziej and Stewart[15] and those extrapolated from Scott.[5] See Fig. 9 for a comparison of the rates versus 1/T. One can see in Fig. 39 that for heat fluxes less than about 29 Btu/ft^2-s the heating for the two sets of recombination rates is comparable, however, at higher heat fluxes the Kolodziej and Stewart rates result in lower heating. The high heat flux using Scott's rates is due to the extrapolation outside the range for which his data was measured. In Fig. 40 we see the ratio of q to q_{FC} for both sets of recombination rates as function of time in the AFE trajectory. Note that because of the maximum in Kolodziej and Stewart's rates as function of temperature, there is a minimum in the heat flux ratio at peak heating.

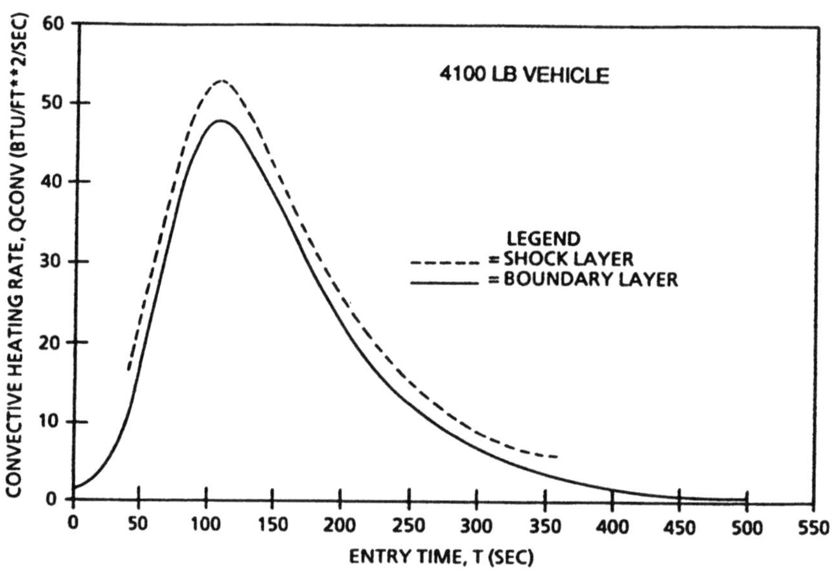

Figure 38.- 4100 lb AFE convective heating rates at stagnation point. BLIMPK S/L & B/L fully catalytic wall, $R_N = 8.3$ ft.

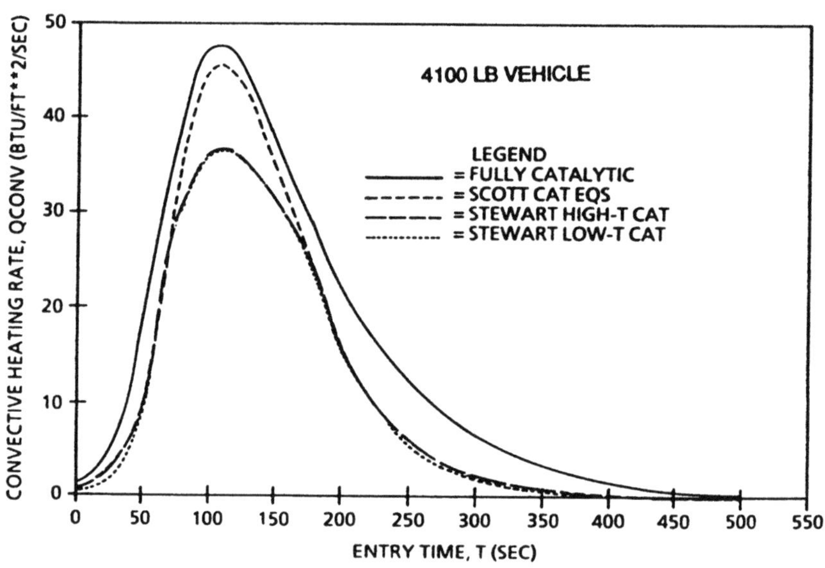

Figure 39.- 4100 lb AFE convective heating rates at stagnation point. BLIMPK B/L fully & partially catalytic wall, $R_N = 8.3$ ft.

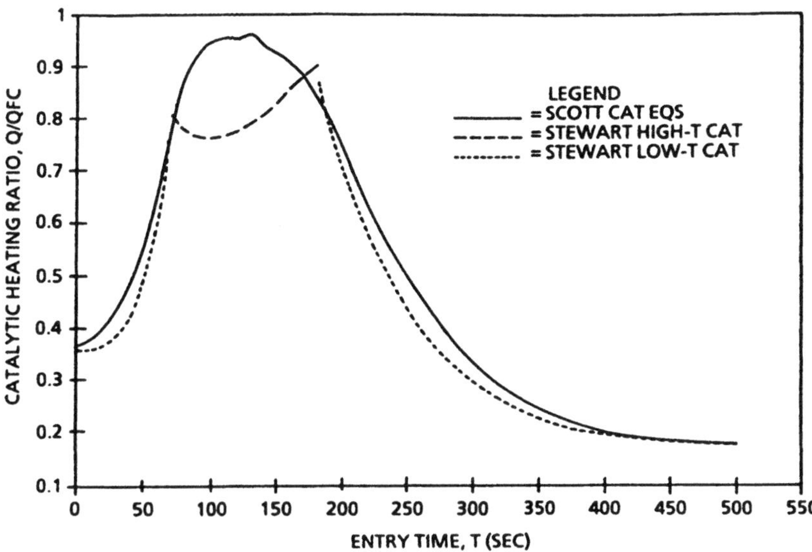

Figure 40.- Ratio of partially catalytic to fully catalytic heating rate at stagnation point. 4100 lb AFE BLIMPK B/L, R_N = 8.3 ft.

The distribution in heat flux for three different assumptions of the wall catalysis is shown in Fig. 41. In the stagnation region on the ellipsoid there is a significant reduction in the heat flux for finite rate catalytic recombination as well as for a non-catalytic surface. However, on the cone and skirt (S/L in the range from about 0.2 to 0.7), where the temperature of the surface is lower, there is not much reduction in heating due to finite rate catalysis. It is also seen that if the surface were non-catalytic there would be a significant reduction in heating everywhere.

The heating distribution on the cone-side of the aerobrake is shown in Fig. 42 for 5 different times in the AFE trajectory. The shape of the curves vary with time due to the effects of catalysis and levels of surface temperature.

As the AFE develops and the methodology evolves there will be refinements in the heating predictions used for design. The design of the AFE thermal protection system will be assessed from time to time to determine the effects of the updated methodology. Finally, when the AFE flies and the measurements are analyzed we will have a benchmark by which to assess the assumptions and methods used in the design predictions and for CFD validation.

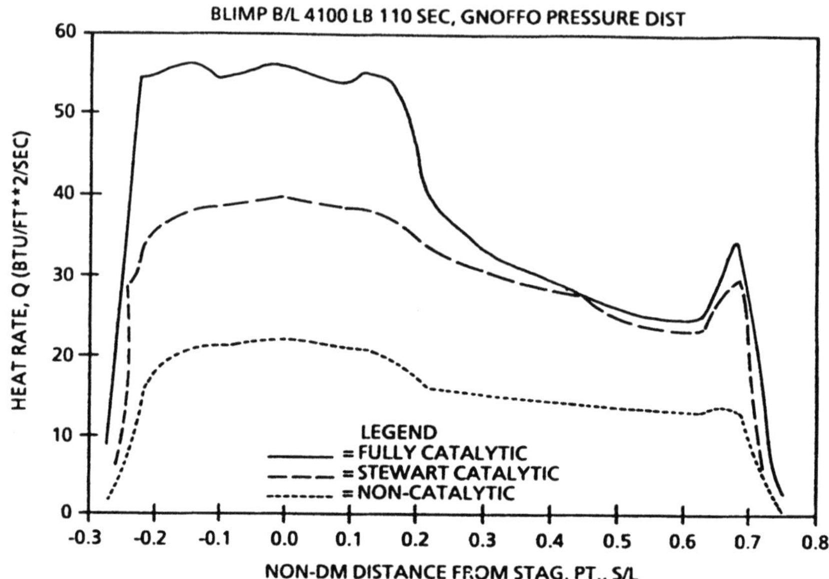

Figure 41.- Heating rate distribution around AFE pitch plane.
BLIMP B/L 4100 lb 110 sec, GNOFFO pressure distribution.

HYPERSONIC TRANSATMOSPHERIC VEHICLES AND SLENDER BODIES. Up to this point we have been discussing primarily blunt body flows where the flow has come through a near normal shock, and in which a large amount of dissociation has occurred. The AOTV's described and the AFE certainly fall in this category, whereas, even the Shuttle Orbiter exhibits many of the characteristics of a blunt body since it flies at the fairly large angle of attack of 40 degrees. In the case of sharp nosed vehicles such as the National Aerospace Plane (NASP) or other hypersonic cruise vehicles that would fly at high Mach number, the significance of the nonequilibrium and catalysis may not be as great. This is especially true if the vehicle is a sharp nosed-narrow cone or wedge, for which the shock wave is very oblique. The temperature behind a very oblique shock is not nearly as high as for normal shocks and therefore the air may not be dissociated very much. If such is the case, then there will not be much chemical energy to reject; hence, a noncatalytic surface would see as much heating as a fully catalytic surface. Of course there is the possibility that there might be dissociation in the boundary layer due to viscous dissipation. The energy equation contains a term involving the shear stress

$$\text{viscous heating} = \mathbf{u} \cdot \tau$$

238

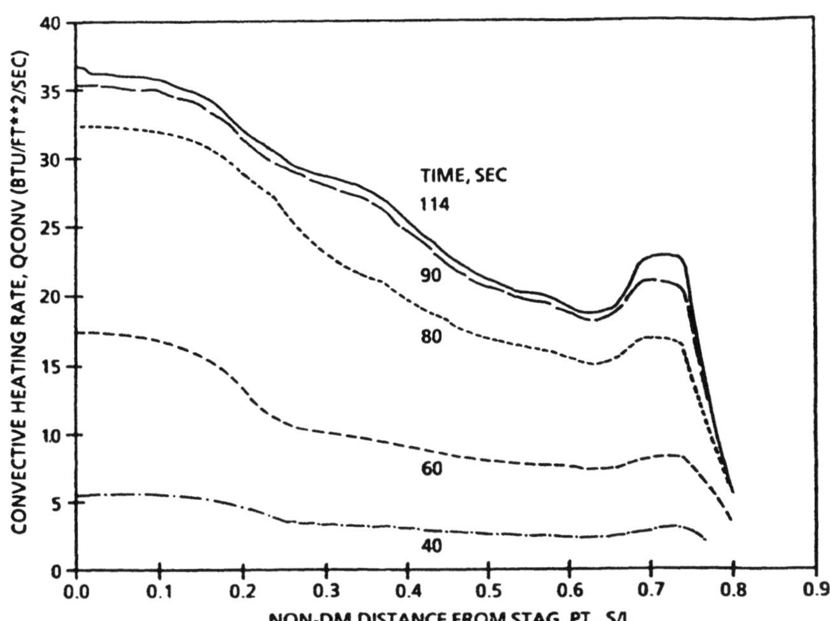

Figure 42.- Convective heating rate distribution around AFE.
Kolodzie's and Stewart's catalysis, Hamilton's pressures,
baseline V trajectory.

which results in a high temperature in the boundary layer. If the viscous heating in the boundary layer is great enough and the dissociation occurs fast enough, then there may be some noncatalytic effects. However, this would require very long running lengths which might preclude the effect to occur and might also allow enough time for the atoms to recombine in the boundary layer near the surface. The principle effect of catalysis down a slender vehicle is probably due to the flow from normal shock in the vicinity of the nose continuing around the body and flowing into the boundary layer of the slender cone. This effect is seen to a certain extent on the Orbiter as was indicated in Fig. 17.

A few results for slender blunted cones recently were presented June 1988 at the AIAA Thermophysics, Plasmadynamics and Lasers Conference in San Antonio. Stewart, et al.[81] calculated the heat flux to the windward surface of a sphere-cone TAV at angle of attack. Their method, developed by Rakich and Bailey,[82] involved the Ames Method of Integral Relations and a method of characteristics code to determine pressure distributions. These pressures were then input in a boundary layer code. Heating distributions along the centerline were calculated for angles of attack of 11°, representing ascent, and for 33°, representing entry.

239

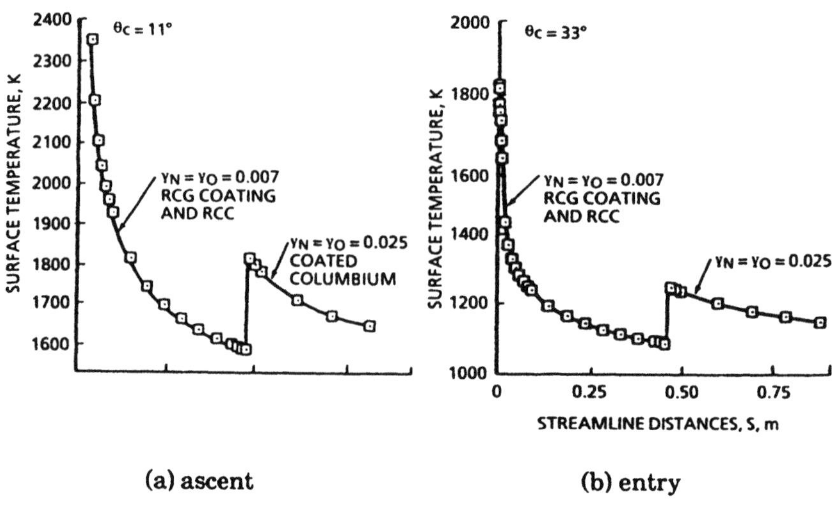

Figure 43.- Effect of surface catalysis on the heating distribution over a TAV during ascent and atmospheric entry. From Ref. 79.

The trajectory points for two cases are given in Table 4. The results of their calculations are shown in Fig. 43, where the radiation equilibrium surface temperature is given. Note that there is a change in surface catalysis at $s = 0.45$ m which corresponds to a change in the material from RCG or RCC to coated columbium. Note also that the ascent condition is much more severe than the entry condition.

Table 4 Trajectory Points for TAV Heating Calculation

	Alt., km	V_∞, km/s	α
Ascent	55.0	8.0	11°
Entry	72.8	6.56	33°

The values of the recombination coefficients used in the calculation were estimated from arc jet measurements by Stewart, et al.[81]

Viscous shock layer solutions for slender sphere cones were recently reported by Zoby, et al.[83] and by Shih, Zwan, and Kelley.[84]

Zoby, et al.[83] presented the heat flux distribution to slender cones of varying cone angles and nose radii for two trajectory conditions. They also investigated the binary scaling law assumed to be applicable for reacting flows. It can be seen in their results for an altitude of 175 000 ft and Mach = 25 that the nose radius has a significant effect on the extent of the low noncatalytic heating. See Fig. 44. The effect of cone angle is seen in Fig. 45. As is consistent with the opening remarks in this section, the larger the cone angle or the larger the nose radius the larger the extent of low q_{NC}/q_{FC}. The authors also concluded from stagnation point calculations of q_{NC}/q_{FC} at various densities and nose radii, that the binary scaling law does not apply. However, it is my opinion based on their data that binary scaling is adequate for fairly good scaling, although the error in heat flux may be on the order of 10-15 percent at the low altitudes or large nose radii.

Shih, Zwan, and Kelley[84] investigated the thermal systems of a transatmospheric vehicle shown in Fig. 46. They showed heating rate ratios along 10° sphere cones having nose radii of 0.25, 0.5, 1.0, and 3.0 ft. Their results, which are similar to those of Zoby, et al.[83], are shown in Fig. 47. Their results are for altitudes of 180 kft and 240 kft and a velocity of 21 ft/s.

From these sets of viscous shock layer calculations and the two-layer calculations of Stewart, et al.[81] it can be concluded that reduced heating will occur near the nose, and that vehicle designers should try to use materials having low catalytic recombination coefficients in that region. Downstream on slender cones it not as important to use noncatalytic materials since the payoff would not be as great.

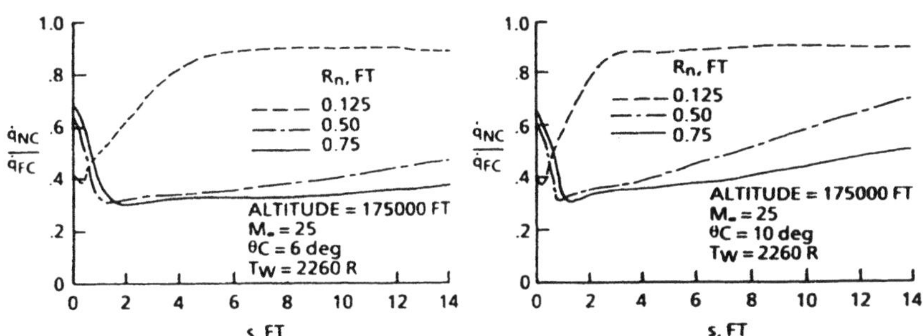

Figure 44.- Nose-radius effect on nonequilibrium heat-transfer distributions. From Ref. 83.

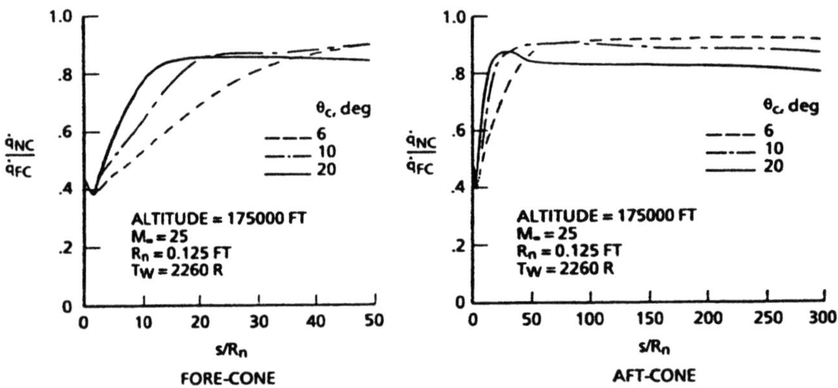

Figure 45.- Cone-angle effect on nonequilibrium heat-transfer distributions, $R_N = 0.125$ ft. From Ref. 83.

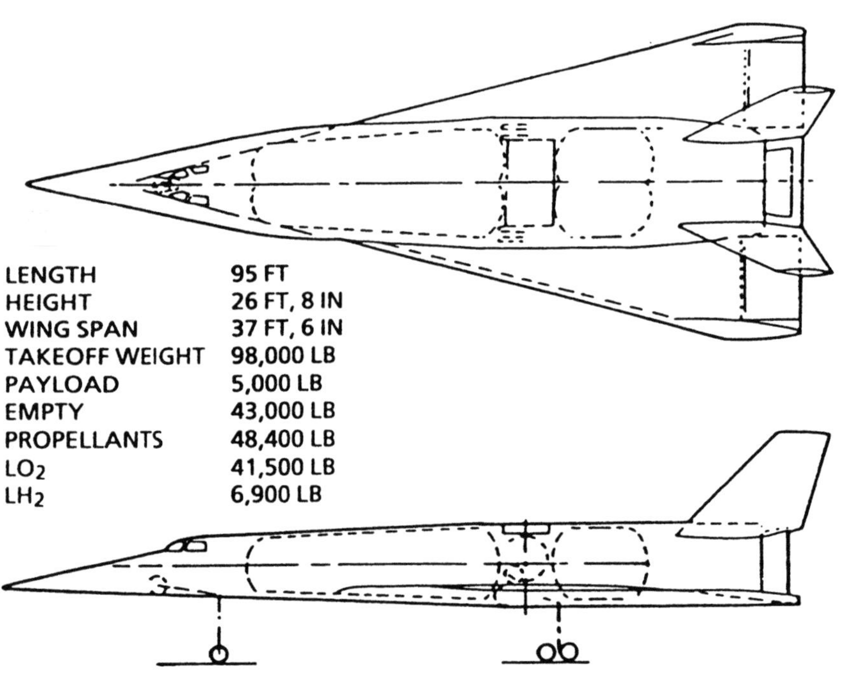

Figure 46.- Hypersonic aerospace vehicle. From Ref. 84.

242

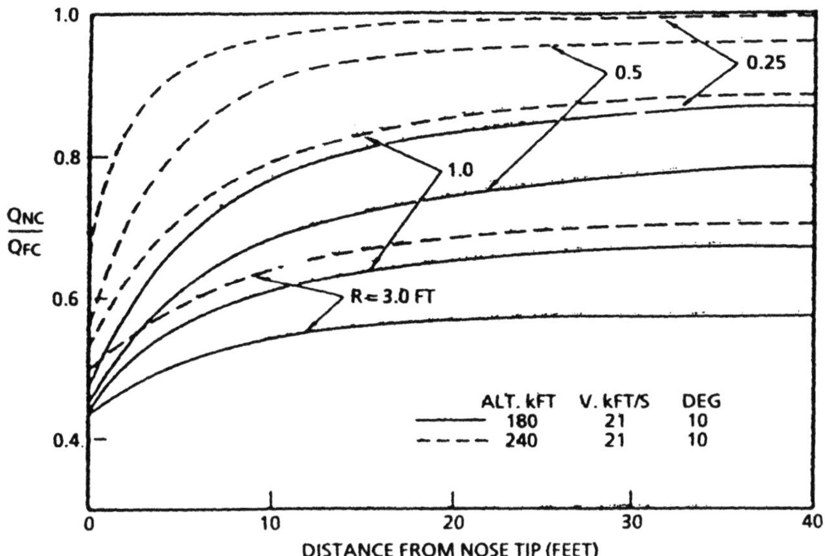

Figure 47.- Windward surface heating ratio. From Ref. 84.

REFERENCES

[1]Scott, C. D., "Effects of Thermochemistry, Nonequilibrium, and Surface Catalysis on the Design of Hypersonic Vehicles," in the notes for the 1st Joint Europe-US Short Course on Hypersonics, Dec. 7-11, 1987, Paris, France, edited by R. Glowinski, J. Bertin, and J. Périaux, Birkhauser Boston, Inc., 1988.

[2]Melin, G. A. and Madix, R. J., "Energy Accommodation During Oxygen Atom Recombination on Metal Surfaces," *Faraday Society Transactions*, Vol. 67, 1971, pp.198-211.

[3]Halpern, B. and Rosner, D. E., "Chemical Energy Accommodation at Catalyst Surfaces," *Chemical Society of London, Faraday Transactions I. Physical Chemistry*, Vol. 74, part 8, 1978, pp. 1883-1912.

[4]Breen, J. , Rosner, D. E., et al, "Catalysis Study for Space Shuttle Vehicle Thermal Protection Systems," NASA CR-134124, 1973.

[5]Scott, C. D., "Catalytic Recombination of Nitrogen and Oxygen on High Temperature Reusable Surface Insulation," *Progress in Astronautics and Aeronautics*, Vol. 77, edited by A. L. Crosbie, AIAA, 1981, pp. 192-212.

[6]Rosner, D. E. and Feng, H. H., "Energy Transfer Effects of Excited Molecule Production by Surface Catalyzed Atom Recombination," *J. of the Chemical Society, Faraday Transactions I*, Vol. 70, 1974, pp. 884-907.

[7]Scott C. D., "Wall Boundary Equations with Slip and Catalysis for a Multicomponent Nonequilibrium Gas," NASA TMX 58111, 1973.

[8]Scott, C. D., "Reacting Shock Layers with Slip and Catalytic Boundary Conditions," *AIAA J.*, Vol. 13, 1975, pp. 1271-1278.

[9]Gupta, R. N. and Simmonds, A. L., "Hypersonic Low Density Solutions of the Navier-Stokes Equations with Chemical Nonequilibrium and Multicomponent Surface Slip," AIAA Paper 86-1349, 1986.

[10]Gupta, R. N., Scott, C. D., and Moss, J. N., "Surface Slip Equations for Low Reynolds Number Multicomponent Air Flows," in *Progress in Astronautics and Aeronautics*, Vol. 96, edited by H. F. Nelson,1985.

[11]Scott, C. D., "Effects of Nonequilibrium and Wall Catalysis on Space Shuttle Heat Transfer," *J. of Spacecraft and Rockets*, Vol. 22, 1985, pp. 489-499.

[12]Jumper, E. W., R. G. Wilkins, and B. L. Preppernau, "Wall-Catalytic Fluorine Recombination in an HF Laser Nozzle," *AIAA J.*, Vol. 26, No. 1, Jan. 1988, pp.57-64.

[13]Jumper, E. W., C. J. Ultee, and E. A. Dorko, "A Model for Fluorine Atom Recombination on a Nickel Surface," *Journal of Physical Chemistry*, Vol. 84, 1980, pp. 41-50.

[14]Seward, W. A., "A Model for Oxygen Atom recombination on a Silicon Dioxide Surface," Ph. D. dissertation submitted to the Air Force Institute of Technology, 1985.

[15]Kolodziej, Paul and Stewart, D. A., "Nitrogen Recombination on High-Temperature Reusable Surface Insulation and the Analysis of its Effect on Surface Catalysis," AIAA-87-1637, 1987.

[16]Seward, W. A., and Jumper, E. J., "Oxygen Recombination on Space Shuttle Thermal-Protection-Tile Like Surfaces," AIAA-90-0054, 1990.

[17]Willey, R. J., "Mechanistic Model for Catalytic Recombination During Aerobraking Maneuvers," NASA-CR-185611, 1989.

[18]Swaminathan, P. K., B. C. Garrett, C. S. Murthy, and M. J. Redmon, "Formation and Quenching of Electronically Excited Molecules on Surfaces," Chemical Dynamics Corp. SBIR Phase I Final Technical Report to NASA-Ames Research Center, Sept. 25, 1986.

[19]Myerson, A. L., "Exposure-Dependent Surface Recombination Efficiencies of Atomic Oxygen," *J. Chem. Phys.*, Vol. 50, 1969, pp. 1228-1234.

[20]Marinelli, W. J. and Campbell, J. P., "Spacecraft-Metastable Energy Transfer Studies," Physical Sciences, Inc., Andover, Mass., final Report PSI-G565/TR-595, Contract No. NAS9-17565, 31 July 1986.

[21]Anderson, L. A., "Effect of Surface Catalytic Activity on Stagnation-Point Heat Transfer Rates," *AIAA J.*, Vol. 11, 1973, pp. 649-656.

[22]Pope, R. B., "Stagnation-Point Convective Heat Transfer in Frozen Boundary Layers, *AIAA J.*, Vol. 6, 1968, pp. 619-626.

[23]Scott, C. D., "Catalytic Recombination of Nitrogen and Oxygen on Iron-Cobalt-Chromia Spinel," AIAA Paper 83-0585, 1983.

[24]Rakich, J. V., Stewart, D. A. and Lanfranco, M. J., "Catalytic Efficiency of the Space Shuttle Heat Shield," *Progress in Astronautics and Aeronautics,* Vol. 85, edited by P. E. Bauer and H. E. Collicutt, AIAA, 1983, pp. 97-122.

[25]Zoby, E. V., Gupta, R. N., and Simmonds, A. L., "Temperature-Dependent Reaction Rate Expressions for Oxygen Recombination," *Progress in Astronautics and Aeronautics*, Vol. 96, edited by H. F. Nelson, 1985. pp. 445-464.

[26]Greaves, J. C. and Linnett, J. W., "The Recombination of Oxygen Atoms at Surfaces," *Transactions of the Faraday Society*, Vol. 54, 1958, pp. 1323-1330.

[27]Greaves, J. C. and Linnett, J. W., "Recombination of Atoms at Surfaces, Part 5 - Oxygen Atoms at Oxide surfaces," *Transactions of the Faraday Society,* Vol. 55, Part 8, 1959, pp. 1346-1354.

[28]Rahman, M. L. and Linnett, J. W., "Recombination of Atoms at Surfaces Part 10 - Nitrogen Atoms at Pyrex Surfaces," *Transactions of the Faraday Society,* Vol. 67, Part 1, 1971, pp. 170-198.

[29]Wood, B. J. and Wise, H., "The Interaction of Atoms with Solid Surfaces," *Rarefied Gas Dynamics Supplement 1.* edited by L. Talbot, Academic Press, 1961, pp. 51-59.

[30]Dickens, P. G. and Suttcliffe, M. D., "Recombination of Oxygen Atoms at Oxide Surfaces Part 1 Activation Energies of Recombination," *Transactions of the Faraday Society*, Vol. 60, 1964, pp. 1185-2308.

[31]Smith, *J. Chem. Phys.*, Vol. 11, 1943, pp. 110.

[32]Marinelli, W. J., "Collisional Quenching of Atoms and Molecules on Spacecraft Thermal Protection Surfaces," AIAA-88-2667 (1988).

[33]Prok, G. M., Effect of Surface Preparation and Gas Flow on Nitrogen Atom Surface Recombination," NASA TN D-1090, 1961.

[34]Myerson, A. L., "Mechanisms of Surface Recombination form Step-Function Flows of Atomic Oxygen over Noble Metals," *J. Chem. Phys.* Vol. 42, No. 9, 1965, pp. 3270-3276.

[35]Hartunian, W. P. Thompson, and S. Safron, "Measurements of Catalytic Efficiency of Silver for Oxygen Atoms and the O-O2 Diffusion Coefficient," *J. Chem. Phys.* Vol. 43, No. 11, (1965), pp. 4003-4006.

[36]Goulard, R. J., "On Catalytic Recombination Rates in Hypersonic Stagnation Heat Transfer," *Jet Propulsion*, Vol. 28, 1958, pp. 737-745.

[37]Scott, C. D., "Measured Catalycities of Various Candidate Space Shuttle Thermal Protection System Coatings at Low Temperature," NASA TN D-7113, 1973.

[38]Stewart, D. A., Rakich, J. V., and Lanfranco, M. J., "Catalytic Surface Effects Experiment on the Space Shuttle," *Progress in Astronautics and Aeronautics*, Vol. 82, edited by T. E. Horton, AIAA, 1982, pp. 248-272.

[39]Zwan, A. D., Crooks, R. S., and Whatley, W. J., "Arcjet Validation of Surface Catalycity Using a Viscous Shock-Layer Approach," in *Validation of Computational Fluid Dynamics*, AGARD-CP-437, pp. 24-1—24-13, 1988.

[40]Arepalli, S., Yuen, E. H., and Scott, C. D., "Application of Laser Induced Fluorescence for Flow Diagnostics in Arc Jets," AIAA-90-1763, 1990.

[41]Cunnington, G. R., Robinson, J. C., and Clark, R. K., "Non-Catalytic Coatings for Hypersonic Vehicle Applications," AIAA-90-1742, 1990.

[42]McCaffrey, B. J., and East, R. A., "Non Equilibrium Stagnation Point Heat Transfer Measurements to Catalytic Surfaces in Shock Heated Air," in the Proceedings of the 10th International Shock Tube Symposium, *Modern Developments in Shock Tube Research*, Edited by G. Kamimoto, Shock Tube Research Society, Japan, 1975.

[43]Fay, J. A. and Riddell, F. R., "Theory of Stagnation Point Heat Transfer in Dissociated Air," *J. Aeronautical Sciences*, Vol. 25, 1958, pp. 73-85.

[44]Bird, G. A., *Molecular Gas Dynamics*, Clarendon Press, Oxford, 1976.

[45]Inger, G., "Nonequilibrium Recombination-Dissociation Boundary-Layer Flows Along Arbitrary-Catalytic Hypersonic Vehicles," AIAA 90-0055, 1990.

[46]Blottner, F. G., "Viscous Shock Layer at the Stagnation Point with Nonequilibrium Air Chemistry," *AIAA J.*, Vol. 7, 1969, pp. 2281-2287.

[47]Davis, R. T., "Viscous Shock Layer at the Stagnation Point with Nonequilibrium Air Chemistry," AIAA Paper 70-805, 1970.

[48]Miner, E. W., and Lewis, C. H., "Hypersonic Ionizing Air Viscous Shock-Layer Flows Over Nonanalytical Blunt Bodies," NASA CR-2550, 1975.

[49]Moss, J.N., "Reacting Viscous-Shock-Layer Solutions with Multicomponent Diffusion and Mass Injection," NASA TR R-411, June 1974.

[50]Kim, M. D., Swaminathan, S., and Lewis, C. H., "Three Dimensional Nonequilibrium Viscous Flow over the Space Shuttle Orbiter," *Journal of Spacecraft and Rockets*, Vol. 21, 1984, pp. 29-35.

[51]Kim, M. D., Swaminathan, S., and Lewis, C. H., "Three Dimensional Viscous Flow over the Shuttle with Surface Catalytic Effects," AIAA Paper 83-1426, 1983.

[52]Thompson, R. A., "Comparison of Nonequilibrium Viscous-Shock-Layer Solutions with Windward Surface Shuttle Heating Data," AIAA-87-1473, 1987.

[53]Bhutta, B. A., C. H. Lewis, and F. A. Kautz, "A Fast Fully Iterative Parabolized Navier-Stokes Scheme for Chemically-Reacting Reentry Flows," AIAA-85-0926, 1985.

[54]Prabhu, D. K., J. C. Tannehill, and J. G. Marvin, "A New PNS Code for Chemical Nonequilibrium Flows," *AIAA J.* Vol. 26, No. 7, July 1988, pp. 808-815.

[55]Prabhu, D. K., J. C. Tannehill, and J. G. Marvin, "A New PNS Code for Three-Dimensional Chemically Reacting Flows," AIAA-87-1472, 1987.

[56]Tannehill, J., J. Ievalts, D. Prabhu, and S. Lawrence, "An Upwind Parabolized Navier-Stokes Code for Chemically Reacting Flows," AIAA-88-2614, 1988.

[57]Inger, G. R., "Nonequilibrium Hypersonic Stagnation Flow with Arbitrary Surface Catalycity Including Low Reynolds Number Effects,", *Int. J. Heat and Mass Transfer*, Vol. 9, 1966, pp. 755-772.

[58]Scott, C. D., Ried, R. C., Maraia, R. J., Li, C.-P., and Derry, S. M., "An AOTV Aerobraking and Thermal Protection Study," *Progress in Astronautics and Aeronautics,* Vol. 96, edited by H. F. Nelson, AIAA, 1985, pp. 309-337.

[59]Scott, C. D. "Space Shuttle Laminar Heating with Finite Rate Catalytic Recombination," *Progress in Astronautics and Aeronautics*, Vol. 82, edited by T. E. Horton, AIAA, 1982, pp. 273-289.

[60]Rakich, J. V. and Lanfranco, M. J., "Numerical Computation of Space Shuttle Laminar Heating and Surface Streamlines," *J. of Spacecraft and Rockets,*Vol. 14, 1977, pp. 265-272.

[61]Li, C.-P. "Development of Hypersonic Flow Models and Prediction Methods," Notes for the 3rd Joint Europe-US Short Course in Hypersonics, Oct 1-5, 1990, Aachen, FRG.

[62]Rakich, J. V., Stewart, D. A., and Lanfranco, M. J., "Catalytic Surface Effects of Space Shuttle Thermal Protection System During Earth Entry of Flights STS-2 Through STS-5," Paper presented at Langley Conference on Shuttle Performance: Lessons Learned, Hampton VA., March 1983.

[63]Williams, S. D. and Curry, D. M., "An Analytical and Experimental Study for Surface Heat Flux Determination," *Journal of Spacecraft and Rockets,* Vol. 14, 1977, pp. 632-637.

[64]Scott, C. D. "Space Shuttle Laminar Heating with Finite Rate Catalytic Recombination," *Progress in Astronautics and Aeronautics*, Vol. 82, edited by T. E. Horton, AIAA, 1982, pp. 273-289.

[65]Scott, C. D. and Derry, S. M., "Catalytic Recombination and the Space Shuttle Heat Shield," *Progress in Astronautics and Aeronautics,* Vol. 85, edited by P. E. Bauer and H. E. Collicutt, AIAA, 1983, pp. 123-148.

[66]Rakich, J. V. and Lanfranco, M. J., "Numerical Computation of Space Shuttle Laminar Heating and Surface Streamlines," *J. of Spacecraft and Rockets,*Vol. 14, 1977, pp. 265-272.

[67]Tong, H., Buckingham, A. D., and Morse, H. L., "Nonequilibrium Chemistry Boundary Layer Integral Matrix Procedure," NASA CR 134039, 1973.

[68]Shinn, J. L., Moss, J. N., and Simmonds, A. L., "Viscous- Shock-Layer Heating Analysis for the Shuttle Windward Plane with Surface Finite Catalytic Recombination Rates," *Progress in Astronautics and Aeronautics,* Vol. 85, edited by P. E. Bauer and H. E. Collicutt, AIAA, 1983, pp. 149-180.

[69]Goodrich, W. D., Li, C.-P., Houston, C. K., Chiu, P. B., and Olmedo, L., "Numerical Computations of Orbiter Flow fields and Laminar Heating Rates," *J. of Spacecraft and Rockets,* Vol. 14, 1977, pp. 257-264.

[70]Zoby, E. V., R. N. Gupta, and A. L. Simmonds, "Temperature Dependent Reaction Rate Expressions for Oxygen Recombination," *Thermal Design of Aeroassisted Orbital Transfer Vehicles,* edited by H. F. Nelson, Vol. 96 of Progress in Astronautics and Aeronautics, 1985, pp. 445-465.

[71]Curry, D. M., W. C. Rochelle, D. C. Chao, and P. C.Ting, "Space Shuttle Orbiter Nose Cap Thermal Analysis," AIAA 86-0388, 1986.

[72]Ting, P. C., W. C. Rochelle, and Curry, D. M., "Comparison of Viscous Shock Layer and Boundary Layer Reentry Heating Techniques for Orbiter Nose Cap," AIAA 86-1350, 1986.

[73]Cheng, H. K., "The Blunt Body Problem in Hypersonic Flow at Low Reynolds Number," Cornell Aeronautical Lab Rept. AF-1284-A-10, June 1963.

[74]Siemers, P. M., H. Wolf, and M. W. Henry, "Shuttle Entry Air Data System (SEADS) - Flight Verification of an Advanced Air DATA System Concept," AIAA 88-2104, May 1988.

[75]Ting, P. C., W. C. Rochelle, and D. M. Curry, "Prediction of Aerodynamic Heating and Pressures on Shuttle Entry Air Data System (SEADS) Nose Cap and Comparison with STS-61C Flight Data," presented at the 1st International Conference on Hypersonic Flight in the 21st Century, Sept. 20-23, 1988, Grand Forks, ND.

[76]Blackwell, H. E., Scott, C. D., Hoffman, J. A., Mende, S. B., and Swenson, G. R., "Spectral Measurements of the Space Shuttle Leeside Shock Layer and Wake," AIAA Paper 86-1262, 1986.

[77]Shinn, J. L, and J. J. Jones, "Chemical Nonequilibrium Effects on Flowfields for Aeroassist Orbital Transfer Vehicles," AIAA-83-0214, Jan. 1983.

[78]Scott, C. D., Ried, R. C., Maraia, R. J., Li, C.-P., and Derry, S. M., "An AOTV Aerobraking and Thermal Protection Study," *Progress in Astronautics and Aeronautics,* Vol. 96, edited by H. F. Nelson, AIAA, 1985, pp. 309-337.

[79]Stewart, D. A. and P. Kolodziej, "Wall Catalysis Experiment on AFE," AIAA 88-2674, June, 1988.

[80]Ting, P. C.,W. C. Rochelle, S. R. Mueller, J. E. Colovin, C. D. Scott, and D. M. Curry, "Development of AFE Aerobrake Aerothermodynamic Data Book," AIAA-89-1734, June, 1989.

[81]Stewart, D. A., W. D. Henline, P. Kolodziej, and E. M. W. Pincha, "Effect of Surface Catalysis on Heating to Ceramic Coated Thermal Protection Systems for Transatmospheric Vehicles," AIAA-88-2706, June 1988.

[82]Rakich, J. V. and H. E. Bailey, "Computation of Nonequilibrium, Supersonic Three-Dimensional Inviscid Flow Over Blunt Bodies," *AIAA J.,* Vol 21, No. 6, June 1983, pp. 834-841.

[83]Zoby, E. V., K. P. Lee, R. N. Gupta, R. A. Thompson, and A. L. Simmonds, "Viscous Shock-Layer Solutions with Nonequilibrium Chemistry for Hypersonic Flows Past Slender Bodies," AIAA-88-2709, June, 1988.

[84]Shih, P. K., A. D. Zwan, and M. N. Kelley, "Thermal Protection System Optimization for a Hypersonic Aerospace Vehicle," AIAA-88-2739, June, 1988.

PHYSICAL ASPECTS OF HYPERSONIC FLOW:
FLUID DYNAMICS AND NON-EQUILIBRIUM PHENOMENA

Maurizio Pandolfi
Dipartimento di Ingegneria Aeronautica e Spaziale
Politecnico di Torino, Torino.

1 Preliminaries

The fluid dynamics in the hypersonic regime is characterized by many features.

Discontinuities, as shock waves and contact surfaces, develop over the flow field. Shocks are generated because of geometrical conditions and may present complex patterns. Interaction of shocks initiate contact surfaces. The Euler equations provide the description of these aspects.

Transport phenomena are of primary importance. The diffusion of molecules, as momentum and energy are concerned, leads to viscosity and thermal conductivity. The additional terms of the Navier Stokes equations account for these effects. Moreover, for a wide range of flying conditions, flows are not laminar and turbulence becomes fundamental.

The high speeds in hypersonics make available a large amount of energy, that prompts variations in the physical nature of the gas. In the lower part of the regime (high supersonic) energy is transfered to the vibrational excitation of biatomic molecules. At higher speeds, these molecules dissociate and generate new different species. On the upper limit of hypersonics, ionization can occur. Each of these phenomena presents a typical time of relaxation and the hypothesis of equilibrium flow can not be accepted. Therefore, proper equations have to be considered to describe these non-equilibrium phenomena.

The existence of different species in the flow field with a non uniform distribution (finite gradient of concentrations) requires some modelling about their diffusion; for example, the Fick's law of diffusion can be introduced.

In the present contribution, we would like to focus the attention on the interaction between fluid dynamics and non-equilibrium phenomenology. Since the above picture is rather complex, we assume a crude modelling of the actual physics, by neglecting some of the previous features. The transport phenomena are disregarded; therefore viscosity and thermal conductivity are not considered. Then, we suppose the vibrational energy excited to some fixed level (half or full excitation), or even in equilibrium conditions, so that its relaxation is neglected. Moreover, ionization is ignored.

The only non-equilibrium here considered is the chemical one. A five species (O, N, NO, O_2 and N_2) model is taken into account. In addition, the diffusion of the species is ignored.

After all these drastic assumptions, the resulting flow is described by the Euler equations and by the finite rate equations for the chemical species.

Hereafter, we will cover the following points :

- governing equations for the **non-equilibrium** flow

- models of **frozen and equilibrium** flows

- **speeds of sound** through reacting media

- interaction between fluid dynamics and non-equilibrium: the **Damkhöler** number

- **quasi linear** form of the equations for reacting flows

- non-equilibrium flow **tending** to frozen and equilibrium configurations

- the **propagation of waves** in reacting media

Let me remind the reader interested on this matter the basic references [1] and [2].

2   Equations for Non-Equilibrium Flow

I consider the Euler equations for the 1-D unsteady flow problem. These equations can be written in two different forms : laws of conservation and quasi-linear differential equations. The first one, also known as divergence form, expresses straightly the physical principles of conservation of mass, momentum and energy :

$$\rho_t + (\rho\, u)_x = 0$$

$$(\rho\, u)_t + (p + \rho\, u^2)_x = 0 \tag{1}$$

$$e_t + [u(p + e)]_x = 0$$

Methods that aim to provide a correct numerical capturing of discontinuities are based on the integration of Eqs. 1. The quantities differentiated with respect to time $[\rho, \rho u, e]$ are called the conservative variables and those differentiated with respect to the space coordinate $[\rho u, p + \rho\, u^2, u(p + e)]$ are the fluxes.

The second way of writing the Euler equations, always related to the same physical principles, is the quasi linear form :

$$\rho_t + u\, \rho_x + \rho\, u_x = 0$$

$$u_t + u\, u_x + p_x/\rho = 0 \tag{2}$$

$$h_t - p_t/\rho + u\, (h_x - p_x/\rho) = 0$$

Here we have written the energy equation (the last of Eqs. 2) by introducing the enthalpy, rather than giving evidence to the entropy. Indeed the latter does not play the significant role in non-equilibrium flow, as it does for the inert gas.

The quasi linear form is very meaningful, since it is here that we recognize the hyperbolic nature of the problem and work out the definition of characteristic rays and compatibility equations. The propagation of waves is described on the basis of these elements. The quantities $[\rho, p, h, u]$ are called the primitives variables.

The above forms, in which the Euler equations are presented, are equally important. They emphasize two concepts, the conservation principles and the propagation of waves. Both are fundamental for the development of efficient numerical methods in hypersonics. The conservative form (Eqs. 1) is used, in the integration, to obtain the numerical approximation of the weak solution (*numerical capturing*). The quasi linear form (Eqs. 2) is needed to conceive and develop **upwind formulations** (*flux vector* or *flux difference splittings*), that provide the more significant information of the fluxes to be introduced in the integration procedure.

These two sets of equations (Eqs. 1 and 2) describe the flow evolution for an inert gas as well as for a reacting one. In addition, a state equation and proper definitions are required to relate the different thermodynamical properties with the concentrations of species.

The finite rate equations are written on the basis of the rate of production of species. we assume a five species (O, N, NO, O_2 and N_2) model and, therefore, we can only consider the production of three species (for instance O, N, and NO), since the others (O_2 and N_2) follow from the conservation of the atomic species, oxygen and nitrogen. By neglecting the diffusion of species, the finite rate equation for the i-species is given by :

$$Y_{it} + u\, Y_{ix} = \mu_i\, \omega_i \tag{3}$$

where $i = 1, 2, 3$, respectively for O, N and NO and μ_i represents the molecular mass. The concentration of the species (Y_i) is here expressed as the mass concentration. This equation states that the local variation in time of a species (Y_{it}) is balanced by the convection term ($u\, Y_{ix}$) and by the source term that represents the reaction rate (ω_i).

The conservative form of Eq. 3 is given by :

$$\rho_{it} + (u\, \rho_i)_x = \rho\, \mu_i\, \omega_i \tag{4}$$

where $\rho_i = \rho\, Y_i$ is the partial density of the i-species. Eq. 4 will be coupled with Eqs.1, in the integration procedure.

I point out that, in the non-equilibrium modelling of the physics, discontinuities are only generated by the fluid dynamics. The partial density ρ_i presents a jump through shocks, because of ρ, but the transition of the concentration Y_i is continuous.

The variation of the concentrations of species is related to the amount of energy h_{for}, defined as it follows :

$$h_{for} = \sum_{i=1}^{3} Y_i\, h^{\circ}{}_i$$

where $h^{\circ}{}_i$ represents the heat of formation of the i-species.

Also, we can define the energy related to the vibrational excitation for the biatomic molecules. Such a value can be taken at a fixed level of excitation or in local equilibrium, since we are not considering any relaxation for it. Here we suppose this energy given by equilibrium conditions. So, the energy h_{vib} is given by:

$$h_{vib} = \sum_{i=3}^{5} Y_i\, (e_{vib})_i$$

where $(e_{vib})_i$ is determined by the characteristic vibrational temperature and by the translational and rotational temperature (T).

The static enthalpy is defined by :

$$h = \sum_{i=1}^{5} Y_i\, c_{p_i}\, T + h_{for} + h_{vib} \tag{5}$$

where the specific heat at constant pressure is given by:

$$c_{p_i} = \frac{5}{2}\mathcal{R}/\mu_i \qquad (i = 1, 2)$$

$$c_{p_i} = \frac{7}{2}\mathcal{R}/\mu_i \qquad (i = 3, 4, 5)$$

The universal constant of gases is denoted by $\mathcal{R}$.

Finally, the pressure is given by the equation of state, written for a mixture of perfect gases :

$$p = \rho\,\mathcal{R}\,T \sum_{i=1}^{5} \frac{Y_i}{\mu_i} \tag{6}$$

The rate of production (ω_i) of the i-species is given by the contribution of all those reactions taken into account in the model. A model with seventeen reactions, as the one proposed in [3], can be used in applications. We have:

$$\omega_i = \omega_i(p, \rho, Y_j) \tag{7}$$

The value of ω_i depends strongly on the temperature T, proportional to p/ρ (Eq. 6). Such a dependence appears in the forward and reverse reactions constants (or one of the two and their ratio, that represents the equilibrium constant). Moreover ω_i depends on the concentrations Y_j of all the species. Note that ω_i is vanishing when the concentrations tend to the equilibrium values (Y_i^*), and is increasing as the concentrations deviate from them. Indeed, the equilibrium concentrations follow from :

$$\omega_i(p, \rho, Y_j^*) = 0 \tag{8}$$

3 Frozen and Equilibrium Flows

The frozen and equilibrium flows represent two alternative models to describe reacting flows in particular extreme conditions. For both of them, the equations of the fluid dynamics seen above (Eqs 1 and 2) do not change. On the contrary, the equations that provide the production of species (Eqs. 3 and 4) are replaced by completely different conditions.

The frozen flow is a very particular model of reacting flow, where no reactions occur at all. The differential equations (Eqs. 3 and 4) are replaced by the conditions that the concentrations Y_i remain constant along the path of any mass element. If all of them come from the same upstream conditions, the concentrations Y_i are uniform all over the flow field. The model looks equivalent to the non-equilibrium flow in the case

the source term in Eqs. 3 or 4 is set equal to zero, but such an interpretation could
be misleading. Later, we will come back to this point. Since the concentrations are
constant, the enthalpy, defined in Eq. 5, is now depending only on the temperature,
that is on pressure and density :

$$h = h(p, \rho) \tag{9}$$

In the model of **equilibrium flow**, the differential equations (Eqs. 3 and 4) which
provide the description of the concentrations are replaced by algebraic relationships :

$$Y_i = Y_i^*(p, \rho) \tag{10}$$

These relationships are obtained by setting the rate of reactions equal to zero (Eq. 8).
Often, we are brought to say that this model is equivalent to a non-equilibrium flow
where the concentrations adjust immediately to any variation of pressure or density due
to fluid dynamics. Such an interpretation may be misleading, because it gives the feeling
of high speeds of reaction, whilst the ω_i values are zero in the equilibrium flow. We will
reconsider this point later. From Eq. 10, one sees that the enthalpy is dependent, once
more, on pressure and density only, part explicitly and part through the equilibrium
concentrations (Y_i^*) :

$$h = h(p, \rho, Y_i^*(p, \rho)) \tag{11}$$

4   Speeds of Sound

At this point, it is convenient to introduce the definition of the speed of sound in reacting
media.

Consider a perturbation which propagates in a gas at rest. Let the perturbation be
characterized by small jumps in pressure (dp) and density $(d\rho)$. Assuming the propagation phenomenon as governed by the continuity and momentum equations and isentropic
the transition through the perturbation, the velocity of propagation of its front is given
by :

$$a^2 = \left(\frac{\partial p}{\partial \rho}\right)_{S = const} \tag{12}$$

Such a velocity is called the speed of sound.

In order to define a speed of sound in a reacting flow, we have to verify wheter the
transition through the front of the perturbation is isentropic. It can be shown that
this condition is satisfied only in the limiting cases of frozen and equilibrium flows.
For a non-equilibrium flow, any variation of concentrations leads to the production of
entropy. Moreover, the definition of a speed of sound is achieved only for a front that
separates two uniform flow regions. This is the case of the frozen and equilibrium flow,

but not of the non-equilibrium flow, where relaxations (and related non uniform flow) are occurring over a certain distance, behind the front.

So, we can only define speeds of sound in the frozen and equilibrium flow.

First, we consider the frozen flow. The variation of enthalpy (Eq. 9) of a mass element, that goes through the front, is depending on the jumps dp and $d\rho$:

$$dh = h_p\, dp + h_\rho\, d\rho$$

Since the energy equation (the last of Eqs. 2) requires, for such an element:

$$dh - \frac{dp}{\rho} = 0 \tag{13}$$

we have :

$$(h_p - \frac{1}{\rho})\, dp + h_\rho\, d\rho = 0$$

From Eq. 12, we have the speed of sound in a frozen flow, defined as :

$$a_f^2 = \frac{h_\rho}{1/\rho - h_p} \tag{14}$$

Let me now consider the equilibrium flow. In this case the variation of enthalpy through the front is obtained from Eq. 11 :

$$dh = h_p\, dp + h_\rho\, d\rho + \sum_{i=1}^{5} h_{Y_i}(Y_{i\,p}^{*}\, dp + Y_{i\,\rho}^{*}\, d\rho)$$

Once more from Eq. 13, we have :

$$(h_p - \frac{1}{\rho} + \sum_{i=1}^{5} h_{Y_i} Y_{i\,p}^{*})\, dp + (h_\rho + \sum_{i=1}^{5} h_{Y_i} Y_{i\,\rho}^{*})\, d\rho = 0$$

The speed of sound in the equilibrium flow is then defined as :

$$a_e^2 = \frac{h_\rho + \sum_{i=1}^{5} h_{Y_i} Y_{i\,\rho}^{*}}{1/\rho - h_p - \sum_{i=1}^{5} h_{Y_i} Y_{i\,p}^{*}} \tag{15}$$

The equilibrium speed of sound is different from the frozen one, because of the reactions that develop through the front.

However, at low temperature, the two speeds practically present the same values, since the equilibrium concentrations Y_i^* do not depend on p and ρ ($Y_{i\,p}^{*}$ and $Y_{i\,\rho}^{*}$ are vanishing) and remain constant. The same is true at very high temperature. This is due to the fact that the values of Y_i^* do not change anymore, because the dissociation process has fully developed. In the intermediate range of temperature, the difference between the two can be noticeable. It can be shown that the ratio $(a_f/a_e) > 1$. Indeed, let me assume that the perturbation front is characterized by the same $dp > 0$, for both the frozen and equilibrium flow. The jump of density is positive, and $(d\rho)_e > (d\rho)_f$ because the energy required by the dissociation tends to cool the equilibrium flow. Therefore, because of Eq. 12, we have $a_f > a_e$.

5 The Damkhöler number

Let me consider the classical problem of a supersonic flow about a blunt body, for instance a cylinder with a prescribed radius r.

The qualitative description of the flow field is well known. A bow shock is located upstream of the body, at a stand-off distance that depends on the radius r. The mass element crossing the shock is raised up to very high temperature, specially on the symmetry line. Chemical relaxations are triggered by this temperature. First, molecules of oxygen dissociate and generate atoms of oxygen. Meanwhile, molecules of nitrogen dissociate too, at much lower rate. The few available atoms of nitrogen combine with atoms of oxygen and nitric oxide is generated. Later on, the nitric oxyde dissociates. Therefore, along a streamline we observe a maximum of concentration of NO, somewhere behind the shock. The mass element that goes through the region confined by the bow shock and the body experiences changes of pressure and temperature due to the fluid dynamics induced by the body. Therefore chemical reactions develop inside the mass element because the local concentration Y_i are, in general, different from the equilibrium values Y_i^*, predicted by the local values of pressure and density (see Eq. 8).

Such a physical phenomenon is described on the basis of two reference lengths. One is related to the fluid dynamics (l_{FD}) and the other is dictated by the rate of chemical reactions (l_{CH}). The length l_{FD} represents a typical physical distance over which variations in the fluid dynamics take place. Clearly, we can assume the radius of the cylinder r as l_{FD}. The other length, l_{CH}, represents the distance over which a significant production of species can be appreciated. Since the shock triggers the chemical relaxation, this distance may be given by the thickness of the chemical layer behind a normal shock in the 1-D problem. For a prescribed set of upstream conditions (M_∞, p_∞, ρ_∞), an estimation of l_{CH} is easily obtained.

Therefore, taking the radius of the cylinder as l_{FD} and evaluating l_{CH} from the upstream conditions, we define the Damkhöler number ($\mathcal{D}$) :

$$\mathcal{D} = \frac{l_{FD}}{l_{CH}}$$

The physical significance of this number follows immediately from its definition. It represents the parameter that gives an idea of the interaction between fluid dynamics and non-equilibrium and identifies the degree of non-equilibrium. Its definition and relevance are not only confined to chemical non-equilibrium, but can be extended to any kind of non-equilibrium.

Let me now consider the effects of $\mathcal{D}$ on the structure of the flow field. we prescribe upstream conditions (∞), so that the chemical reference length l_{CH} is fixed and consider different radius of the cylinder, therefore different l_{FD}.

First, we take into account very small cylinders ($r \rightarrow 0$), so that $\mathcal{D} \rightarrow 0$. In this case, the residence time of mass elements in the region of interest is very small. Therefore, no production of species takes place because of lack of time, even if the reaction rates are rather large. The picture looks very similar to the frozen flow model. However, the equivalence of the non-equilibrium flow for $\mathcal{D} \rightarrow 0$ with the frozen flow is not complete. The velocity of particles, that approach the stagnation point, is going down to zero, linearly with the distance from it. This implies that an infinite time is required for particles flowing on the symmetry line to reach the stagnation point. Moreover, once the particles leave the stagnation point and flow away, downstream along the wall, pressure and temperature drop down in a very short distance, proportional to the radius r. The recombination expected to follow the dissociation, that took place ahead of the stagnation, does not develop. This happens because the low temperature brings down the reaction rates and because the time available for this relaxation is vanishing, with $r \rightarrow 0$. So, the flow field appears everywhere frozen, with respect to the upstream conditions (∞), except:

- at the stagnation point, where we have equilibrium conditions, and

- on the wall of the cylinder, where the flow is still frozen, but now with reference to the equilibrium conditions of the stagnation.

At this point, we are going to consider a very large cylinder ($r \rightarrow \infty$), that corresponds to the case of $\mathcal{D} \rightarrow \infty$. The stand-off distance, related to r, is very large, but the thickness of the chemical layer behind the shock remains finite. Therefore such a layer tends to vanish if compared with any significant length in the field. Moreover, the distance over which we observe finite differences of pressure and density induced by the fluid dynamics is so large that any mass element has enough time to develop all the reactions, for readjusting the concentrations to the local equilibrium values Y_i^*. Therefore the non-equilibrium flow at very high $\mathcal{D}$ looks equivalent to the equilibrium flow. The only difference is the finite thickness of the chemical layer behind the shock. However, note that such a layer becomes undectable and invisible on the large scale of the cylinder.

In conclusion, the model of non-equilibrium is different from the models of frozen and equilibrium flows, since it accounts for finite rate reactions, not included in the others. Nevertheless, except some differences, the non-equilbrium flow field, at $\mathcal{D} \to 0$ or $\mathcal{D} \to \infty$, tends to be equal to the frozen and to the equilibrium configuration respectively.

6 Equations in Quasi Linear Form for Reacting Flows

The quasi linear form of the equations that describe reacting flows show three thermodynamical properties (ρ, p, h), only two of them are independent. Indeed, for the non-equilibrium flow, we can eliminate one of the them, by using the definition of the enthalpy (Eq. 5) and the values of Y_i provided by Eq. 3. For frozen and equilibrium flows, we consider the enthalpy defined in Eqs. 9 and 11 respectivly, with uniform concentrations or the equilibrium values (Y_i^*) provided by Eq. 10.

It is convenient to eliminate the density.

For non-equilibrium flow, the total derivative ($D\ /Dt = \partial\ /\partial t + u\,\partial\ /\partial x$) of the enthalpy is given by :

$$\frac{D\,h}{D\,t} = h_p \frac{D\,p}{D\,t} + h_\rho \frac{D\,\rho}{D\,t} + \sum_{i=1}^{5} h_{Y_i} \frac{D\,Y_i}{D\,t}$$

Since the energy equation (the last of Eqs.2) gives :

$$\frac{D\,h}{D\,t} - \frac{1}{\rho}\frac{D\,p}{D\,t} = 0$$

we obtain :

$$\frac{D\,\rho}{D\,t} = \frac{1}{h_\rho}(\frac{1}{\rho} - h_p)\frac{D\,p}{D\,t} - \frac{1}{h_\rho}\sum_{i=1}^{5} h_{Y_i} \frac{D\,Y_i}{D\,t}$$

The term that multiplies the total derivative of the pressure represents the frozen speed of sound, defined in Eq. 14. Also, we replace the total derivative of the concentration with the rate of production of the species, according to Eq. 3. Therefore, we obtain:

$$\frac{D\,\rho}{D\,t} = \frac{1}{a_f^2}\frac{D\,p}{D\,t} - \psi$$

where :

$$\psi = \frac{1}{h_\rho} \sum_{i=1}^{5} h_{Y_i}\, \mu_i\, \omega_i$$

The total derivative of the density which appears in the continuity equation, the first of Eq. 2, is replaced by the relationship just written. Then the continuity equation, in the quasi linear form, becomes :

$$\frac{D\,p}{D\,t} + \rho\, a_f^2\, u_x = a_f^2\, \psi \tag{16}$$

Now, we replace the density in the continuity equation considered for the frozen flow. Since in this case the enthalpy is given by Eq. 9, I have :

$$\frac{D\,\rho}{D\,t} = \frac{1}{a_f^2}\frac{D\,p}{D\,t}$$

and the continuity equation for the frozen flow becomes :

$$\frac{D\,p}{D\,t} + \rho\, a_f^2\, u_x = 0 \tag{17}$$

For the equilibrium flow, the enthalpy is defined in Eq. 11. Its total derivative is given by :

$$\frac{D\,h}{D\,t} = (h_p + \sum_{i=1}^{5} h_{Y_i}\, Y_{i\,p}^*)\frac{D\,p}{D\,t} + (h_\rho + \sum_{i=1}^{5} h_{Y_i}\, Y_{i\,\rho}^*)\frac{D\,\rho}{D\,t}$$

Therefore, we have :

$$\frac{D\,\rho}{D\,t} = \frac{1/\rho - h_p - \sum_{i=1}^{5} h_{Y_i}\, Y_{i\,p}^*}{h_\rho + \sum_{i=1}^{5} h_{Y_i}\, Y_{i\,\rho}^*}\,\frac{D\,p}{D\,t}$$

The term that multiplies the total derivative of pressure represents the equilibrium speed of sound, defined in Eq. 15. Therefore, the continuity equation for the equilibrium flow is given by :

$$\frac{D\,p}{D\,t} + \rho\, a_e^2\, u_x = 0 \tag{18}$$

The continuity equations for the frozen (Eq. 17) and for the equilibrium (Eq. 18) flows are almost identical, except the different definition of the speed of sound. For the non-equilibrium flow, the continuity equation (Eq. 16) is identical to the frozen flow equation (Eq. 17) as the left hand side is concerned, but a source term $(a_f\,\psi)$ appears on its right, straightly related to the rates of the chemical reactions.

7 Non-Equilibrium Tending to Frozen and to Equilibrium Flows

Now, we would like to show how the non-equilibrium problem tends to the frozen problem, for $\mathcal{D} \to 0$, and to the equilibrium one, for $\mathcal{D} \to \infty$. We assume steady state configurations, so that $\partial\ /\partial t = 0$. Then we look at the two more interesting equations, the rate (Eq. 3) and the continuity equation (Eq. 16). As done previously, we consider the flow about a cylinder. The upstream conditions are kept constant, so that l_{CH} is fixed, and we assume different values for the radius r of the cylinder, that is of l_{FD}.

First, we analize the case with $\mathcal{D} \to 0$, corresponding to an infinitesimal body. From Eq. 3, written in the dimensional form, I have:

$$Y_{i_x} = \frac{\mu_i\,\omega_i}{u} \tag{19}$$

Let me interpret any x-derivative as the ratio of the difference of flow property (ΔY_i) over the distance (Δx) on which we consider such a difference. Because ω_i is finite, the derivative Y_{i_x} is finite. Since any distance here is infinitesimal ($\Delta x \to 0$), also the difference of concentrations is very small ($\Delta Y_i \to 0$), in order to allow Y_{i_x} be finite. So, despite high rates of reaction, the concentrations Y_i do not change because the particles have not enough time to develop any appreciable reaction, over such a small region.

Looking at the continuity equation (Eq. 16), the derivatives p_x and u_x tend to infinity, being finite differences of pressure Δp and velocity Δu occuring over infinitesimal distances Δx. The two terms, where they appear, balance each other, since the source term ψ is infinitesimal with respect to each of them.

Obviously, the same conclusinos can be drawn if we consider the equations written in dimensionless form, with lengths normalized with respect to r. The dimensionless form is important because numerical solution are obtained by carrying on the integration on the equations written in this form. Now the dimensionless source terms are multiplied by r and tend to zero. The derivative of the concentrations vanishes and this implies that no variations of Y_i occur over finite normalized distances. In the continuity equation, the derivatives of pressure and velocity are finite and the terms, where they appear, cancel each other, since the source term is infinitesimal.

At this point, let me consider the case for $\mathcal{D} \to \infty$, where the radius of the cylinder is very large. First we are looking at the dimensional form of the equations. Being the flow close to the equilibrium, the rate ω_i vanishes and the derivative Y_{i_x} goes to zero. However, the differences of the concentrations (ΔY_i) over the field are not zero. On the contrary, the concentrations Y_i coincide almost with the equilibrium values (Y_i^*), which in turn vary due to changes of pressure and density (Eq. 8). Therefore, the derivative $Y_{i_x} \to 0$, since the finite difference of the concentration is spread over an infinite distance $(\Delta x \to \infty)$. In the continuity equation, finite differences of pressure and velocity are also spread over infinite distances, and the derivatives p_x and u_x tend to zero, as well as the source term $(\omega_i \to 0)$.

Now, we consider the dimensionless form of the equations. The normalized source terms tend to finite values, since they are products of infinitesimal chemical rates multiplied by the reference length that tends up to infinity. The conclusions are the same as in the dimensional analysis. The derivatives of the concentration, of pressure and velocity are finite, since the distances (Δx) have been normalized and balance the finite values of the normalized source terms.

We note that for $\mathcal{D} \to 0$, the non-equilibrium equations tend to coincide with those of the frozen flow. However, in the case of $\mathcal{D} \to \infty$, we do not observe the same peculiarity. Let me consider the dimensionless equation of the continuity for both the non-equilibrium and equilibrium models and point out two differences. First, the speed of sound which multiplies the divergence of the velocity (u_x in the 1-D problem) is the frozen speed for the non-equilibrium and the equilibrium speed of sound for the equilibrium model. Then, the finite source term in the non-equilibrium case has no corresponding term in the equilibrium flow. A contradiction seems to exist, since we have to expect that the same field of pressure and velocity has to satisfy the non-equilibrium equations, in the limit of $\mathcal{D} \to \infty$, and the equilibrium equations. Because two are the differences, we could expect that they cancel each other.

In the next paragraph we analyse this point and show that the contradiction is only apparent.

8 Propagation of wave in reacting media

By combining properly the continuity equation with the momentum equation (the second of Eqs. 2), compatibility equations are obtained. These equations describe the propagation of signals along characteristic rays.

For the non-equilibrium flow, the continuity equation is given by Eq. 16. The compatibility equations yield :

$$\frac{\partial R_{1,3}}{\partial t} + \lambda_{1,3}\,\frac{\partial R_{1,3}}{\partial x} = a_f^2\,\psi \tag{20}$$

where the signals $dR_{1,3}$ and the slope of the characteristics are given by :

$$dR_{1,3} = dp \mp \rho\, a_f\, du \qquad (21)$$

$$\lambda_{1,3} = u \mp a_f \qquad (22)$$

For the frozen flow, the continuity equation is given by Eq. 17. Then, we have :

$$\frac{\partial (R_f)_{1,3}}{\partial t} + (\lambda_f)_{1,3}\, \frac{\partial (R_f)_{1,3}}{\partial x} = 0 \qquad (23)$$

where :

$$d(R_f)_{1,3} = dR_{1,3} = dp \mp \rho\, a_f\, du \qquad (24)$$

$$(\lambda_f)_{1,3} = \lambda_{1,3} = u \mp a_f \qquad (25)$$

Finally, for the equilibrium flow, the continuity equation is given by Eq. 18. The compatibility equations become :

$$\frac{\partial (R_e)_{1,3}}{\partial t} + (\lambda_e)_{1,3}\, \frac{\partial (R_e)_{1,3}}{\partial x} = 0 \qquad (26)$$

where :

$$d(R_e)_{1,3} = dp \mp \rho\, a_e\, du \qquad (27)$$

$$(\lambda_e)_{1,3} = u \mp a_e \qquad (28)$$

The above equations put in evidence the features of the non-equilibrium flow, when it is considered in the limiting cases of $\mathcal{D} \to 0$ and $\mathcal{D} \to \infty$.

First, we note that the slope of the characteristics for the non-equilibrium (Eq. 22) is the same as in the frozen flow (Eq. 25). Even the signals which propagate along the charateristic lines present the same definitions (Eqs. 21 and 24). However, the source term (ψ) related to the rate of the chemical reactions and appearing on the right hand side of Eq. 20, contributes in changing the non-equilibrium signal along its propagation. On the other hand, the compatibility equation for the equilibrium flow (Eq. 26) is homogeneous, as for the frozen flow, but both signals and characteristics (Eqs. 27 and 28) are defined with the equilibrium speed of sound. In both the frozen and the equilibrium flow, we can define Riemann Invariants, whereas in the non-equilibrium flow the signal is varying along the characteristics.

Looking at the wave propagation, the limit of the non-equilibrium flow to the frozen one, is rather clear. On the contrary, the limit to the equilibrium flow presents some difficulties, as already previously seen for the continuity equations (Eq.16 and 18). The problem can be understood by working out a linearized approach.

Let me write the non-equilibrium equations in the simple case of biatomic molecules and atoms of only one species (A_2 and A). I proceed to the linearization, by considering perturbations with respect to a gas at rest and in equilibrium, with conditions denoted by overlining the symbols. The perturbations are denoted by $\rho, p, h, u, Y_i, Y_i^*$. Hence, the governing equations are :

$$\rho_t + \bar{\rho}\, u_x = 0 \tag{29}$$

$$u_t + p_x/\bar{\rho} = 0 \tag{30}$$

$$h_t - p_t/\bar{\rho} = 0 \tag{31}$$

$$Y_t = (Y - Y^*)/\bar{\tau} \tag{32}$$

These equations are the linearized counterpart of the continuity, momentum and energy equations (Eqs. 2) and of the rate equation (Eq. 3, being Y the concentration of the atomic species A). Note that the rate of reaction depends on the departure of the concentration Y form the equilibrium value Y^* and on a relaxation time $\bar{\tau}$ of the reaction ($A_2 \rightleftharpoons 2\,A$). In addition, we define the enthalpy as :

$$h = h_p\, p + h_\rho\, \rho + h_Y\, Y \tag{33}$$

Let me differentiate Eq. 33 with respect to time . After Eqs. 14, 29 and 31, we have:

$$Y_t = \frac{h_\rho}{h_Y} \left(\frac{1}{\bar{a}_f^2}\, p_t + \bar{\rho}\, u_x \right) \tag{34}$$

The perturbation of the equilibrium concentration is given by :

$$Y^* = Y_p^*\, p + Y_\rho^*\, \rho$$

By recalling Eq. 29, the derivative with respect to time is :

$$Y_t^* = Y_p^* \, p_t - \bar{\rho} \, Y_\rho^* \, u_x \tag{35}$$

From Eqs. 34 and 35, we have :

$$(Y - Y^*)_t \;=\; \frac{h_\rho + h_Y \, Y_\rho^*}{h_Y} \; \left(\frac{1}{\overline{a_e^2}} \, p_t + \bar{\rho} \, u_x \right) \tag{36}$$

Note that the frozen speed of sound, which appears in Eq. 34, is transformed in the equilibrium value in Eq. 36, because of the contribution of equilibrium terms in Eq 35. Let me point out that the terms in Eq. 36 are proportional to the derivative in time of the rate of reaction (Eq. 32).

After this preliminary work, we eliminate the derivative of the density in Eq. 29, just as previuosly done in Eq. 16. After Eqs. 29 and 33, we obtain the linearized continuity equation for the non-equilibrium flow:

$$p_t + \overline{\rho \, a_f^2} \, u_x \;=\; \overline{a_f^2} \, \frac{h_Y}{h_\rho} \, \frac{(Y - Y^*)}{\overline{\tau}} \tag{37}$$

By combining Eq. 37 with Eq. 30, we get the compatibilty equations :

$$\frac{D^{ne} p}{Dt} \mp \overline{\rho \, a_f} \, \frac{D^{ne} u}{Dt} \;=\; \overline{a_f^2} \, \frac{h_Y}{h_\rho} \, \frac{(Y - Y^*)}{\overline{\tau}} \tag{38}$$

where the non-equilibrium operator is:

$$\frac{D^{ne}}{Dt} = \frac{\partial}{\partial t} \mp \overline{a_f} \frac{\partial}{\partial x}$$

Note that the compatibility equations for the frozen flow are given by :

$$\frac{D^f p}{Dt} \mp \overline{\rho \, a_f} \, \frac{D^f u}{Dt} \;=\; 0 \tag{39}$$

where the frozen flow operator is equal to the non-equilibrium one:

$$\frac{D^f}{Dt} = \frac{D^{ne}}{Dt} = \frac{\partial}{\partial t} \mp \overline{a_f} \frac{\partial}{\partial x}$$

Let me now differentiate Eq. 38 with respect to time :

$$\frac{\partial}{\partial t} \left(\frac{D^{ne} p}{Dt} \mp \overline{\rho \, a_f} \, \frac{D^{ne} u}{Dt} \right) \;=\; \frac{\overline{a_f^2}}{\overline{\tau}} \, \frac{h_Y}{h_\rho} \, (Y - Y^*)_t \tag{40}$$

On the other hand, by combining Eq. 30 with Eq. 36, I have :

$$\frac{D^e p}{Dt} \mp \overline{\rho\, a_e}\, \frac{D^e u}{Dt} \;=\; \overline{a_e^2}\, \frac{h_Y}{h_\rho + h_Y\, Y_\rho^*}\, (Y - Y^*)_t \tag{41}$$

where :

$$\frac{D^e}{Dt} \;=\; \frac{\partial}{\partial t} \mp \overline{a_e}\, \frac{\partial}{\partial x}$$

By eliminating the derivative in time $(Y - Y^*)_t$ from Eqs. 40 and 41, we obtain:

$$\frac{\partial}{\partial t}\Big(\frac{D^f p}{Dt} \mp \overline{\rho\, a_f}\, \frac{D^f u}{Dt}\Big) \;=\; \frac{\overline{a_f^2}}{\overline{a_e^2}}\, \frac{h_\rho + h_Y\, Y_\rho^*}{h_\rho\, \overline{\tau}}\, \Big(\frac{D^e p}{Dt} \mp \overline{\rho\, a_e}\, \frac{D^e u}{Dt}\Big)$$

This equation is very meaningful. If we consider its dimensionless form, it may be shown that :

$$\frac{\partial}{\partial t}\Big(\frac{D^f p}{Dt} \mp \overline{\rho\, a_f}\, \frac{D^f u}{Dt}\Big) \;\propto\; \mathcal{D}\, \Big(\frac{D^e p}{Dt} \mp \overline{\rho\, a_e}\, \frac{D^e u}{Dt}\Big) \tag{42}$$

Recall that Eq. 42 comes from the compatibility equation for the non-equilibrium flow.

For $\mathcal{D} \to 0$, this equation reduces to the compatibility equation for the frozen flow :

$$\frac{D^f p}{Dt} \mp \overline{\rho\, a_f}\, \frac{D^f u}{Dt} = 0$$

For $\mathcal{D} \to \infty$, Eq. 42 provides the compatibility equation for the equilibrium flow :

$$\frac{D^e p}{Dt} \mp \overline{\rho\, a_e}\, \frac{D^e u}{Dt} = 0$$

It is clear that the contradiction resulting from the limit of the non-equilibrium flow to the equilibrium one is only apparent.

In conclusion, features of both the frozen and equilibrium compatibility flow are embedded in the compatibility equation for the non-equilibrium flow. The frozen flow appears explicitely through the frozen speed of sound that is used in defining signal and slope of the characteristic. The equilibrium flow is hidden in the source term, where the rate of reaction, which represents the non-equibrium phenomenology, contains the elements related to the equilibrium.

References

[1] Vincenti,W.G. and Kruger,C.H., "Introduction to Physical Gas Dynamics", John Wiley & Sons, New York,1965.

[2] Clarke,J.F. and McChesney,M. ,"The Dynamics of Real Gas", Butterworths,London,1964.

[3] Park,C., "On Convergence of Computation of Chemically Reacting Flows", AIAA Paper-85-0247, Jan. 1985.

PERMISSIONS

In addition to the credit lines on appropriate figures, Birkhäuser Boston thanks the following publishers for permission to reproduce their copyrighted material in this volume.

Chapter 4, Chul Park:

Fig. 3. Reprinted with permission of Cambridge University Press from Hornung H (1974): Non-equilibrium dissociating nitrogen flow over spheres and circular cylinders. *J Fluid Mech* 64:149-176.

Fig. 4. Copyright © AIAA 1987. Used with permission. From Macrossan MN, Stalker RJ (1987): Afterbody flow of a dissociating gas downstream of a blunt nose. AIAA Paper 87-0407.

Fig. 5. Reprinted with permission of Pergamon Press from Lobb RK (1964): Experimental measurement of shock detachment distance on spheres fired in air at hypervelocities. In: *The High Temperature Aspects of Hypersonic Flow, AGARDograph 68*, Nelson WC, ed. Copyright Pergamon Press, Ltd.

Fig. 7. Copyright © AIAA 1968. Used with permission. From Hillje ER, Savage R (1968): Status of aerodynamic characteristics of the Apollo entry configuration. AIAA Paper 68-1143.

Figs. 10, 11. Copyright © AIAA 1988. Used with permission. From Sharma SP, Huo WM, Park C (1988): The rate parameters for coupled vibration-dissociation in a generalized SSH approximation. AIAA Paper 88-2714.

Chapter 6, Carl D. Scott:

Fig. 10a. Reprinted with permission of The Royal Society of Chemistry from Greaves JC, Linnett JW (1958): The recombination of oxygen atoms at surfaces. *Transactions of the Faraday Society* 54:1323-1330.

Fig. 10b. Reprinted with permission of The Royal Society of Chemistry from Halpern B, Rosner DE (1978): Chemical energy accommodation at catalyst surfaces. *Chemical Society of London, Faraday Transactions I. Physical Chemistry* 74(8):1883-1912.

Figs. 13 and 14. Reprinted with permission of Pergamon Press from Inger GR (1966): Nonequilibrium hypersonic stagnation flow with arbitrary surface catalycity including low reynolds number effects. *Int J Heat and Mass Transfer* 9:755-772. Copyright Pergamon Press Ltd.

Figs. 23–26. Copyright © AIAA 1987. Used with permission. From Thompson RA (1987): Comparison of nonequilibrium viscous-shock-layer solutions with windward surface shuttle heating data. AIAA Paper 87-1473.